FARM POWER AND MACHINERY MANAGEMENT

Tenth Edition

FARM POWER AND MACHINERY MANAGEMENT

DONNELL HUNT

Iowa State Press
A Blackwell Publishing Company

Donnell Hunt is a retired assistant dean of engineering and professor of agricultural engineering, University of Illinois, Urbana.

©2001 Iowa State Press
A Blackwell Publishing Company
All rights reserved

©1983, 1977, 1973, 1968, 1964, 1960, 1965, 1954
Iowa State University Press

Blackwell Publishing Professional
2121 State Avenue, Ames, Iowa 50014

Orders: 1-800-862-6657
Office: 1-515-292-0140
Fax: 1-515-292-3348
Web site: www.blackwellprofessional.com

Printed on acid-free paper in the United States of America
Tenth edition, 2001

Library of Congress Cataloging-in-Publication Data

Hunt, Donnell.
    Farm power machinery management/Donnel Hunt.
—10th ed.
    p. cm.
Includes bibliographical references (p.) and index.
ISBN-13: 978-0-1756-9 (alk. paper)
ISBN-10: 0-8138-1756-0 (alk. paper)
1. Agricultural machinery. 2. Farm mechanization.
I. Title

S676. H83 2001
631.3—dc21   2001024374

Last number is the printing: 10 9 8 7 6 5 4

CONTENTS

PREFACE

Farm Machinery Management is the section of farm management that deals with the optimization of the equipment phases of agricultural production. It is concerned with the efficient selection, operation, repair and maintenance, and replacement of machinery.

As North American agriculture has developed, it has become increasingly dependent on mechanization. With the technology available today, one farm provides the food and fiber needs of 128 people. Additionally, these farms exported more than 25% of the total value of the production of corn, wheat, and soybeans. Such performance would be impossible without augmenting human strength and energy with farm machinery.

The costs of owning and operating machinery is often exceeded only by the cost of land use. A typical Corn Belt grain farm, 400 ha [1000 a], can have machinery (no labor) costs of $200/ha [$80/a]. Machinery investment costs can be $1200/ha [$500/a]. Potentially, there is opportunity to improve farm profits by reducing machinery costs.

The measure of the worth of a management decision is assumed to be dollar profit. While it is recognized that many farmers use personal preference, comfort, and convenience as deciding factors in machinery management, such emotional responses are ignored in this presentation unless there is an apparent economic benefit. The philosophy adopted is that a farm is a factory that markets several products, and the management goal is to maximize profits. The machines on the farm are merely tools of production and have costs that subtract from gross income. But present and future concerns for the environment and for conservation of natural resources cause activity for farm machinery that may not be strictly economic. In 1997 more than 13,000,000 ha [33,000,000 a] were enrolled in U.S. government conservation programs.

The objectives of this textbook are to analyze the factors that comprise machinery management, to explain the function of the various machines and mechanisms as they affect economic operation, and to indicate approaches and procedures for making management decisions. Farm machinery operation practices are ever changing and are related to both the crop and the geographic area. It is impractical to completely cover the subject in a textbook. It is hoped that instructors using this text can augment the book with material specific for students' needs.

An understanding of agricultural practices, college algebra, and trigonometry should be adequate preparation for using this text. Concepts of economics and physics are developed as needed.

Metric units are used to conform with policy pronounced by the Congress. Customary units are retained in brackets following the SI (Système Internationale d'Unitès) units since it is anticipated that people engaged in agriculture will be using both systems. Units such as litres, kilometres per hour, hectares, and revolutions per minute are not recommended by SI but are used because of their particular utility for machinery management. The units have been chosen to reflect the precision of the measurement. For example, tillage depth is reported in centimetres since a field measurement can seldom be determined more closely than a whole centimetre. A description of the SI system of units, abbreviations, and a table of common conversions is included in Appendix J.

The subject matter may be of direct value only to the present or potential farm operator or farm manager. However, the management principles and the machinery operating details should be useful to students preparing for careers in agricultural education, agricultural mechanization, or agricultural business and to the agricultural engineer who might serve as a machinery manager on a large commercial farm.

Students of machinery management are encouraged to procure and study the many excellent informational sales brochures produced by the equipment manufactures that are available at dealerships. Detailed explanations and illustrations, usually in color, show the equipment in much greater detail than can be included in this text.

ACKNOWLEDGMENTS

A compilation of machinery management facts would be impossible without the cooperation of many farm equipment companies, governmental bodies, and technical associations who have supplied data and illustrations. These groups contributed information for this volume.

Aerovent Fan and Equip. Co.
Ag Leader Technology
Agco Corporation
ASAE (American Society of Agricultural Engineers)
Behlen Manufacturing Co.
Robert Bosch, Inc.
Burch Plow Co.
Campbell Industries, Inc.
Cane Machinery and Engineering Co.
Case IH
Caterpillar, Inc.
Century Engineering Corp.
Deere and Co.
Delco Remy Div., General Motors Corp.
Ever-Tite Manufacturing Co.
Farmhand, Inc

Flexi-Coil
FMS/Harvest
Gandy Co.
Gehl Co.
Great Plains Manufacturing, Inc.
Herd Co.
Hypro Div., Lear Sigler
John Blue Co., Inc.
Kuhn Farm Machinery, Ltd.
Lilliston Corp.
LML Corp.
M & W Gear Co.
Nebraska Tractor Test Laboratory
New Holland North America Inc.
Portable Elevator
Prairie Agricultural Machinery Inst.
Schumacher Company, LC
The Snow Co.
Society of Automotive Engineers
Spraying Systems Co.
Texico Inc.
Tire and Rim Association
Transland Aircraft
Vegors Enterprises
U.S. Department of Agriculture
U.S. Internal Revenue Service

ECONOMIC PERFORMANCE

Optimum farm machinery management occurs when the economic performance of the total machine system has been maximized. Admittedly many farm machines are operated for tradition's sake, for pleasure, and even for therapeutic value; however, the successful commercial farm, composed as it is of many enterprises for which machines are only tools of production, will operate its machinery in a businesslike manner to produce goods at a profit.

The performance of a machine system is profitable only when it can add value to products and processes beyond the system's cost of operation. A minimum cost would appear to be an optimum economic goal, but overall profit maximization is the true goal of business and on the farm this does not necessarily occur with a minimum cost system of operations. Profit for the total business should also be foremost at the individual machine level. This may dictate that an individual machine operate at other than its possible minimum cost. Good machinery management, then, requires that the individual operations in a machine system must be adjusted and combined in a manner so that their overall performance returns the greatest profit to the farm business.

The economic performance of a machinery system is measured in terms of dollars per unit of output. Examples are $120 machinery cost/ha of corn harvested, $40/t of soybeans grown, 75¢/kg of beef marketed, etc. (See Appendix L for abbreviations of units.) In these terms maximum system performance occurs when the production cost per unit is lowest.

The three components of economic performance are

1. Machine performance
2. Power performance
3. Operator performance

Sometimes these items are referred to inaccurately as "efficiencies," as if there existed an ultimate value on which some fractional performance could be based. It should be manifest that zero cost is the only theoretical limit to the potential economic performance of machine systems, and there is no way to express the efficiency of a machine system in percentages since the output-input ratio is in terms of economic measures to physical factors.

The dimensional units of machine, power, and labor performances are in quantity per time. These three performance figures add to become an economic performance figure when the quantity per time is divided into the cost per time of each component. For example, a machine system produces 5t of forage/hr with a machine cost of $10/hr. The system requires 1.5 work-hours and 1.1 tractor-hours at $8 and $7,

respectively. The economic performance of the system is:

$$5.94\frac{\$}{t}=\frac{\$10}{hr}\times\frac{1hr}{5t}+\frac{\$8}{hr}\times\frac{1.5hr}{5t}+\frac{\$7}{hr}\times\frac{1.1hr}{5t}$$

See Laboratory Exercise 1, Problem Solving, to become familiar with the unit factor procedure for making calculations.

Hardware, energy, and human labor and management are the components of farm machinery systems. The economic performance of these components are examined in the next three sections.

Chapter 1

MACHINE PERFORMANCE

Measures of agricultural machine performance are the rate and quality at which the operations are accomplished. Rate is an important measure because few industries require such timely operations as agriculture with its sensitivity to season and to bad weather. Completeness is that portion of quality describing a machine's ability to operate without wasted product. As most agricultural materials are fragile and many are perishable, the amount of product damage or reduction in product quality due to a machine's operation is another important measure of machine performance. Farm operators are very aware of the need for complete and speedy operations but they often ignore the economic penalties resulting from crop and soil damage. Quality as well as quantity must be considered when evaluating machine performance.

A rate of machine performance is reported in terms of quantity per time. Most agricultural field machine performance is reported as area per hour. Harvesting machine performance is sometimes quoted as bushels per hour; quintals per hour; tons or tonnes per hour; and in the case of balers, bales per hour. Processing equipment performance is usually expressed as bushels or tonnes per hour. Such performance figures are properly called *machine capacity*.

Capacities

Capacity, when expressed only as area per time, is usually not a sufficient indicator of a machine's true performance, particularly with harvesting machines. Differences in crop yields and crop conditions can mean that one machine may have a low area per hour capacity but a high mass per hour capacity when compared with an identical machine in a different field. In this case a valid comparative capacity would be mass per hour.

The concepts of weight and mass must be understood for confidence in expressing machine capacities and crop yields in both customary and SI units. *Mass* is to be thought of as the substance of a body that resists acceleration and is attracted to the mass of the earth. A body will accelerate rapidly toward the center of the earth unless restrained. This restraining force is equal to the body's *weight*.

The relationship between mass and weight is

$$f = m\, ac$$

where f = force acting on the body,
m = mass of the body, and
ac = resulting acceleration in units of distance/s²

When the acceleration is caused by the earth's gravitational attraction, the term, ac, is labeled g and the force, f, is called weight. At sea level, g is considered to be 32.2 ft/s² in customary units and 9.807 m/s² in SI units. The value of g decreases slightly with elevation above sea level.

Unless measured with a beam balance scales, the weight of a body will vary at different places over the earth; therefore, the SI system uses mass units rather than weight units for measuring quantities of agricultural products. Kilograms, quintals (100 kg), and tonnes (1000 kg—also megagram) are common commercial trade units. These masses are commonly determined, however, by measuring their weights. Balance beam scales (no springs) are to be used for legal trade.

The customary system traditionally determined grain quantities with a volume term, the bushel. In recent years the bushel has come to mean a quantity of produce weighing a designated number of pounds depending on the crop and its moisture content. Some produce and most forages are measured in hundred-weight, cwt (100 lb in the United States, 112 lb in England), and in tons, T (2000 lb in the United States, 2240 lb in England).

Combines, potato harvesters, and similar machines that separate desired material from undesirable material need a special capacity comparison term. Rather than a report on the weight of material harvested, the weight of material handled is the proper capacity measure. The term *throughput* has come to mean the time rate of processing a total mass of material through a machine. As an example, the kilograms per hour throughput of a combine is assumed to include the total mass of grain, chaff, straw, and weeds that enter the header. Even throughput is not always a constant base for comparison, as it varies with crop moisture conditions. Throughput capacity ratings should be accompanied by a material moisture report.

Calculations of machine capacity involve measuring areas or masses and times. The computations are relatively easy if attention is paid to units. The units for area in this text are hectares, abbreviated as ha, and acres, abbreviated as a.

As an example case, three types of machine capacity are computed below—field capacity, material capacity, and throughput capacity.

It is determined that a 5-m [16.4-ft] width-of-cut combine is traveling 1.5 m/s [4.9 ft/s] In one minute's time 50 kg [110 lb] of grain are collected in the grain tank and 60 kg [132 lb] of material are discharged out of the rear of the machine.

Machine Capacities

1. Field capacity

$$2.7\frac{ha}{hr} = \frac{1.5\,m}{s} \times \frac{5\,m}{1} \times \frac{1\,ha}{10,000\,m^2} \times \frac{3600\,s}{1\,hr}$$

$$\left[6.64\frac{a}{hr} = \frac{4.9\,ft}{s} \times \frac{16.4\,ft}{1} \times \frac{a}{43,560\,ft^2} \times \frac{3600\,s}{hr} \right]$$

2. Material capacity

$$3000\frac{kg}{hr} = \frac{50kg}{min} \times \frac{60min}{hr}$$

$$\left[6600\frac{lb}{hr} = \frac{110\,lb}{min} \times \frac{60\,min}{hr} \right]$$

3. Throughput capacity

$$6.6\frac{t}{hr} = \frac{110\,kg}{min} \times \frac{t}{1000\,kg} \times \frac{60\,min}{hr}$$

$$\left[7.26\frac{T}{hr} = \frac{242\,lb}{min} \times \frac{T}{2000\,lb} \times \frac{60\,min}{hr} \right]$$

The capacities just calculated are *theoretical capacities* as distinguished from *effective capacities*. It is usually impossible to operate machines continuously or at their rated width of action; therefore, their effective or actual capacities will be substantially less than their theoretical or potential capacities.

Time Efficiency

Time efficiency is a percentage reporting the ratio of the time a machine is effectively operating to the total time the machine is committed to the operation. Any time the machine is not actually processing the field is counted as time waste. Rather strict definitions are required as to what should really be counted as time waste chargeable against the machine. The following list describes the time elements that involve labor, that are associated with typical field operations, and that should be included when computing the capacities or costs of machinery related to the various farm enterprises.

1. Machine preparation time at the farmstead (includes removal from and preparation for storage, also shop work)
2. Travel time to and from the field
3. Machine preparation time in the field both before and after operations (includes daily servicing, preparation for towing, etc.)
4. Theoretical field time (the time the machine is operating in the crop at an optimum forward speed and performing over its full width of action)
5. Turning time and time crossing grass waterways (machine mechanisms are operating)
6. Time to load or unload the machine's containers if not done on-the-go
7. Machine adjustment time if not done on-the-go (includes unplugging)
8. Maintenance time (refueling, lubrication, chain tightening, etc., if not done on-the-go—does not include daily servicing)
9. Repair time (the time spent in the field to replace or renew parts that have become inoperative)
10. Operator's personal time

Not all of these time elements are commonly charged against machine operations. The operator's personal time, 10, is a highly variable quantity and is usually unrelated to the operating efficiency of the machine; consequently, it is often ignored as a time waste charged against the machine. Similarly, 1, 2, and 3 are often excluded from consideration. The remaining elements, 4-9, are the items included in the term *field efficiency*.

Specifically, field efficiency is the ratio of the time in 4 to the total time spent in the field (4-9). Field efficiencies are not constant values for specific machines but vary widely. Table 1.1 lists the ranges for some of the common farm machines.

TABLE 1.1. Range in Typical Field Efficiencies and Implement Operating Speeds

Operation	Equipment	Field Efficiencies, %	Operating Speeds km/hr	[MPH]
Tillage	moldboard plow	88-74	5-9	[3.1-5.6]
	disk harrow	90-77	6-10	[3.7-6.2]
	spring-tooth or spike-tooth harrow	83-65	6-12	[3.7-7.5]
	field cultivator, chisel plow	90-75	6-9	[3.7-5.6]
Cultivation	row crop cultivator	90-68	3-9	[1.9-5.6]
	rotary hoe	88-80	9-20	[5.6-12.4]
Seeding	row planter with fertilizer	78-55	7-10	[4.3-6.2]
	grain drill with fertilizer	80-65	5-10	[3.0-6.2]
	broadcaster	70-65	7-10	[4.3-6.2]
	potato planter	80-55	9-12	[5.6-7.5]
Harvesting	mower-conditioner	95-80	5-9	[3.0-5.6]
	rake	89-62	6-9	[3.7-5.6]
	baler, rectangular	80-65	5-10	[3.0-6.2]
	baler, round	50-40	5-19	[3.0-12.0]
	forage harvester, shear bar	76-50	6-10	[3.7-6.2]
	combine	90-63	3-8	[1.9-5.0]
	corn picker	70-55	3-6	[1.9-3.7]
	windrower, swather	85-75	6-10	[3.7-6.2]
	potato harvester	90-50	3-6	[1.9-3.7]
	cotton, spindle picker	90-65	3-5	[1.9-3.1]
Miscellaneous	sprayer	80-55	7-10	[4.3-6.2]
	anhydrous ammonia applicator	65-55	6-9	[3.7-5.6]
	rotary stalk chopper, mower	85-65	6-10	[3.7-6.2]
	fertilizer spreader	90-60	6-10	[3.7-6.2]

With a field efficiency established, an equation for effective field capacity can be determined:

$$C = \frac{S\,w\,e}{c} \qquad (1.1)$$

where C = effective capacity, ha/hr [a/hr]
  S = speed, km/hr [mi/hr]
  w = rated width of implement, m [ft]
  e = field efficiency as a decimal
  c = constant, 10 [8.25]

Because traditional land measure is not expected to change, American farmers may be faced with operating metric machines in fields having areas expressed in acres. An approximate equation in the same form as Eq. 1.1 gives the customary capacity of a metric machine as a/hr = S w e/4.

An expression for effective material capacity can be derived from Eq. 1.1 by incorporating a yield per area term in the numerator. If this term should be tons per hectare, the machine capacity will be expressed in tons per hour, etc.

$$M = \frac{S\,w\,e\,y}{c} \qquad (1.2)$$

where M = material capacity, units/hr
  y = yield, units/area

The suggested time increment in the SI system is the second, and velocities are to be reported as metres per second. However, farm machine speeds have traditionally been reported on an hourly basis as in the customary unit, miles per hour. It is anticipated that the hour unit will persist as the more practical unit for agriculture and thus the field speeds reported in this book will be kilometres per hour. To calculate speeds in metres per second, the kilometres per hour speed is divided by 3.6.

A more precise mathematical statement of field capacity for normally operated machines (without breakdowns or unexpected stoppages) can be developed without reference to such an all-inclusive and general term as field efficiency. Eq. 1.3 applies when only turning time, swath overlap, and such area-related times as filling seed boxes, unloading grain tanks, or unhitching yield-collecting wagons detract from machine performance. A rectangular field with headlands is assumed.

$$C = \frac{S\,w\,L\,E_w}{(c1)\,L + D\,S\,w\,L\,E_w + (c2)\,S\,t} \qquad (1.3)$$

where C, S, w have the same units as in Eq. 1.1
  E_w = effective swath coverage, decimal of rated width
  D = unproductive time, hr/ha [hr/a]
  L = length of field, m [ft]
  t = turning time, s/turn
  c1 = constant, 10 [8.25]
  c2 = constant, 2.7778 [12.1]

As can be intuitively deduced, Eq. 1.3 shows that long fields, quick turns, wide machines, fast speeds, and short loading and unloading times all contribute to high machine capacities.

Factors affecting field efficiency are

1. Theoretical capacity of the machine
2. Machine maneuverability
3. Field patterns
4. Field shape
5. Field size
6. Yield (if a harvesting operation)
7. Soil and crop conditions
8. System limitations

Each of these factors is discussed in turn.

Theoretical Capacity

Field efficiency decreases with increases in theoretical capacity. One can feel intuitively that a minute wasted with a large machine represents more loss in potential production than the same minute lost with a smaller machine. An example case was the study by K. K. Barnes, T. W. Casselman, and D. C. Link (Iowa State University). They found that increasing the width of a 4-row implement by 50% will increase the effective field capacities by only 35% for corn planters and by only 40% for cultivators.

Field capacity of an implement also depends on its travel speed, and one can expect a drop in field efficiency when the implement's operating speed is increased. Increased field speeds will decrease the actual working time required; but if the time losses remain essentially the same, mathematically the field efficiency will be less. Such a result suggests that as far as speed considerations go, it is not good management to try to maximize field efficiency—i.e., one should not use slow speeds to keep the numerical value of field efficiency high. As high field and material capacities are possible with fast ground speeds, the good operator will test and judge the crop and soil conditions and then operate as fast as possible without having the quality of the operation suffer to any great extent.

Field speeds may be limited by any of the following factors:

1. Overloading the machine's functional units
2. Inability of the operator to steer the machine accurately
3. Loss of function and structural damage to the machine due to rough ground surface
4. Need to handle materials gently (slowly)

Machine Maneuverability

Farm machines need to be easily maneuvered both in the field and on the road to the field. Field machines need to be designed to permit them to make short turns at the ends of the field and while following crop rows planted on the contour and in curves.

Considerable time and space are required to turn large machines at a headland (Fig. 1.1). Yet the total field turning time for large machines is slightly less than for smaller machines if the large machine can negotiate a turn within the same multiple or fraction of its width as the smaller machine, because the large machine makes fewer turns than the smaller machine in equivalent fields.

The turning radius of implements is an important factor affecting the time lost in end travel and at corners. The automotive and farm equipment industries generally define turning radius as the radius of the circle within which the vehicle can make its shortest turn. Such a definition really reports the radius of the path of an extreme part of the vehicle, R, and usually does not describe the outer radius, r_0, of the

Fig. 1.1. Maneuverability is important for efficient use of large machines.

machine's working head (as illustrated for a self-propelled combine in Fig. 1.3).

In this presentation the term *turning radius* will refer to the outer radius of the effective path of the implement when in its sharpest turn, and it shall be designated r with the implement engaged and r_0 with the implement disengaged. The former is usually substantially larger than the latter.

The radius, r, is of most concern to the machine operator in irregular or contoured fields. The radius, r, may be greater than R for a row planter if the effective width of the planter is wider than its frame, or it may be much less as in the case of a side-mounted tractor mower.

Square Corners

Few tillage or seeding machines can make square turns. With most cutter bar mowers the turning radius is short enough to permit square corners. The succeeding raking, windrowing, and baling operations, however, usually follow a rounded corner pattern.

Trailed harvesters can make an essentially square corner if the hitch will clear the tractor wheels in a tight turn. Fig. 1.2 indicates the paths of various points of a tractor and a trailed, offset harvester as they negotiate a corner. This diagram assumes that the operator can start the turn by instantly pivoting the tractor about its right rear wheel; hold this pivot until impending fouling occurs between the tire and the implement tongue, point A; continue the turn from point A to point B holding a near-rub position between the tire and tongue; pivot about the left rear wheel of the tractor from points B to C; and then continue in a straight line at right angles from the original direction of travel. For the implement shown, the turn must start about 0.6 m [2 ft] before the cutter bar cuts out of the crop. Note that the corner remaining is not exactly square and a small area is missed by the left divider of the cutter bar. Also note that the tractor front wheels are driven to the edge of the standing crop at B and as the outer edge of the cutter bar swings from A to C it will back over a portion of the uncut crop.

Diagrams such as Fig. 1.2 and 1.3 are about the only way of predicting the turning ability of farm implements. One may assume that turns are applied instantly. Such an assumption is not unrealistic for slow-moving implements and implements equipped with quick-acting power steering.

Each part of an implement in a turn instantaneously rotates about a common point called a *turning center*.

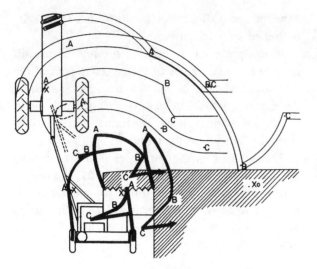

Fig. 1.2. Turning diagram of trailed, offset harvester.

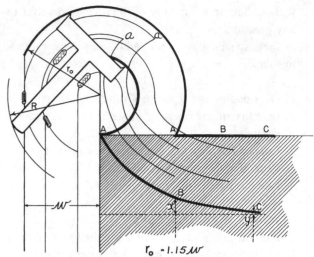

Fig. 1.3. Self-propelled combine at p/2-rad [90°] corner.

The location of this turning center may change quite rapidly as the turn progresses. In Fig. 1.2 the turning center for the implement at the start of its turn is located by the intersection of a line along the implement's axle and a line from the pivot point of the tractor through the drawbar pin. As the turn progresses to point A the turning center gradually moves to the left. It goes off the diagram and goes to infinity when the drawbar pin is about halfway to point X. As the turn progresses farther the turning center comes from infinity on the right and is located at point X_0 when the drawbar pin is at point X. From points A to B the tractor and implement have a fixed angular arrangement; consequently, the turning center is a fixed point at the intersection of lines along the axles of the tractor and the implement. This fixed turning center is located approximately at the position of the left end of the cutter bar when it first starts the turn.

When the turning center is not fixed, very small incremental plottings of the position of the points of interest are necessary to obtain accuracy. The locations of these points are determined by compass arcs from the instantaneous turning center and the arbitrarily advanced position of a significant point. In Fig. 1.2 the drawbar pin was used as the significant point from which all other points were determined.

Self-propelled (SP) implements, particularly those with rear wheel steering, have some unique turning problems. Some self-propelled windrowers have the ability to pivot about the midpoint of their drive axles, as the direction of rotation of their drive wheels can be reversed. A square corner is easily produced.

Self-propelled combines with a minimum turning radius equal to 1.15 their width of action are not capable of maintaining a square corner, as shown in Fig. 1.3. The paths of the left and right divider points of the cutter bar are indicated by heavy, dark lines. (It is assumed that any part of the crop bracketed by these points will be harvested.) After the combine has cut through the corner of the crop and pivoted until the divider points are at points A and A, the operator must instantly turn the rear wheels at points a and a from full right to full left. The remainder of the turn is accomplished holding the left divider along the edge of the standing grain while the right divider pivots into the standing grain. Even with such an extreme maneuver the cutter bar is unable to assume a full-width cut until it is beyond point C. On the next round the combine will have to proceed the additional distance x before making the pivot and will be short by the distance 2y of taking the required width of cut at the new point C. Consequently, the corner becomes progressively sharper each round and additional time will be required when finishing the field.

A comparison of the time lost between the $3\pi/2$-rad [270°] left turn and the $\pi/2$-rad [90°] right turn shows that the $\pi/2$-rad [90°] turn of Fig. 1.3 will be the most efficient. As given in the development of Eq. 1.13, the unproductive distance traveled by the outside divider point for a $3\pi/2$-rad [270°] turn is $2r_0$ - w + $3\pi/2r_0$. The same distance for the $\pi/2$-rad [90°] turn is w + x + (n-1)y + πr_0 where n equals the number of the round. As in Fig. 1.3 where $r_0 = 1.15w$, the distance for the $3\pi/2$-rad [270°] turn becomes 6.7w and for the $\pi/2$-rad [90°] turn becomes 4.84w when n = 2, x = 0.2w, and y = 0.03w. If the travel speeds for the two turns are essentially the same, one could make approximately 64 rounds before the time for the π/2-rad [90°] turn would be as much as that for the 3π/2-rad [270°] left turn. At such a time only three or four extra passes would be required to even up the corner. Remember that the x and y distances in Fig. 1.3 are the shortest possible. If the operator is unable to produce a turning radius of 1.15w, the efficiency for the π/2-rad [90°] turn will descend to that of the 3π/2-rad [270°] turn.

Curved rows and fields planted on contour lines restrict machine use. Fig. 1.4 shows a vehicle with wheel spacings twice the row spacing, s, negotiating a curve keeping a clearance, c, between the row and the center of the wheel. The vehicle wheelbase, L, must be no greater than $(s^2 + 2\,Rs - 2sc - 4Rc)^{1/2}$ to make the curve successfully. Tractors or self-propelled implements with tricycle configuration are slightly more adept in these turns.

Transport of machines over roads should be rapid, convenient, and safe. Some wide implements are made more maneuverable by folding to smaller widths (Fig. 1.5). Very large and heavy implements use the tractor's hydraulic power to do the folding. Other designs readjust the hitch to permit trailing wide implements in a lengthwise manner (Fig. 7.3).

Fig. 1.5. Implement folded for transport.

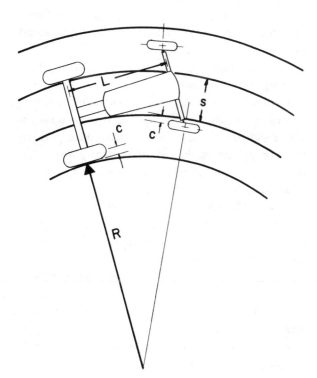

Fig. 1.4. Limiting-row curvature for vehicle with 2s wheel spacing.

Fig. 1.6. Twelve-bottom moldboard plow's width reduced by one-third for road travel.

Field Patterns

Substantial improvements in field efficiency can be made by analyzing and varying the pattern of field operations. Of course the pattern of operations is closely related to the size and shape of the field, but some pattern considerations can be studied independently of field configurations.

The primary objective in establishing an efficient field pattern is to minimize the amount of field travel. The number of nonworking turns, the travel distance in a turn, and the amount of nonworking travel in the interior of a field are all nonproductive users of valuable time and energy and should be eliminated if possible.

Objectives other than time minimization modify the choice of a field pattern. In flood-irrigated fields especially, but in other fields too, a tillage pattern should produce a level surface to eliminate water ponding. Repeated machine travel over a particular area of the field, usually an end of field condition, may cause compaction in some soils. Efficient planting patterns may be sacrificed occasionally to permit more efficient harvesting patterns. Soil conservation measures are probably the most important modifier of time-efficient field patterns; however, the economic benefits of soil conservation are not easily ascertained and the decision as to an optimum balance between time conservation and soil conservation is uncertain.

Terminology

Some terms in common use and some definitions relating to field patterns include the following.

1. A *round* refers to the travel of the machine across or around a field from a selected starting point to a point adjacent to the first. A *trip* is a half round or the travel from one end of the field to the other.
2. Operating in *lands* describes the practice of dividing the field into subareas and operating on these subareas individually.
3. *Turn strips* are unprocessed areas that provide room for making turns. These strips are processed either previously or at a later time. When the turn strip is at the end of a field it is called a *headland*.
4. A *headland pattern* has trips that parallel each other, are incremented successively by the operating width of the implement, and initiate at one boundary and terminate at the opposite. Turns are π-rad [180°] over the headlands.
5. A *circuitous pattern* describes the operation of the implement operation paralleling each land's boundaries and is commonly described as "going around the field." If the land is straight sided, the operations may start from the center of the field; otherwise, operations start at the outside boundaries. Circuitous patterns having π/2-rad [90°] turns are: (a) rounded corners, (b) square corners, and (c) diagonal turn strips. A circuitous pattern with 3π/2-rad [270°] turns may start either at the center or at the boundaries of a land.
6. An *alternation pattern* is sometimes used in processing established row crops. To provide a turn that is easy to negotiate, the trips are not adjacent. The alternation pattern may be a modification of the continuous pattern or the whole land may be alternated. Turns are π-rad [180°].

Fig. 1.7 illustrates the common patterns for straightsided fields.

Several farm machines have rather rigid pattern requirements because they are one-way oriented; i.e., they are commonly said to be either right- or left-handed. Moldboard plows; pull-type mowers, rakes, and windrowers; and most pull-type harvesters require a definite operating position with respect to the unprocessed portion of the field. Most tillage implements and most self-propelled implements are more liberal in their pattern operating requirements. In fact, if all farm implements were self-propelled or front-mounted on tractors the field pattern problems would be reduced considerably.

Moldboard plowing involves the most complex field patterns. Common plow bottoms throw to the right and require consecutive, adjacent, same-direction trips or undesirable *dead furrows* and *back furrows* will result. Dead furrows occur when consecutive plowing trips are adjacent on the left, back furrows when adjacent on the right. Some dead furrows and back furrows are practically unavoidable; they are not to be considered disastrous unless it is imperative to maintain a uniformly level surface on the field. A good practice is to alternate the positions of dead furrows and back furrows each season.

Comparison of patterns is straightforward with rectangular fields. It is relatively easy to calculate the maximum efficiency when the implement operates in a predictable manner. Patterns common to moldboard plowing serve as examples for comparing pattern efficiencies.

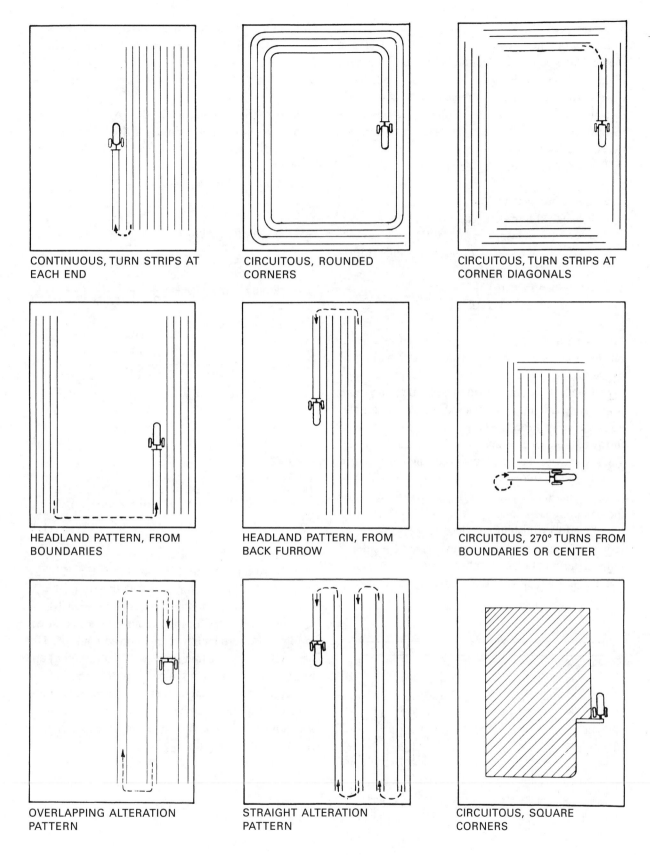

CONTINUOUS, TURN STRIPS AT
EACH END

CIRCUITOUS, ROUNDED
CORNERS

CIRCUITOUS, TURN STRIPS AT
CORNER DIAGONALS

HEADLAND PATTERN, FROM
BOUNDARIES

HEADLAND PATTERN, FROM
BACK FURROW

CIRCUITOUS, 270° TURNS FROM
BOUNDARIES OR CENTER

OVERLAPPING ALTERATION
PATTERN

STRAIGHT ALTERATION
PATTERN

CIRCUITOUS, SQUARE
CORNERS

Fig. 1.7. Common field machine patterns for rectangular fields.

Headland Pattern

Laying out the most efficient number of lands for the headland pattern is a common plowing problem. If the lands are made too large, excessive headland travel results. If the lands are made too small, excessive time is used to finish off the numerous dead furrows. Maximum efficiency will lie somewhere between these extremes.

The variables to be considered are

w = effective width of plow, m [ft]
f = length of furrow, m [ft]
S_p = speed of plowing, km/hr [MPH]
S_e = effective speed of headland travel, km/hr [MPH]
W = width of field, m [ft]
L = length of field, m [ft]
$c3$ = constant, 1000 [5280]

It is assumed that both back-furrowed and dead-furrowed lands are of equal size and one additional trip across the field is required to finish a dead-furrowed land. (Finishing a land includes plowing some of the soil back into the exposed furrow to partially level the soil surface.) The procedure for a 2-back-furrowed field or a 3-land field is illustrated in Fig. 1.8. As shown, the back-furrowed lands have been completed and the dead-furrowed land remains.

Computation of pattern efficiency (PE) requires the calculation of time spent for plowing, for idle travel, and for finishing the dead furrows. The time required for plowing the headlands is not considered in the determination of number of lands, as this time is constant regardless of the number of lands used.

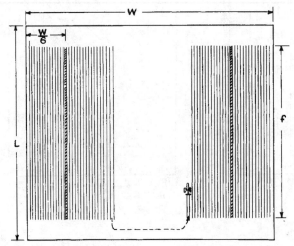

Fig. 1.8. Headland pattern. Average headland travel is W/6 for 2-back-furrowed, 3-land pattern.

Number of back-furrowed lands = n, number of dead-furrowed lands = $n - 1$, and total number of lands = $2n - 1$
Time in hours for one trip (also time to finish a dead furrow) = $f/[(c3) S_p]$
Number of trips to complete lands = W/w
Total hours for plowing lands = $f W/[(c3) S_p w]$
Average distance for turning one end of land = 1/2 land width
Width of land = $W/(2n - 1)$

$$\text{Average time for turning end} = \frac{W}{2(2n-1)(c3)S_e}$$

$$\text{Total turning time for lands} = \frac{W^2}{2(2n-1)(c3)S_e w}$$

$$\text{PE for lands} = \frac{\dfrac{fW}{(c3)S_p w}}{\dfrac{fW}{(c3)S_p w} + \dfrac{W^2}{2(2n-1)(c3)S_e w} + \dfrac{(n-1)f}{(c3)S_p}}$$

Rearranging, the pattern efficiency (PE) for the lands becomes

$$PE = \frac{(4n-2)\,fW\,S_e}{(4n-2)\,fW\,S_e + W^2 S_p + (4n^2 - 6n + 2)\,fwS_e} \tag{1.4}$$

The last term in the denominator of Eq. 1.4 may be omitted with an error of less than 2% for $n = 5$, but the error rises rapidly if n is greater than 10.

A plot of Eq. 1.4 for the example values is shown in Fig. 1.9 where length of the plowed land is f. The peak efficiency for this field of 12 ha [29.65 a] appears to be about 5 back furrows or 9 lands.

Rather than plotting a curve for each field situation, the optimum number of lands, as specified by the number of back furrows, can be expressed by Eq. 1.5. (See Appendix B.)

$$n = 0.5 + \sqrt{\frac{W^2 S_p}{4\,fw\,S_e}} \tag{1.5}$$

Using Eq. 1.5 on the example data in Fig. 1.9 gives the optimum number of back furrows as 5.24. Yet, because of the flatness of the curve near its peak, little loss in pattern efficiency exists with significant variation in the number of back-furrowed lands. Eq. 1.6 gives the range in back-furrowed lands for a permissible reduction in pattern efficiency.

$$n_{1,2} = \frac{W^2}{2w}\left(\frac{1}{Z}-1\right)+0.75$$

$$\pm\sqrt{\frac{W^2}{4w^2}\left(\frac{1}{Z}-1\right)^2+\frac{W}{4w}\left(\frac{1}{Z}-1\right)-\frac{W^2 S_p}{4\,fw\,S_e}+0.0625}$$

$$(1.6)$$

where the new symbols are

$n_{1,2}$ = double answer defining the range

Z = acceptable field efficiency, less than the optimum

In the example data of Fig. 1.9, the pattern efficiency is optimum at $n = 5.24$ and is computed to be approximately 0.95. If a 2% drop in efficiency is acceptable, then $Z = 0.93$ and the range in the number of back-furrowed lands is 3.0–10 according to Eq. 1.6. For this example it is seen that the number of lands is not critical if the extreme values are avoided. Excessive wheel traffic on the soil in the headlands, especially near the ends of the back furrows, can be considered at this point. The upper end of the range should be used if soil compaction is troublesome.

An easier computation is for the optimum width of the lands, Eq. 1.7,

$$\text{opt. land width} = \sqrt{fw\,S_e\,/\,S_p} \qquad (1.7)$$

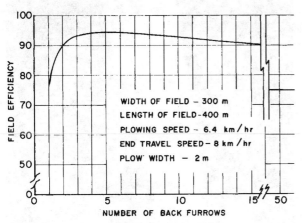

Fig. 1.9. Headland plowing pattern efficiency as function of number of lands.

For the data in Fig. 1.9, the optimum land width is found by Eq. 1.7 to be about 32 m. This width needs to be adjusted slightly to 33 m so that a whole number of lands, 9, will be scheduled. On the other hand, 32 m is exactly 16 trips by the plow. The best management scheme would have a whole number of plow trips in a whole number of lands.

One-way oriented implements other than moldboard plows, such as side-mounted cutterbar mowers and offset pull-type harvesters, can be modeled with some modification by Eq. 1.5 and 1.6. For harvesting implements there is no dead-furrow finishing time; but extra time (and crop damage) may occur as the lands are opened with the initial travel to break-through the standing crop. If one can equate the extra time to break-through to that gained by not requiring a finish trip, Eq. 1.5, 1.6, and 1.7 are applicable without modification. In such an instance, n is the number of break-throughs required.

Headland patterns are also used for self-propelled machines. The width of the lands is determined by the length of an easy turn on the headland and is shown in Fig.1.7 as an overlapping alternation pattern.

To enable comparisons of patterns, the time for plowing the headlands must be added to Eq. 1.4. The headlands are really 2 additional lands and are treated as back-furrowed lands with the average end travel equal to $(1/2)(L-f)/2$. Since the idle travel distance is small, the end travel speed might be slower than the assumed headland travel speed S_e

$$\text{Plowing time to finish headlands} = \frac{(L-f)\,W}{(c3)\,w\,S_p}$$

$$\text{Total time turning at headland ends} = \frac{(L-f)^2}{4\,(c3)\,w\,S_e}$$

Adding these two items to Eq. 1.4, the complete pattern efficiency, PE, becomes

$$(1.8)$$

$$PE = \frac{4(2n-2)\,LWS_p}{4(2n-1)\,LWS_p+2W^2 S_p+4(2n^2-3n+1)fwS_e+(2n-1)(L-f)^2 S_p}$$

Circuitous Pattern, Diagonal Turn Strips

The circuitous pattern with diagonal turn strips is another popular pattern for plowing. For proper operation the turn strips must be exact bisectors of the square corners of the field. A strip equal to the width of the turn strips, s, is left in the center and is finished with the turn strips. It is assumed that turning speed, finishing speed, and plowing speed are equal.

The total plowing time for this pattern is computed by dividing the width of the plow, w, into the net area of the field to get a total furrow distance:

$$\text{land plowing time} = \left(\frac{LW}{w}\right)\frac{1}{(c3)S_p}$$

$$\text{total time lost in turns} = \frac{4\,s}{(c3)S_p}\left(\frac{W-s}{2\,w}\right)$$

$$\text{time to turn 6 turn strip ends} = 3\frac{s}{2}\left(\frac{s}{w}\right)\frac{1}{(c3)S_p}$$

$$\text{time to finish dead furrows} = \frac{4(1.414)W/2+L-W}{(c3)S_p}$$

$$PE = \frac{L\,W}{L\,W+2s\,W-s^2/2+1.828\,w\,W+w\,L} \qquad (1.9)$$

It is assumed that the turn strip, s, will be a whole multiple of w to avoid inefficient finishing travel. The time to lay out the diagonal turn strips is omitted. These turn strips should be marked off with a shallow furrow before starting to plow the field. Potentially, the time for this marking process is about the same as the dead-furrow finishing time.

Continuous Pattern

The efficiency of the continuous pattern depends greatly on quick turns at the headlands. E. S. Renoll (Auburn University) suggests that headlands should be smooth and wide enough to permit an easy turn in a few seconds. Narrow headlands will increase the turning time if the forward speed must be slowed to accomplish short turns, but they will require less time to process when field operations are being concluded. A headland of 2w width provides adequate room for high-speed turns yet limits finishing travel to two passes.

Wide maneuverable machines can usually make a pivot turn on a headland (Fig. 1.1). Unlike other vehicles, farm tractors and self-propelled implements have individual wheel brakes to assist in making sharp turns.

The time loss for a continuous pattern, using a two-way plow, involves only the turning time, t, at the headland. Since this turn requires little travel and

may involve stopping and backing, the time loss is best expressed as seconds per turn. The headlands are finished in a continuous pattern also.

$$\text{Total time for turning} = \frac{t}{3600}\left(\frac{W}{w}+\frac{L-f}{w}\right)$$

$$\text{Total time for plowing} = \frac{L\,W}{w\,(c3)\,S_p}$$

$$PE = \frac{L\,W}{L\,W+(c3)\,t\,S_p(W+L-f)/3600} \qquad (1.10)$$

Notice that the pattern efficiency is not affected by the width of the plow, w.

Circuitous Pattern, Rounded Corners

The circuitous pattern would appear to have the greatest field efficiency as the operation is continuous (Fig. 1.10). However, an unprocessed, crescent-shaped area is left at each turn. Fig. 1.11 illustrates

Fig. 1.10. Aerial view of continuous field pattern.

the effect of a large turning radius, r, of an implement of width, w, as it processes a π/2-rad [90°] corner. Note that the turn must begin w units earlier each round. The shorter the turning radius and the wider the implement, the more able it is to negotiate the turn. But even if the implement were able to pivot about one side of its effective cut, an area would still be missed.

The unprocessed areas left by this pattern may be unimportant considerations for some tillage operations but for plowing their effect is considerable. The area of one of these missed crescent-shaped areas is given by Eq. 1.11.

$$\text{crescent area} = 0.215 \, w \, (2r - w) \text{ units}^2 \qquad (1.11)$$

In a rectangular field with dimensions L by W, there are W/(2w) rounds; and with 4 wasted areas per round the total area wasted is 0.43 W (2r − w). For example, if W = 1000, r = 16, and w = 4, the area lost is 12,040 units². Note that the length of the field is not a factor in the amount of area lost but does have an effect on the proportion of area wasted.

If L equals	% of area wasted is
1000	1.20
2000	0.60
4000	0.30
8000	0.15

To eliminate the unprocessed area in a rounded corner pattern, extra passes with the implement must be made in a manner such that the actual outside radius of the turn is equal to or greater than the minimum turning radius of the implement. Fig. 1.12 shows the necessary pattern for accomplishing the turn with no missed areas. It is interesting that neither the size of the implement, w, nor its turning radius, r, affects the need for additional passes—only the angle of the corner is significant. Variations in crescent width and the ratio of extra passes for common angles are:

Corner angle in			Ratio of extra
rad	deg	Crescent widths	passes to rounds
2.36	135	0.082 w	5 to 61
1.57	90	0.414 w	12 to 29
1.05	60	1.0 w	1 to 1
0.79	45	1.60 w	8 to 5
0.52	30	2.86 w	20 to 7

In Fig. 1.12, a ratio of minimum turning radius to width of implement has been set at 4 and the required extra-pass-to-round relationship has been established for the first 5 rounds. For π/2-rad [90°] corners the ratio of 12 to 29 (or 0.414 to 1) is a critical ratio. As soon as the ratio drops below 0.414 to 1, an extra pass is required. When the corner has been turned on round 5, the ratio of extra passes to rounds is 2 to 5 or 0.400. Since this ratio is less than

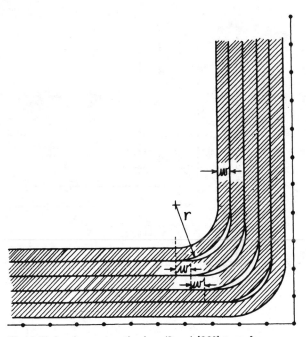

Fig. 1.11. Implement paths in p/2-rad [90°] turn for rounded corner pattern; r/w = 4.

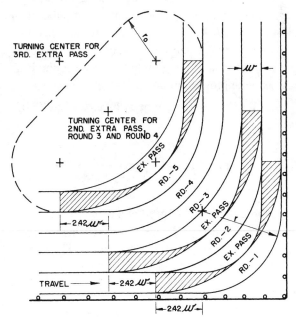

Fig. 1.12. Extra pass requirements for p/2-rad [90°] rounded corners; r/w = 4.

0.414, an extra pass is required before continuing round 5 from this corner. Note that the extra passes must start their curvature a distance of 2.42w prior to the previous point where curvature commenced. The extra pass will overlap some portion of the previous round except at the center of the crescent where the full width of the implement will be utilized.

The time wasted in the circuitous rounded corner pattern will be considered to be only that lost in making the extra passes at the corners. If the implement is capable of a shorter turning radius when disengaged, as with a moldboard plow, the time for extra passes may be shortened. Fig. 1.12 indicates r_0 as a short turning radius with the implement disengaged.

Assuming the average speed of travel for the extra pass is S_p, the total time lost for $\pi/2$-rad [90°] corners equals

$$\left[\frac{270}{360}(2\pi\, r_0) + \frac{r + 2.42w - r_0}{0.707} + \frac{2\pi}{4}(r + 2.42w) \right] \frac{0.414\,W4}{2\,w\,(c3)\,S_p}$$

The pattern efficiency for the rounded corner pattern is approximated by Eq. 1.12:

$$PE = \frac{L}{L + 2.73\, r_0 + 4.48\, r + 6\, w} \qquad (1.12)$$

Circuitous Pattern, 3π/2-rad [270°] Turns

The circuitous pattern with $3\pi/2$-rad [270°] turns is usually started at the center of a land when plowing so that the turns will be on unplowed rather than plowed ground. As far as field efficiency is concerned it is immaterial where the operation commences if one assumes that turning on the plowed ground is as fast as on unplowed ground.

Fig. 1.13 shows the geometry of a $3\pi/2$-rad [270°] turn.

$$\text{Time for a turn} = \left[2r_0 - w + \frac{3}{4}(2\pi) \right] \frac{1}{(c3)S_p}$$

In this case S_p is the speed of the outermost part of the width of action. It should be realized that the average speed of the implement is less than S_p. The pattern efficiency for the $3\pi/2$-rad [270°] turn pattern is given by Eq. 1.13.

$$PE = \frac{L}{L - 2w + 13.42 r_0} \qquad (1.13)$$

Obviously some additional time will be required to locate the position and length of the initial back furrow. While these times are not included in Eq. 1.13, adequate time spent initially will reduce the finishing time at the edges of the field. This pattern is a good plowing pattern to alternate with the circuitous types that throw soil away from the center of the field.

Table 1.2 compares the plowing pattern efficiencies for a square field of 16 ha [40 a] in area. Patterns that require the least turning time exhibit higher field efficiencies. The alternation pattern and the circuitous pattern with square corners are omitted as they do not apply to moldboard plowing. The continuous pattern can be realized for moldboard plowing with a two-way plow.

Tillage implements other than moldboard plows make good use of the continuous pattern if their turning radii are no greater than the effective width of the implement. Good use is also made of the rounded corner, circuitous pattern, as the extra pass problem occurring in plowing may be replaced with diagonal finishing trips. With these implements the efficiency for the rounded corner pattern can be greater than for plowing. The only wasted time is that spent finishing the diagonals where previously processed ground is reworked. (Fig. 1.11.)

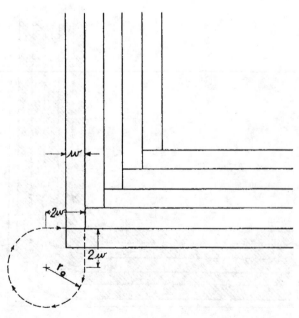

Fig. 1.13. Geometry of 3 π/2-rad [270°] turn; r_0 = 2w.

TABLE 1.2. Plowing Pattern Comparisons, Square Fields

Pattern	Comments	Pattern Eff., %
Headlands	f = 394 m [1292 ft] 6 m [19.6 ft] headlands n = 5 (optimum) S_e = 8.0 km/hr [5 MPH]	93.0
Continuous	t = 10 s f = 394 m [1292 ft]	95.7
Circuitous,	s = 10 m [32.8 ft] diagonal turn strips	94.8
Circuitous, rounded corners	$\pi/2$-rad [90°] corners r_0 = 3 m [9.8 ft] r = 6 m [19.6 ft]	91.9
Circuitous,	r_0 = 6 m [19.6 ft] $3\pi/2$-rad [270°] turns	84.0

Note: L = W = 400 m [1312 ft]; S_p = 6.4 km/hr [4 MPH]; and w = 2 m [6.56 ft].

Row crops may be planted in any of the plowing patterns, but the continuous pattern is most used because of the need to supply the planter at the headlands and because the rows are commonly spaced from a boundary. If the rounded corner pattern is used for planting, the pattern efficiency is 100%, but the missed crescent-shaped areas will reduce the effective area of the field.

Both row crop cultivation and harvesting patterns are dependent on the prior planting pattern. The alternation patterns are possible only with tractor-mounted or self-propelled equipment and continuous or headlands planting patterns. The alternation pattern efficiency can surpass other types of headland patterns when the easier turn permitted by processing alternate sets of rows results in less time for turning.

When the yield of a field is large, consideration for the problems in harvesting should override planting pattern considerations. Ideally the harvesting pattern should be such that the first half of the travel is away from the field gate and the last half toward the gate. Thus each binful or wagon load is released at a point requiring the least field travel for transportation equipment.

One optimum harvesting pattern follows planting patterns that use a headland. If the size of the field and the crop yield are balanced, the harvester makes a whole number of rounds per unloading and always empties a full load on the headland nearest the gate. Such ideal conditions are rare.

For large fields, consideration should be given to dividing the fields into lands. For example, if corn is

yielding 8 t/ha [127 bu/a], a 4-row corn combine with a 1.6-t [63-bu] grain tank is filled after only 500 m [1640 ft] of travel. Lands 250 m [820 ft] in length are needed. Fig. 1.14 illustrates the division of a field into lands of efficient length. The turn strips may be left open, planted to an early maturing crop, or planted to the main crop and harvested first. An alternative to dividing the fields into lands is to use larger wagons and/or use extension sides on the grain tank.

The circuitous pattern with diagonal turn strips should not be overlooked as a possible planting pattern for row crops because of its potential benefits for harvesting. After the diagonal turn strips are harvested the field is opened to effective use of transport equipment. The harvesting starts in the center of the field. The transport equipment is thus able to go to the center of the field and proceed directly to a meet with the harvester. Such a pattern eliminates traveling around the field to the far side to pick up a load, which is a 1 in 4 chance if harvesting starts from the boundary. Crops having high tonnage yields such as silage crops and sugar beets will profit most from this pattern.

For solidly planted crops such as small grains and hay crops, the circuitous pattern with square corners is worked from the outside in. The ability to keep the

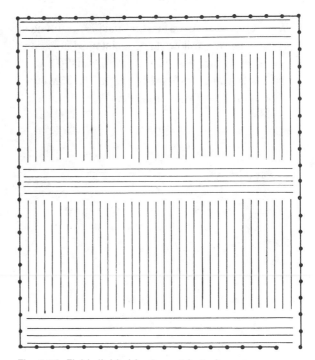

Fig. 1.14. Field divided by turn strip to improve harvesting pattern efficiency.

corners square is of prime importance in keeping the pattern efficiency high; otherwise, the field will not finish out evenly and extra time will be spent in processing the odd-shaped areas remaining.

Field Shape

The field efficiency for irregularly shaped fields is expected to be significantly less than for rectangular fields because of excessive turning time. Even if the irregular fields are straight-sided the ratio of turning time to operating time will be high.

All the patterns illustrated in Fig. 1.7 are possible patterns for plowing irregular fields. The continuous pattern used with a two-way plow has the greatest efficiency for contour work and other irregular shapes. If the headlands pattern is used, the headlands should be selected to be as perpendicular to the back furrows as possible. The $3\pi/2$-rad [270°] circuitous type pattern is almost impossible to start from the center of an irregular field, but the pattern can be used if started from the boundaries. The actual degrees of turn will probably range widely on each side of $3\pi/2$-rad [270°].

Figure 1.15 indicates a combination of the turn strip and rounded corner patterns for moldboard plowing a very irregular field. This pattern follows principles listed here that were established by Leo Ahart (Dow City, Iowa) in 1922.

1. When corners are more than $\pi/2$-rad [90°], a rounded corner pattern is followed and maintained unless the corner angle sharpens to less than $\pi/2$-rad [90°], as in the lower right-hand corner of Fig. 1.15, whereupon a turn strip is established.

2. Uniform-width turn strips are marked off with a shallow plow furrow on each side of the bisector of the corner angles. In corners where the angles change rapidly, only short mark furrows are plowed at a time.

3. It is necessary to continue the pattern even when short furrows result, as indicated near the center of the field in Fig. 1.15. Subsequently, the turn strips may be plowed as lands with either back furrows or dead furrows. If the turn strips have been carefully laid out with a constant width, they will finish out evenly with little time loss.

Ahart sold a device to aid in accurately bisecting the angles. A small length of cardboard was held out at a distance determined by a taut string knotted through the cardboard and held in the teeth. The width of the angle was measured by the scale marked on the cardboard and the bisector determined by dividing the measured width by two. A prominent feature of the landscape was noted to indicate the direction of the bisector. The mark rows were then laid out about 4.5 m [15 ft] on each side of the bisector—all without leaving the tractor seat.

Long fields can improve pattern time efficiencies significantly. In a study of two-row cultivation, E. S. Renoll (Auburn University) found sizable differences in the proportion of turning time required for irregular fields with different row lengths:

	Field A	Field B
Range in row length, ft	400-165	1060-1000
Turning time, % total time	20	3
Field capacity, a/hr	1.9	2.9

The effective field capacity of the cultivator increased over 50% when operated on longer row lengths. Carried to a ridiculous extreme, the most efficient shape is the field only as wide as the implement; thus no time is lost in turning.

Fields of irregular shapes have inefficiencies related to headlands at an angle to machine travel (Fig. 1.16). A machine having a swath width, w, approaches the headland at angle A. To completely cover all the ground, the implement must cover all the cross-hatched areas too. In addition, the length of the turn is increased by w/tan A over that of the π-rad [180°] turn. This extra travel causes losses in time and effort—and seed and fertilizer in the case of seeding machines. In row crop production this loss is called *point-row loss*.

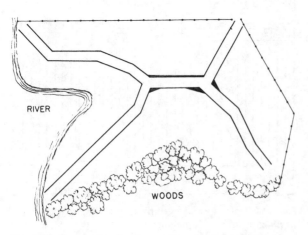

Fig. 1.15. Turn strip plowing pattern for irregular field.

The cross-hatched area is processed needlessly again when the headland is processed. The total area of this loss for a complete turn is w²/tan A. This relationship shows

1. The wasted area approaches zero as A approaches a right angle, but is extremely high if A is π/6-rad [30°] or less.
2. The loss varies as the square of machine width. The loss for a 4-row machine is 4 times as great, and the loss for an 8-row machine is 16 times as great, as that for a 2-row machine.

The total area of angled headland loss depends on the relative size of the implement to the field and on the field pattern. For example, a triangular 16-ha [40-a] field has a length of 800 m [2624 ft] and a width of 400 m [1312 ft], Fig. 1.17. Should the rows or swaths be parallel to side L, the total number of trips across the field would be W/w and the total wasted area would be the number of angled turns, W/(2w), times the wasted area per turn, w²/tan A. Eq. 1.14 shows the total area of wasted angled headland.

$$AHL = \frac{w\,W}{2\tan A} \qquad (1.14)$$

where AHL = area wasted because of angled headland, m² [ft²]

w = width of machine, m [ft]

W = dimension of field perpendicular to machine travel, m [ft]

A = angle of headland (Fig. 1.16), rad [deg]

L = length of field, m [ft]

Substituting the field dimensions and a machine width of 4 m [13.12 ft] into Eq. 1.14 gives

$$\frac{4\times400}{2\times0.5}=1600\ \text{m}^2\ [17{,}222\text{ft}^2]\ \text{wasted area}$$

This loss, as a fraction of field area, is 1600/160,000 or 1%; that is, a 4-m machine working in this field will operate 1% more for every crop operation every year than in the same size field without angled headlands. Also, 1% of the seed, fertilizer, and chemicals applied to the field will be wasted every year.

In a more general case, angled headlands are found at both ends of a field as in Fig. 1.18 where

the machine swaths are parallel to the hypotenuse. The wasted areas at each end equal w²/tan A, and the angles are arc tan (L/W) for one end and W/L for the other. The number of trips across the field will be approximately X/w where X is W L/√(L² + w²) . The number of turns at each end equals X/(2w); therefore, the total wasted area is w √(L² + W²/2) .

For the 16-ha [40-a] field and 4-m machine used previously, the total wasted area is

$$(4/2)\sqrt{800^2+400^2}=1788\text{m}^2\,[19{,}255\,\text{ft}^2]$$

or 1.1% of the field.

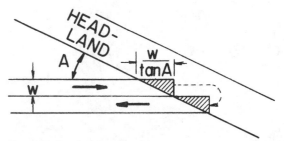

Fig. 1.16. Angled headland loss.

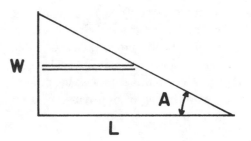

Fig. 1.17. Pattern parallel to perpendicular side.

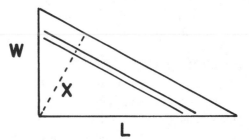

Fig. 1.18 Pattern parallel to angled side.

This comparison illustrates a general truth. If two sides of a field are perpendicular, the machine pattern should be operated parallel to the longer of these two sides for the greater efficiency as only one angled headland is encountered. If a triangular field has no right angles, the machine pattern should be parallel to the longest side to reduce the number of angled headland turns.

The increased travel of machines in making an angled headland turn (Fig. 1.16) increases the total time for turns over that for conventional π-rad [180°] turns. At headland travel speeds of 4.8 km/hr [3 MPH], the increased turning times for a 4-m machine at various headland angles are

Angle		Increased
rad	[deg]	turning time, s
1.57 ($\pi/2$)	[90]	0.00
1.31	[75]	0.80
1.05	[60]	1.73
0.79 ($\pi/4$)	[45]	3.00
0.52 ($\pi/6$)	[30]	5.20
0.35	[20]	8.24
0.17	[10]	17.01

Field Size

Large fields do not necessarily have greater field efficiencies than small fields. If the dimension of the field is doubled and the width of a tillage implement is also doubled, the pattern efficiency remains the same if the forward speed remains the same and the turning radii are identically proportional to width. But if the width of implement remains unchanged, the field efficiency will improve because the proportion of implement operating time increases with respect to its turning time.

Yields

The yield of a field affects the field efficiency of harvesting machines. Heavy yields will usually slow the forward speed. If the yield is caught in wagons or it is necessary to stop to unload the machine's container, sizable time losses can result.

Crop and Soil Conditions

When crop or soil conditions are poor for machine operations, forward speed must usually be reduced. Mathematically, this condition will improve field efficiency but it is not, of course, a desirable operating condition. An extreme time loss can occur

in harvesting if the crop has been windblown to such a position that it can be approached from only one direction.

System Limitations

Few field operations are completely independent of other production operations. Typically, production farming involves the operation of a *system* of machines. As a result the field efficiency of any single machine may be limited by the capacity of other operations in the system. Harvesting operations furnish the best example of a system of machines. The harvester, the transport wagons or trucks, and the unloading equipment comprise a system operation in which the individual machine efficiencies depend on the performance of the system as a whole. Seedbed preparation and planting is a system any time seeding is required immediately after the soil preparation. Even primary tillage can have system considerations when plowing is preceded by stalk chopping and fertilizer application. Hay harvesting, which involves at least partial drying of the hay in the field, is a very important and complex production system.

The timing of one individual operation with respect to another is the feature that defines a machine system. The timing is merely sequential when tillage must precede seeding. It is parallel when two operations must proceed simultaneously, as in the field curing of hay where various implement operations must match the drying capacity. Many midwestern farm managers have found that ignoring the system aspects of production can result in a situation where increasing the size of the corn combine does not increase the corn harvesting rate because the dryer capacity was limiting.

System performance is not as easily determined as individual machine performance. As with individual machine operations, actual system performance may be quite variable. The machinery manager is usually content to estimate only the steady-state operation of the system and does not try to predict the exact state of the system at any instant. Algebra suffices to predict individual machine capacities in Eq. 1.1—1.3, but it is not completely adequate for system estimates. The necessity for comparative timing of individual machines requires a diagrammatic approach for an estimation of system performance.

A cycle diagram aids in field machinery system analysis. Such diagrams indicate the activity of the whole machine system. Fig. 1.19 is a complete cycle

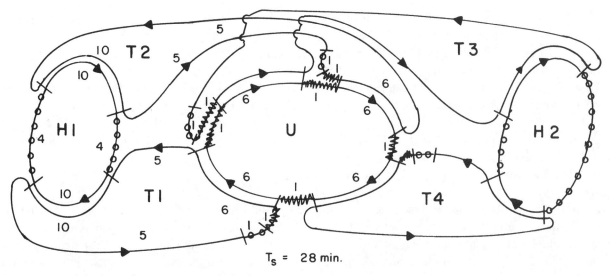

Fig. 1.19. Cycle diagram for forage harvesting system using truck transports.

diagram for an operation composed of two harvesters, H1 and H2; four trucks, T1, T2, T3, and T4; and one unloader, U. The activity of each machine is represented by a closed loop making up a complete time cycle. A zigzag line denotes support time (time necessary to the operation but not actually productive time). A line with superimposed circles indicates idle time. The arrows indicate the progression of clock time and the sequence of events. Fig. 1.19 models a system in which 10 min are required to load a truck that is driven alongside (accompanying, not attached to) the harvester. Field to unloader travel time is 5 min each way, and the loader requires 6 min to unload the truck with 1 min of support time required to position the truck before commencing the unloading operation. The numbers between the hash marks indicate the time in minutes spent on that portion of the cycle. No attempt is made to make the length of the intervals proportional to the time. The system cycle time is T_s.

Development of a cycle diagram is guided by the following principles.

1. All machine cycles must add up to the same total time, T_s.
2. Idle times are permitted only after the completion of the immediate specific operation.
3. Cycle times and diagrams need to be computed only for a representative cycle of each machine type used in the system for a steady-state solution.
4. Transport times must be determined for average travel distance and speeds to get an average system performance.
5. The system capacity is limited to that of the cycle with zero idle time.

Definite steps are followed in the development of a cycle diagram.

1. Sketch the cycles to show the proper machine interrelationships.
2. Mark the productive and support times along the cycles.
3. Sum the times *required* for each cycle. For example:

   Unloader, $6 + 1 + 6 + 1 + 6 + 1 + 6 + 1 = 28$ min
   Harvester, $10 + 10 = 20$ min
   Transport, $10 + 5 + 1 + 6 + 5 = 27$ min

   The longest cycle time (unloader) defines the system time, T_s, of 28 min.

4. Add idle time to the other cycles to bring their total times up to T_s. The harvester will have $28 - 20$ or 8 min of idle time in its cycle (4 min between each truck). The trucks will have $28 - 27$ or 1 min idle time in their cycles.

Four loads are delivered during one cycle of Fig. 1.19. The steady-state system capacity would be predicted as 8.57 loads/hr. The labor efficiency, with one worker required for each machine, is 8.57/7 or

1.22 loads/work · hr. Should the field efficiency of the harvester normally have been 80%, the 10 min for loading in the H1 and H2 cycles could have been broken down into 8 min of theoretical field time and 2 min for turns, adjustments, field maintenance, and repair. When the system operation is considered, the 4 min of idle time must be included, which means the actual field efficiency is 8/(8 + 2 + 4) or 57%.

The cycle diagram is more helpful when the analysis considers an attached wagon and shuttle tractor scheme since the actual system time required is not obvious. The attached transport system requires support times of 2 min for hitching a wagon to the harvester and 1 min for hitching a wagon to a shuttle tractor. Fig. 1.20 shows only half of the complete cycle diagram. The portion of the figure relating to H2 merely repeats H1 and need not be shown. Note the requirement that the shuttle tractor must return an empty wagon to the harvester before it can haul in a loaded one.

Analysis of cycle times shows that the shuttle tractor is the limiting operation in Fig. 1.20.

H = 10 +2 + 10 + 2 = 24 min
U = 28 as in Fig. 1.19
W = 10 + 1 + 5 + 1 + 6 + 5 + 2 = 30 min
T = 6 + 5 + 1 + 5 + 1 + 6 + 5 + 1 + 5 + 1 = 36 min

System cycle time is 36 min and system capacity is 6.66 loads/hr. The labor efficiency in this scheme is higher than for the scheme in Fig. 1.19 since only 5 workers are needed. The labor efficiency is 6.66/5 or 1.33 loads/ work · hr. The field efficiency of the harvester is 8/(8 + 2 + 8) or 44.4%.

Field efficiencies can be estimated more accurately when the complete machinery system is considered.

Breakdowns

Breakdowns are field stoppages due to sudden failure of a part. The expected repair time for breakdowns is not usually included in the calculation of a predicted field efficiency, but such time losses do interfere with machine performance. The probability for and the lost time due to breakdowns can be considerable. A probability number is the decimal ratio of the number of times a breakdown is observed to the total number of observations.

Table 1.3 lists some data obtained from a 1967 survey of more than 1500 midwestern farms. With the possible exception of cultivation, the probability for breakdowns and time loss increases with the size of the farm operation. The complex, many-component, harvesting machines and the heavily stressed tillage machines have the greater probability for breakdown.

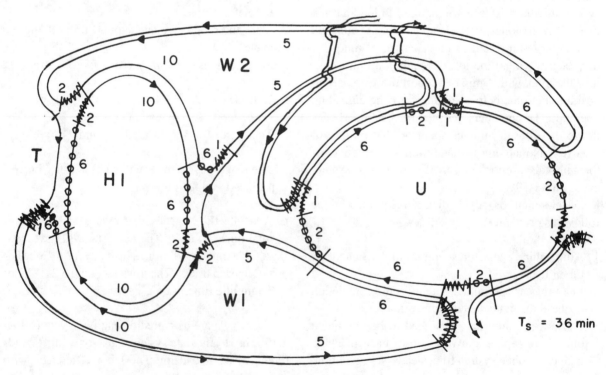

Fig. 1.20. Cycle diagram for system in Fig. 1.19 using wagon transports.

TABLE 1.3. Midwestern Equipment Breakdown Data (probability, downtime, expected occurrence)

		Planting			Harvesting	
Size of Farm	Tillage	Corn	Soybeans	Cultivating	Corn	Soybeans
More than 40 ha[100 ac]						
annual probability	0.196	0.157	0.068	0.081	0.379	0.239
hr/breakdown	10.7	4.4	3.7	7.0	10.7	7.7
More than 120 ha [300 ac]						
annual probability	0.238	0.199	0.089	0.100	0.483	0.308
hr/breakdown	11.7	4.6	4.1	6.3	11.9	7.9
More than 200 ha [500 ac]						
annual probability	0.285	0.255	0.085	0.085	0.570	0.320
hr/breakdown	13.6	5.3	3.7	5.6	12.3	8.2
Expected breakdown interval, ha [ac]	372 [920]	304 [752]	397 [980]	899 [2220]	126 [310]	112 [276]

Note that soybean combining is more subject to breakdown than corn combining when these breakdowns are compared on an area basis.

The data in Table 1.3 can be used to estimate the expected breakdown time. Consider a corn planter used on farms of over 120 ha [300 a]. The expected breakdown time would be 0.199 ×4.6, or 0.92 hr/yr. Assuming average use and size of planter from Appendix E, the percent of lost time due to breakdowns each year during corn planting would be 0.92/43 or about 2%.

The minutes lost during field breakdowns may not be many. But these minutes may be very costly if they occur during a timely operation. Most farm operators think of a machine as being *reliable* if it has little or no breakdown time loss.

Many farm machinery managers attempt to maintain high reliability in their machinery system by using only new equipment. The data from the survey above indicate that such a practice is only partially effective. Fig. 1.21 shows the implied probabilities for breakdowns of specific machines as they vary with age and accumulated use. Old age alone does not seem to mean that a machine is unreliable, but there seems to be a slight increase in the probability of breakdown for planters and plows as their hours of accumulated use increase. The effect is more pronounced for self-propelled combines. During 3500 hr of combine use the probability for breakdown increases 25%. A combine used for both corn and soybean harvesting has a high potential for breakdown at least once during a season. While this implies that breakdowns occur pretty much at random, logic leads one to think that care in operation and maintenance would cause a decrease in breakdowns, an increase in machine reliability, and a higher rate of machine performance.

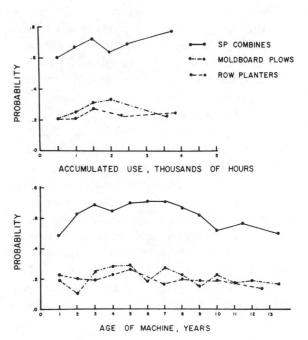

Fig. 1.21. Probabilities for at least one breakdown during season.

Quality Performance

All farm machines handle materials as a part of their performance. Harvesting machines gather and process grain and forage. Seeding machines distribute seed and fertilizer. Tillage machines manipulate soil as a material. The quality of a machine's performance is described by the efficiency with which it handles materials.

A machine's *material efficiency* must describe the completeness with which it handles the material and also the resulting conditions of the processed material. For example, a combine can be reported to have a 90% material efficiency if it is able to place 0.9 t of grain in the tank from an area yielding 1 t. If only quantity is important this report of material efficiency is adequate; but if damaged grain or foreign matter in the tank reduces the value of the harvest, a quantity basis alone is not adequate to define material efficiency. Yet the quality of a harvested product depends only partially on how carefully it was handled by the machine. Furthermore, damaged products or contaminated products are not necessarily worthless. The most realistic measure of material efficiency must be the reduction in the value of a material after being handled by a machine compared to the value it would have with no material losses or deterioration.

Consider the material efficiency of a hay rake. The value of an area of hay before raking might be $10. After hay is windrowed the value might be $15. In some way it is determined that 50¢ worth of material was shattered onto the ground and that a perfectly raked windrow would not have had so many leaves exposed on the windrow surface resulting in a loss in vitamin A worth 30¢. The material efficiency would thus be $5.00/$5.80 or 86%. While this *economic concept* of material efficiency is difficult to apply, one should realize that such a concept is the only true measure of a machine's material efficiency.

Material efficiencies for tillage implements are impossible to calculate unless dollar values can be assigned to different physical conditions of the soil. Seeding machine material efficiencies depend on the value of the seed that is damaged and on the loss associated with not producing a perfect seeding pattern. Cultivator material efficiency must evaluate such intangibles as soil placement, weed kill, soil aeration, etc.

Performance Testing

The prospective user of farm machinery needs to have quantitative measures of the area and/or material capacities of a particular machine, as testing the machine before buying is usually impossible; consequently, the user must depend on the manufacturer's data or on results obtained by experiment stations or testing agencies. In Europe, government testing stations for farm implements are quite common. In the United States, it has been generally agreed that no one central agency could evaluate machine performance with any degree of significance because of the wide variation in crop and soil conditions. The situation would appear to be most unfavorable to the farmer; yet it is a consequence of the complexities of

agriculture and is not a conspiracy against the consumer of farm machinery.

An example of a regional farm machinery testing program is that of the Prairie Agricultural Machinery Institute (PAMI) sponsored by the provinces of Saskatchewan and Manitoba in Canada. PAMI maintains a working relationship with the Alberta Farm Machinery Research Center (AFMRC) to provide machinery performance data to all three prairie provinces. Organized in 1974, the program's objectives were evaluation of existing agricultural machinery and development of new and improved machinery for prairie conditions. A typical evaluation report included a description of the machine; the scope of the tests conducted; results of the functional operation of the implement including capacity, ease and effects of adjustments and power requirements; and a history of mechanical and functional failures and changes required during the testing. The test could last for as long as 100 hours or 1000 ha. Safety for the operator and the adequacy of the operator's manual were also discussed in the report. Institute recommendations for changes in implement design and manufacture were followed by a response to the recommendations by the manufacturer.

PAMI work is helpful to machinery managers in predicting machine performance, fuel consumption and power required. The report for a test of a rectangular baler included photographs of sample bales from over 15,000 produced in baling four forage crops; a plot of both instantaneous and average power requirements (Fig. 1.22) and a mechanical history of the test (Table 1.4).

Recently, PAMI has de-emphasized machine evaluation in favor of research and development responses to machine-related items of interest to the many producer associations in Canada. PAMI now has seven major subject matter divisions: Energy and Processing, Soils and Crops, Hay and Forage, Livestock, Harvesting, Tractors, and Engineering and Testing Services.

PAMI information includes videos, books, and publications "Research Update" and "R&D Report." Considerable information is available from the internet site <www.pami.ca.> The mail address is P. O. Box 1050, Humboldt, Saskatchewan, S0K 2A0 Canada. An annual PAMI subscription entitles one to the Institute's publications. Current subscription fees are $20 for Canadian residents outside Manitoba and Saskatchewan and $25 Canadian for U.S. residents.

Caution should be observed when applying the

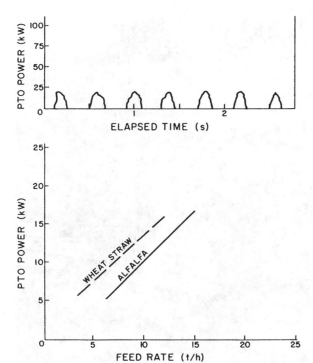

Fig. 1.22. Rectangular baler power measurements by PAMI.

TABLE 1.4. Mechanical History PAMI Test of Rectangular Baler

Item	Operating Hours	Equivalent Bales
Drive train		
Slip clutch began to slip excessively, readjusted to specifications at	58; 88	7,580; 11,900
Feeder assembly		
Feeder carriage roller-to-track clearance checked and adjusted to specifications at	60	8,840
Lost feeder carriage roller cap replaced at	91	12,660
Side play on feeder carriage became excessive; shims added to side wear blocks at	91	12,660
Plunger		
Bent plunger face extension straightened at	9	690
Broken plunger face extension replaced at	32	5,160
Plunger and knife-to-knife clearances checked and adjusted to specifications at	32	5,160
Twine needles		
Broken left twine needle replaced at	9	690
Bale chute		
Broken bale chute chain support bracket replaced at	101	15,460
Cracked bale chute frame at base of deflector arm repaired at	101	15,460

results of machinery tests to specific farm situations. There are wide variations in crops and soils and some variations between individual machines of the same model. But intelligent use of test results such as those from PAMI help reduce the uncertainty facing machinery managers when they seek to select machines for field work.

The USDA National Tillage Machinery Laboratory at Auburn, Alabama, is a facility for research and performance testing of tillage implements, traction, and transport devices. Much of the work is done with an instrumented test car on rails that straddles large bins containing soils ranging from sand to predominately clay. Tests of full-sized tillage tooling (plows, disks, etc.) and of traction devices, such as the crawler tracks in Fig. 1.23, have influenced tractor and implement design since 1936.

Fig. 1.23. Analyzing tractive effort at the USDA National Tillage Machinery Laboratory.

Practice Problems

1.1. Find the theoretical machine capacity in hectares per hour [acres per hour] using the unit factor system described in Laboratory Exercise 1 for a 3-m [9.8-ft] width-of-cut mower observed to travel 100 m [328 ft] in 52 s.

1.2. Show that the constant in the equation for theoretical capacity in acres per hour for metric machines is really 4.047 instead of 4.

1.3. A 5-m [16.4-ft] width-of-cut self-propelled combine makes an average stop of 4 min every time its 2-t [73.3-bu] grain tank is to be unloaded. This stop includes the time for adjustments, lubrication, refueling, and the operator's personal time. The gross yield of the field is 2.1 t/ha [31.17 bu/a]. Material losses are measured as 0.1 t/ha [1.48 bu/a]. The operating speed is 4.8 km/hr [2.98 MPH]. The time for turning on a headland at the ends of the 400-m [1312-ft] field is 15 s. The average actual width of cut is 0.95 of theoretical. Find

a. Theoretical field capacity

b. Effective or actual field capacity

c. Field efficiency

d. Percentage of time loss

e. Material efficiency

1.4. Enter Eq. 1.3 into a programmable calculator or computer for easy and rapid repetitive solutions. Find the most sensitive variable in the equation by determining the largest percentage increase in the effective field capacity upon entering in turn a 20% improvement (increase or decrease in value) for S, w, L, E_w, D, t, and size of grain tank. Use data in problem 1.3 as base values.

1.5. Self-propelled windrowers are capable of reversing their drive wheels independently while their rear wheels swivel. What would be an optimum dimension for the horizontal distance between the cutter bar and the drive axle to permit a perfectly square turn? Express the answer in terms of cutter bar width, w. Would any of the standing grain be backed over by any part of the windrower during this turn?

1.6. A pulled PTO-driven 3-row harvester is used in a row crop planted in 500-mm [19.68-in.] rows. The field is 400 m [1312 ft] long and was planted with 12-row headlands to enable the end turning speeds to equal the field speed of 6.5 km/hr [4 MPH].

a. What is the width of the optimum land for this operation? (Express in terms of the number of implement trips.)

b. If the field is 300 m [984 ft] wide, how many times will it be necessary to break through the field to establish a new land?

c. What is the maximum yield that would just fill a wagon of 4-t [4.4-T] capacity in one trip across the field?

1.7. A round baler spends about 1 min total in stopping forward motion, backing away from the windrow, wrapping the bale with twine, and restarting baling. Assuming a throughput of 9 t/hr [10 T/hr] in a field yielding 2.25 t/ha [1.0 T/a], determine the field efficiency. The windrows have been made from two 3-m [10-ft] swaths and are laid in a circuitous pattern. Neglect any downtime stops. An individual bale weighs 450 kg [1000 lb]. What forward speed is necessary to achieve the throughput capacity?

1.8. The fertilizer hoppers on a planter will hold 90 kg [198 lb] for each row. The required fertilization rate is 340 kg/ha [303 lb/a]. Row spacing

is 1 m [39.4 in.] and the length of the field is 800 m [2625 ft]. Under such conditions, 54.4 kg [120 lb] would have to be added to each hopper at the end of each round. Less time would be wasted if a procedure could be arranged that would allow the fertilizer hoppers to empty further before refilling. Evaluate the following proposals.

a. Go $1^1/_2$ rounds with fertilizer stored at each end of the field.

b. Add extensions to the hoppers to allow 2 complete rounds.

c. Divide the field into lands with row lengths one-fourth the original and place fertilizer supplies at each end of the new lands.

d. Buy more–concentrated fertilizer.

e. Place the fertilizer supply in the middle of the field only.

1.9. Sketch the shape of the cycle diagrams for machine systems made up of

a. One forage harvester, 1 forage blower, and 1 truck

b. One SP baler with bale thrower, 3 wagons, 2 tractors, and an auxiliary engine-driven bale elevator

c. Two SP combines, 3 trucks holding 4 dumps each, 1 grain elevator (show just one transportation loop; indicate others)

1.10. In Fig. 1.19 what unloading time would cause the unloader performance to just match the potential transport performance? What would be the new T_s?

1.11. Suppose Fig. 1.20 is a silage harvesting system. Find the system capacity if the loads carried by the wagons could be increased by 50% and 2 unloaders were used. With the exception of loading and unloading times the other data remain the same.?

POWER PERFORMANCE

A second measure of a machine's economic performance is the effectiveness with which power is applied to accomplish the farm's production objectives. A thorough understanding of the nature of power and its optimum use is essential to good machinery management.

American farmers have purchased many tractors to power their machines. The ratio of the number of people engaged in agriculture to the number of tractors on farms was 11 to 1 in 1930, 2 to 1 in 1950, and approximately 1 to 1 in 1975. In 1990 there were approximately two tractors available for every farm worker in North America.

Tractor power on farms will continue to be an absolute necessity for agricultural production. The total number of workers engaged in farming has dropped to less than 2% of the nation's population, yet total farm production continues to rise. Smaller numbers of tractors and self-propelled implements with greater power ratings will be used in the future if the production from a single farm worker is to continue to increase.

Power is defined as the rate of doing *work*. Work in a technical sense is the application of a force through a distance. The units for mechanical work are force × distance. Power then is work per unit time.

The problem of power units first arose in England with the development of the steam engine. In the latter portion of the eighteenth century, James Watt wished to rate his steam engines in terms of the competition—the horse. He ran a series of tests with average horses and found that a horse could lift 366 lb of coal out of a mine at the rate of 1 ft/s. In other units this was 22,000 ft · lb/min. Watt arbitrarily increased this value by 50% to deliberately underrate his engines. The resulting figure, 33,000 ft·lb/min or 550 ft·lb/s, has been used ever since as the basic unit for horsepower (HP) (Fig. 2.1).

Watt has been memorialized by the SI system, which has named the unit of power a *watt* (W). One watt is the power equivalent of a newton of force expended through a metre of distance in one second. A newton (N) (named after Sir Isaac Newton) is the unit of force required to accelerate one kilogram of mass one metre per second per second. One customary pound of force is approximately 4.448 N. One customary horsepower is the equivalent of 745.7 W, and one kilowatt (kW) is the equivalent of 1.341 HP.

Mechanical power is evident in two forms. *Linear power* occurs when a force is exerted with a linear velocity; *rotary power* is transmitted through rotating bodies. Both forms fit the general relationship of

$$power = \frac{(force \times distance)}{time}$$

An example illustrates the linear form. Suppose a force of 100 N [22.48 lb] is exerted at a velocity of 4 m/s [13.123 ft/s]. Find the power required. In SI units,

$$\underline{400\ W} = 100\ N \times \frac{4m}{1s} \times \frac{1W}{1\ N \cdot m/s}$$

In customary units,

$$\left[\underline{0.536\ HP} = 22.48\ lb \times \frac{13.123\ ft}{1s} \times \frac{1\ HP}{550\ ft \cdot lb/s} \right]$$

Rotary power is a little more complex. The concept of *torque* is often used in these problems. Torque is the product of the length of a pivoted arm (radius of a pulley, pitch radius of a gear, etc.) and a force acting perpendicularly at the free end of that arm. Torque has compound units of newtons times metres, abbreviated as N·m. In customary units the lb·ft is a common torque unit. If a torque is not balanced by an equivalent torque it will cause a body to accelerate its rotation rapidly.

A constant rotary speed has been traditionally reported as revolutions per minute for agricultural machinery. The preferred SI unit is revolutions per second. Machinery managers will need to use both units.

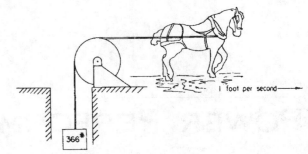

Fig. 2.1. Historical derivation of 1 HP.

Rotary power computation follows the fundamental concept of (force × distance)/time. A force acting on the end of a pivoted arm will travel 2π times the radius of that arm each revolution. Multiplying by the speed of rotation produces the fundamental relationship. Referring to Fig. 2.2, if a belt produces a tangential force, F, of 100 N [22.05 lb] on a pulley having radius, R, of 0.254 m [10 in.] at a speed of 150 rev/min (2.5 rev/s), the power will be, in SI units,

$$\underline{0.4\ kW} = 100\ N \times \frac{2\pi \times 0.254\ m}{rev} \times \frac{2.5\ rev}{s} \times \frac{1\ W}{N \cdot m/s}$$
$$\times \frac{1\ kW}{1000\ W}$$

In customary units,

$$\left[\underline{0.525\ HP} = 22.05\ lb \times \frac{2\pi\ (10/12)\ ft}{rev} \times \frac{150\ rev}{min} \right.$$
$$\left. \times \frac{1\ HP \cdot min}{33,000\ ft \cdot lb} \right]$$

The linear power form is also illustrated in Fig. 2.2. The force, F, produced on the pulley also represents a net tension force in the belt that travels linearly between the two pulleys to transmit power.

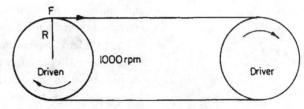

Fig. 2.2. Transmission of power with belt and pulleys.

The linear speed of the belt (distance/time) is $2\pi R \times$ 2.5 rev/s or 4 m/s. Using the linear power form,

$$power = \frac{force \times distance}{time} = \frac{1004}{1} = 0.4 \text{ kW}$$

Fluid power is defined as the product of a weight rate of flow and the resistance to that flow called the *head* of the fluid. Head describes the height of a column of fluid whose mass creates at the bottom of the column a pressure equivalent to that in the flowing system. In general terms,

$$power = \frac{weight \text{ of fluid (force)} \times head \text{ (distance)}}{time}$$

For example, if water weighing 10 kg [22.05 lb] must be pumped up a 100-m [328-ft] hill (through a frictionless pipe) in 10 s, the power required would be, in SI units,

$$0.98 \text{ kW} = \frac{10 \text{ kg}}{10 \text{ s}} \times 100 \text{ m} \times \frac{1 \text{ W} \cdot \text{s}}{1 \text{ N} \cdot \text{m}} \times \frac{1 \text{ kW}}{1000 \text{ W}} \times \frac{9.8 \text{ N}}{1 \text{ kg}}$$

In customary units,

$$\left[1.31 \text{HP} = \frac{22.05 \text{ lb}}{10 \text{ s}} \times 328 \text{ ft} \times \frac{1 \text{HP} \cdot \text{s}}{550 \text{ ft} \cdot \text{lb}} \right]$$

The pressure of the fluid and not its head is used in most machinery management calculations. The preferred units of pressure are pounds per square inch (psi) in customary units and pascals (Pa) in SI units. A pascal is the pressure of one newton of force over an area of one square metre. Since a pascal is a very small unit, the kilopascal (kPa) is commonly used. The kilopascal is also small, being equivalent to 0.145 psi. The bar, 100,000 Pa or 100 kPa, is sometimes used as a convenient unit of pressure as it is numerically close to one atmosphere of pressure.

Tractor Power

Tractors deliver power in several ways. Pulled or towed implements are powered through the traction of drive wheels and the pull or draft from the drawbar. Rotary power is obtained from the power take-off

(PTO) shaft or from a belt pulley. Both linear and rotary power can be produced by a tractor's hydraulic system. Some implements require electric power from tractors.

The tractor power equations (2.1, 2.2, 2.3, and 2.4) are convenient formulas with the necessary unit factor conversions included in the numerical constants.

Drawbar Power (DBP)

$$DBP = \frac{FS}{c} \tag{2.1}$$

where DBP = drawbar power expressed in kW [HP]
F = force measured in kN [lb]
S = forward speed, km/hr [miles/hr or MPH]
c = constant, 3.6 [375]

PTO Power (PTOP)

$$PTOP = \frac{2\pi \text{ FRN}}{c} = \frac{2\pi \text{ TN}}{c} \tag{2.2}$$

where PTOP = PTO power expressed in kW [HP]
F = tangential force, kN [lb]
R = radius of force rotation, m [ft]
N = revolutions per minute (rpm)
T = torque, kN·m [lb·ft]
c = constant, 60 [33,000]

Hydraulic Power (HyP)

$$HyP = \frac{pQ}{c} \tag{2.3}$$

where HyP = hydraulic power, kW [HP]
p = gage pressure, kPa [psi]
Q = flow rate, L/s [gal/min]
c = constant, 1000 [1714]

(Equation 2.3 applies to gaseous and air flows as well as liquid flows. Note that the density of the fluid need not be known.)

Electric Power (EP)

$$EP = IE \tag{2.4}$$

where EP = electric power, W
I = electron flow rate in amperes, A
E = electric pressure in volts, V (DC)

Rolling Resistance

The operation of equipment over typical farm ground involves concepts of rolling resistance and weight components. Rolling resistance is the force required to keep the equipment moving at a constant speed and is proportional to equipment weight. This force is needed to provide the energy required to deflect rubber tires, to compress or push aside soft soil, and to overcome wheel- and axle-bearing friction. The term *coefficient of rolling resistance* is defined as the ratio of the horizontal force (draft) required to pull a loaded wheel over a horizontal surface to the vertical force on that wheel's axle. In customary units this ratio is pull/weight. In the SI system the mass of the load on the wheel must first be converted to force units by multiplying by g, the acceleration of gravity (9.807 m/s^2).

When equipment operates on a slope, the weight no longer acts perpendicularly to the slope through the wheels and axles but vertically at an angle equal to the angle of the slope. Rolling resistance is reduced because the weight on the wheels is reduced.

Slope is defined in this presentation as the tangent of the angle between the soil surface and the horizontal, expressed as a percentage.

Example

A tractor is pulling a 5000-kg [11,000-lb] loaded wagon up a 10% slope at 10 km/hr [6.2 MPH]. The tractor's mass is 3000 kg [6600 lb].

A. Find DBP if the coefficient of rolling resistance is 0.05 for all wheels. Refer to Fig. 2.3 and enter values found.

   1. Angle a = arc tan $0.1 = 0.1$ rad [5.71°]

   2. Components of weight
     a. Perpendicular to the slope

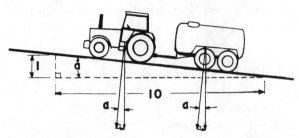

Fig. 2.3. Effects of weight on equipment moving up slope.

       wagon: $5000 \times 9.807 \cos 0.1 = 48.790$ kN
         [$11,000 \cos 5.71° = 10,945$ lb]
       tractor: $3000 \times 9.807 \cos 0.1 = 29.274$ kN
         [$6600 \cos 5.71° = 6567$ lb]
   b. Parallel to the slope
       wagon: $5000 \times 9.807 \sin 0.1 = 4.895$ kN
         [$11,000 \sin 5.71° = 1094.43$ lb]
       tractor:$3000 \times 9.807 \sin 0.1 = 2.937$ kN
         [$6600 \sin 5.71° = 656.7$ lb]
   c. Fill in values in the proper blanks in Fig. 2.3.

   3. Rolling resistance
       wagon: $0.05 \times 48.79 = 2.4395$ kN
         [$0.05 \times 10,945 = 547.25$ lb]

   4. Drawbar pull
     $2.4395 + 4.895 = 7.3345$ kN
     [$547.25 + 1094.43 = 1641.68$ lb]

   5. DBP = 7.3345 kN $\times 10/3.6 = 20.37$ kW
     [$1641.68 \times 6.2/375 = 27.14$ HP]

B. What is the additional power required to move only the tractor up the slope as compared to level ground? Use the data from the folling resistance example and from Fig. 2.3.

   1. For flat surface
     a. Rolling resistance:
       $0.05 \times 3000 \times 9.807 = 1.471$ kN
       [$0.05 \times 6600 = 330$ lb]
     b. Power: $1.471 \times 10/36 = 4.085$ kW
       [$330 \times 6.2/375 = 5.456$ HP]

   2. For slope
     a. Rolling resistance:
       $0.05 \times 29.274 = 1.464$ kN
       [0.05×6567 lb $= 328.35$ lb]
     b. Downhill component of weight: 2.937 kN
       [656.7 lb]
     c. Power:
       $(1.464 + 2.937) \times 10/3.6 = 12.225$ kW
       [$(328.35 + 656.7) \times 6.2/375 = 16.29$ HP]

Increase is 8.14 kW [10.834 HP].

Power Measurement

Power is sometimes defined as energy in motion and as such is invisible. Special apparatus is required to determine the power output from an engine or a tractor. The following sections describe some of the apparatus and the common tests used in power measurement.

Engine Power Definitions

The power output of an engine is a function of the average pressure on the piston head and the engine speed. The average pressure acting during the power stroke of the engine is referred to as the mean effective pressure. *Indicated power* (IP) is the theoretical power an engine should develop from the mean effective pressure existing at the piston head. *Brake power* (BP) is the engine horsepower measured at the flywheel. The numerical difference between IP and BP is the power absorbed by the engine in overcoming friction to run itself (FP).

Engine Testing

Engine testing is accomplished with a dynamometer. The dynamometer may be a *transmission dynamometer* or an *absorption dynamometer*. Most engine tests are conducted with an absorption dynamometer—that is, the dynamometer has within itself the ability to load the engine.

A Prony brake is a simple form of absorption dynamometer that is easily understood. The engine is attached to a large flywheel equipped with an adjustable brake band. The brake band is restrained from turning with the flywheel by a platform scales as shown in Fig. 2.4. Eq. 2.2 may be applied directly to this type of dynamometer, where

F = force read on the scale
R = radius of scale arm
N = rpm

Note that for any one dynamometer all factors in Eq. 2.2 remain constant except F and N. The remaining factors may be combined to form a dynamometer constant, k:

$$k = \frac{2\pi \ R}{c}$$

A simpler power formula may now be written:

BP = kFN   (2.5)

The Prony brake has many disadvantages; an electric or hydraulic dynamometer is a more satisfactory unit.

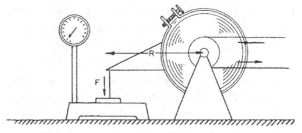

Fig. 2.4. Belt-driven Prony brake dynamometer.

The electric dynamometer may be just a generator. It may be cradle-mounted on bearings or stationary. The armature is connected across a heavy resistance load. The resistance to rotation of the shaft is provided by the electromagnetic forces between the armature and the field coils. The energy is dissipated in the form of heat at the resistance load.

Power may be measured by reading the current output of the armature circuit. In this case the generator characteristics of the dynamometer must be known. When cradled, the reaction force to producing power is measured at the end of the scale arm. In either case the load on the engine is varied by a rheostat placed in the field circuit of the generator.

An eddy-current dynamometer is another type of electric dynamometer that dissipates energy in the form of heat directly into cooling water. The reaction to this dissipation produces the force F contained in Eq. 2.5.

Hydraulic dynamometers are closed-circuit fluid systems that contain a pump, a variable relief valve, and a reservoir. The engine drives the pump and load is applied by closing the relief valve. Either the reaction force is measured or the pressure on the discharge side of the pump is used to estimate the reaction force.

The fuel consumption of an engine, often determined along with power measurement, can be determined in several different ways. One is by measuring the rate at which the fuel is flowing to the engine by some type of flowmeter. Another is by measuring out a volume of fuel and recording the time required to consume this quantity. A third method is to use a measured mass of fuel in the above procedure. In agriculture, fuel consumption is reported on an hourly basis.

Engine fuel efficiency is a ratio of fuel consumption to power output. At least four different efficiencies are computed. The most common and simplest is kW·hr/L [HP·hr/gal]. If different fuels are compared, a kg/(kW·hr) [lb/(HP·hr)] efficiency is most useful.

A ¢/(kW·hr) [¢/(HP·hr)] is helpful in economic analyses. Thermal efficiency is expressed as a percentage indicating the degree of efficiency. The fuel input and the power output are each converted to energy units. The SI system uses the joule (J) as the energy unit while the customary system uses the British thermal unit (Btu). To use Eq. 2.6, the amount of heat energy contained in the fuel must be known.

$$\text{Thermal efficiency, \%} = \frac{P \times c}{FC \times HV} \times 100 \qquad (2.6)$$

where P = power expressed in kW [HP]
      FC = fuel consumption, kg/hr [gal/hr]
      HV = fuel's heating value, kJ/kg [Btu/gal]
      c = constant, 3600 kJ/kW·hr [2545 Btu/Hp·hr]

Tractor Engine Performance

Typical engine performance is shown in Fig. 2.5. The power ratings for the engine alone are shown with dashed lines. The performance is different when the engine is mounted in a tractor and its power determined at the PTO shaft as shown by the solid lines.

The rated speed value of 2200 is the speed point at which the engine governor permits full fuel flow under loaded conditions. Note that the governor action limits the engine's power output above the rated speed even though the engine is capable of greater power output. This arrangement causes a more or less constant speed-power output for a tractor as long as the load remains below a critical value. For ungoverned engine applications, note that the power rating is usually stated as 90% maximum horsepower. Continuous power is rated as 80% maximum. The nearly flat torque curve

indicates the ability of the engine to accept an overload by reducing speed but to retain its torque or twisting power on the crankshaft. Maximum fuel efficiency occurs after the engine load increases beyond the rated load, the load giving maximum power at rated speed.

A PTO dynamometer may be used to indicate engine power and to serve as a load while adjusting the engine. These units may be friction brake type or hydraulic pump type. A hydraulic dynamometer used in tractor service shops is pictured in Fig. 2.6.

Tests of the tractor engine alone are seldom of direct use to machinery managers as the usable power from a tractor comes from the PTO, the drive wheels, the alternator, and the hydraulic pump. Fig. 2.7 indicates the mechanical power transmission efficiency for a tractor on a concrete surface. The PTO rating of tractors has come to be the standard for comparison of performance as it avoids the variables associated with the tractive effort between the wheels and the ground surface.

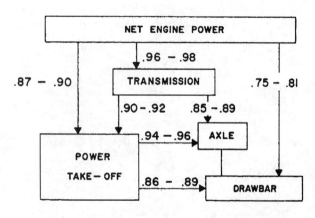

Fig. 2.6. PTO dynamometer.

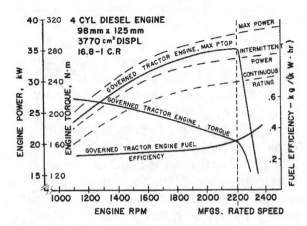

Fig. 2.5. Typical diesel engine performance.

Fig. 2.7. Power efficiencies for tractor on concrete.

Tractor Tests

The power performance of tractors is the most important information item needed by the farm machinery manager. The tractor is the base of the machinery system. It is often the most used machine on the farm and is frequently the most expensive. Economic farm management requires a careful matching of tractor capability to the farm's power needs. Reliable unbiased performance data, based on tests of sample models, is fundamental to good farm machinery management.

The need for reliable performance data arose in the early days of tractor manufacturing. In 1919, W. F. Crozier, a farmer as well as a legislator, introduced mandatory tractor test legislation in the Nebraska House of Representatives to provide farmers with unbiased information about tractor performance. The bill passed and the Nebraska Tractor Test Board was created and given the authority to require tractors offered for sale in Nebraska to be tested before being issued a sales permit. The University of Nebraska in Lincoln was selected as the testing agency. In addition to facilities for PTO testing, an oval concrete track is used for drawbar tests. While the law applied only to Nebraska sales, the industry accepted the results worldwide.

The test procedures have evolved over the years to accommodate changes in tractor technology. The test standards were codified by the American Society of Agricultural Engineers (ASAE) in 1937. A joint standard was issued by the ASAE and the Society of Automotive Engineers (SAE) in 1964. Past tests have helped evaluate such innovations as dual drive wheels, turbochargers, radial tires, new transmissions, and the efficiencies of various fuels.

As tractor manufacturers developed worldwide markets, a need arose for an international tractor test procedure to avoid redundant tests in each national market. The Organization for Economic Cooperation and Development (OECD) approved internationally recognized tractor testing procedures and named the Farm and Industrial Equipment Institute (later renamed the Equipment Manufacturers Institute [EMI]) as a "Designated Authority" for OECD testing. In 1986, the Nebraska Legislature voted to accept the results of either the SAE/ASAE or OECD tractor tests. In 1988, the Nebraska test facility began OECD tests, as well as their traditional ones, in the new role as an approved OECD testing station. OECD testing is global. Fig 2.8 reports results from a test conducted in Italy.

Kenneth Von Bargen, Department of Agricultural Engineering, University of Nebraska-Lincoln, reviewed the provisions for OECD testing at the 1988 Winter Meeting of the ASAE. Five different tractor test codes were listed.

Code I	Standard or full test performance code which consists of PTO, drawbar, hydraulic power, 3-point linkage lift, center of gravity, braking, and sound level tests.
Code II	Restricted standard performance code which has the same PTO test but drawbar tests are conducted without ballast.
Code III	A dynamic protective structure test.
Code IV	A static protective structure test.
Code V	A noise measurement test.

Different testing options are available to meet the requirements of different countries in which the tractor model will be marketed and to meet the marketing needs of the manufacturers. An official test report must provide the basic power and fuel performance characteristics of the tractor, the sound levels, and the 3-point hitch lift capability. The official test results must equal or exceed the manufacturer's claims or representations for power, fuel consumption, and sound levels.

Preparation for Testing

The manufacturer selects the tractor to be tested and certifies that it is a stock model. Each tractor is equipped with the common power-consuming units such as power steering, hydraulics, alternator, etc. An official representative of the company is usually present during the test to see that the tractor gives its optimum performance. Additional weight may be added to the tractor as ballast if desired. The static tire loads and inflation pressures must conform to the limitation established by the Tire and Rim Association. (See Table 2.1).

The engine crankcase is drained and refilled with new oil conforming to the specifications in the manufacturer's operator's manual. This manual is also used as a guide in selecting the proper fuel and for routine lubrication and maintenance.

The tractor is operated for several hours prior to testing to provide a representative test after the pistons and bearings "wear-in." Adjustment of the tractor is permitted at this time. Instrumentation for measuring engine rpm, fan speed, temperatures and pressures, and fuel consumption are installed.

PTO Performance

Engine power performance is tested by connecting a dynamometer to the tractor's PTO shaft. During a preliminary run the manufacturer's representative may make adjustment for injection pump volume and timing. These settings must be maintained for the remaining tests. The manually operated governor control is set to provide the high-idle (no-load) engine speed specified by the manufacturer. During the PTO runs an ambient air temperature of approximately 24°C [75°F] is maintained and the barometer reading should be above 96.6 kPa [28.5 in. Hg].

Power is measured at the rated engine speed specified by the manufacturer. Maximum power may be obtained at other than rated speed and is conducted for 2 hours. When the PTO speed for these test differ from the ASAE and SAE standards, an additional run is made at either the 540 or 1000 rpm standard.

The Varying Power and Fuel Consumption test provides a machinery manager with data on fuel efficiencies at part loads. These tests record the fuel consumption and power developed during six test loadings decreasing from maximum to no load.

Drawbar Performance

Drawbar pull tests are performed in all transmission gears between one gear below the one at which 15% drive wheel slip occurs and that for a maximum speed of 16.1 km/hr [10 MPH]. In each test the governor control is set for maximum speed (high idle) and the horizontal drawbar load increased until the maximum drawbar power is obtained. Measurements are taken of pull, speed, power, slip, fuel consumption, and sound. Sound level readings, db(A), are obtained with a microphone located near the right ear of the tractor operator. A bystander sound reading is taken with a microphone placed 7.5 m [25 ft] from the centerline of the tractor which is accelerating from a lower speed to full speed in its top gear. For tractors with mechanical front-wheel drive (MFWD), the operator ear measurements are made with the front-wheel drive both engaged and disengaged.

A second set of tests investigates part load performance. Drawbar loads of 75% and 50 % of the load at rated engine speed are applied in a gear close to 7.5 km/hr [4.6 MPH] and in the gear where

maximum drawbar power was obtained in the previous tests.

Additional tests may be conducted. The manufacturer may wish a test at reduced engine speed, ballast-aided performance, performance with and without MFWD, and alternate tire configurations.

The procedures specify that distribution and total tractor weight will be in accordance with limits set by the tractor manufacturer, tire manufacturer, and roll-over protection certification. The weight shall include full fuel tanks and an 80-kg [175-lb] operator. Front end ballast can only be provided by a standard weight package and/or front tire ballast supplied or recommended by the manufacturer.

Three-Point Hitch Test

The tractor is tested on the same rear tires used during the drawbar tests. The front tires of two-wheel drive and front-wheel drive assist tractors may be of any size or ply offered by the tractor manufacturer as long as they properly match the rear tires. A quick-attaching coupler is used on all Category III and IV hitches and on any tractors on which it is offered as standard equipment; but the quick coupler is *not* considered part of the load lifted by the hitch.

Hydraulic Lift Capacity and Flow

The hydraulic lift capacity is measured in a special test stand. A frame is fitted to the three-point hitch links and measurements of lift capacity are taken at the hitch points and 61 cm[24 in.] behind the lower, horizontal hitch links. The load is generally applied with a hydraulic cylinder and the links move step-wise through the lift range. The number that is reported is 90% of the load which can be carried throughout the lift range.

A second test determines the pressure/flow relationship and performance of the hydraulic system for supplying power to external cylinders and motors. Measurements reported are pump flow rates at minimum pressure at rated engine speed and the pressure and flow at maximum hydraulic power .

OECD reports are available from the Tractor Test Laboratory, P. O. Box 830832, Lincoln NE 68583.

Some interpretation of the OECD tests is required for farm machinery management use. The PTO performance data can be used directly as indicative of the maximum useful power of the tractor (approximately 90% engine power). Variations from these values would be expected if operating air temperatures and pressures were different from those for the test. Weather combinations of low pressure and high temperature or high pressure and low temperature could cause as much as an 8% power decrease or increase, respectively. Decreases in air pressure with altitude can cause power losses of approximately 3%/300 m [1000 ft] altitude. The decreases are even greater at altitudes above a mile.

The drawbar performance data are not so readily applicable to farm machinery management. The tests for rubber-tired tractors are conducted on a concrete track and include only horizontal pulls. Neither condition is representative of the field tractor with integral equipment. The justification for such a procedure is that a standard test is accomplished permitting fair comparisons of different models of tractors.

Examination of the example OECD tests, Fig.2.8, 2.9, and 2.10 can lead to many comparisons of value to a machinery manager. The test data do lead to comparisons of different tractors; but, they also can indicate the effects of management decisions about different configurations of the same model tractor and about operational adjustments.

Fig. 2.9 allows a judgment of the increased performance from MFWD as the report shows performances with the MFWD engaged and disengaged. Engaging the MFWD in 5th gear increased the maximum drawbar power 7% from 25.91 to 27.73 kW [34.75 to 37.19 HP]. The drawbar pull remained essentially the same, but the % slip dropped from 14.84 to 8.49. Comparisons at just one test gear can mislead. Making the same comparison as above in 6th gear shows that engaging the MFWD provides minimal increase in maximum drawbar power, 27.27 to 27.39 kW [36.57 to 36.73 HP], 2.7% reduction in wheel slip, 7.80 to 5.09, and an actual decrease in drawbar pull, 9.12 to 8.85 kN [2051 to 1990 lb]. Keep in mind that these tests are on a concrete surface and are not always indicative of performances on soils.

Fig. 2.9 (including the supplement) also allows a judgment about engine turbochargers. Adding a turbocharger increased the power by an average of 13% and improved fuel efficiency by an average of 8%. (Note the name change of the manufacturer between the two tests.)

Fig. 2.10 indicates the effect of adding ballast and the benefits from reduced engine speed. Comparing maximum drawbar powers in 9th gear at 2200 rpm, adding 982 kg [2165 lb] of ballast to the front of the tractor produced a 2.9% increase in drawbar power, 145.97 to 150.2 kW [195.75 to 201.43 HP]; a 2% increase in drawbar pull, 70.84 to 72.73 kN [15925 to 16238 lb]; and a 29% reduction in slip, 3.28 to 2.34 %. The relatively small increase in power occurs because adding ballast consumes power as more weight has to be moved over the ground. This tractor's engine characteristics give greater performance at an engine speed of 2000 rpm. Comparing the ballasted tests in ninth gear, tested power increased 11.3%, 150.20 to 167.20 kW [201.43 to 224.21 HP], and fuel efficiency increased 6.5%, 3.06 to 3.26 kW·hr/L [15.55 to 16.57 HP·hr/gal].

The ability of a tractor engine to accept an overload is important in most field operations. Listed below the Varying Power and Fuel Consumption test in the Summary reports is a report of engine torque characteristics. The data in Fig. 2.8 reports that the maximum engine torque is developed at 1249 rpm, well below the rated engine speed of 2350. This maximum torque was 25% greater than at rated speed, 179.2 N·m [132 lb·ft]. An additional torque rise percentage is reported at an engine speed intermediate between maximum power and maximum torque. The industry often refers to this engine characteristic as its lugging ability.

Performance of tractors tested on concrete can be estimated for soil surfaces by curves shown in Fig. 2.11. Compiled by Frank M. Zoz (John Deere Waterloo Tractor Works), this chart considers the effects of engine power, ground speed, rolling resistance, ballast or added weight, and the type of implement hitch on the dynamic weight shift from the front axle to the rear. It predicts drawbar pull, drive-wheel slippage, and actual forward speed for three soil surface conditions and for concrete. It is limited to rear-axle-drive tractors with single-drive tires. It assumes proper ballast distribution to ensure adequate steering control at all times. Approximately 25%, 30%, and 35% of the total tractor mass is required on the front wheels for towed, semimounted, and mounted implements, respectively. The ratio of drawbar height to wheelbase is R. Slippage is defined as

$$\frac{\text{advance under no load} - \text{advance under load}}{\text{advance under no load}}$$

for a specified number of drive-wheel revolutions. Ten revolutions of the drive wheel is a common base for measuring the advance.

Figure 2.8.

SUMMARY OF OECD TEST 1610—NEBRASKA SUMMARY 217
WHITE 6045 DIESEL
16 SPEED

POWER TAKE-OFF PERFORMANCE

Power HP (kW)	Crank shaft speed rpm	Gal/hr (l/h)	lb/hp.hr (kg/kW.h)	Hp.hr/gal (kW.h/l)	Mean Atmospheric Conditions
MAXIMUM POWER AND FUEL CONSUMPTION					
Rated Engine Speed (PTO speed—612 rpm)					
45.7 (34.1)	2354	2.75 (10.42)	0.421 (0.256)	16.60 (3.27)	
Standard Power Take-ff speed (540 rpm)					
44.1 (32.9)	2077	2.55 (9.64)	0.404 (0.246)	17.31 (3.41)	
VARYING POWER AND FUEL CONSUMPTION					
45.7 (34.1)	2354	2.75 (10.42)	0.421 (0.256)	16.60 (3.27)	Air temperature
39.7 (29.6)	2415	2.48 (9.37)	0.436 (0.265)	16.04 (3.16)	66°F (19°C)
30.0 (22.4)	2443	2.04 (7.74)	0.477 (0.290)	14.67 (2.89)	Relative humidity
20.1 (15.0)	2460	1.56 (5.91)	0.542 (0.330)	12.89 (2.54)	62%
10.2 (7.6)	2500	1.18 (4.46)	0.809 (0.492)	8.64 (1.70)	Barometer
....	2552	0.87 (3.31)	.....	.....	29.8" Hg (101.1 kPa)

Maximum Torque 132.1 lb.-ft. *(179.2 Nm)* at 1249 rpm
Maximum Torque Rise 29.5%
Torque rise at 1850 engine rpm 16%

DRAWBAR PERFORMANCE
FUEL CONSUMPTION CHARACTERISTICS

Power Hp (kW)	Drawbar pull lbs (kN)	Speed mph (km/h)	Crank-shaft speed rpm	Slip %	lb/hp.hr (kg/kW.h)	Hp.hr/gal (kW.h/l)	Temp.°F (°C) cool-ing med	Air dry bulb	Barom. inch Hg (kPa)
75% of Pull at Maximum Power—Five Hours 10th (1 FH) Gear									
30.3 (22.6)	2400 (10.67)	4.73 (7.62)	2420	3	0.554 (0.337)	12.62 (2.49)	air cld	46 (8)	29.5 (100.0)
MAXIMUM POWER IN SELECTED GEARS									
7th (4NL) Gear									
32.7 (24.4)	5115 (22.75)	2.40 (3.86)	2397	15	0.546 (0.332)	12.80 (2.52)	air cld	41 (5)	29.6 (100.2)
8th (4NH) Gear									
36.5 (27.2)	4340 (19.30)	3.15 (5.07)	2350	8	0.528 (0.321)	13.24 (2.61)	air cld	41 (5)	29.6 (100.1)
9th (1FL) Gear									
38.8 (28.9)	4100 (18.24)	3.55 (5.70)	2350	8	0.500 (0.304)	14.01 (2.76)	air cld	41 (5)	29.6 (100.2)
10th (1FH) Gear									
39.0 (29.1)	3210 (14.27)	4.56 (7.34)	2350	4	0.494 (0.301)	14.14 (2.79)	air cld	43 (6)	29.5 (100.0)
11th (2FL) Gear									
38.0 (28.3)	2455 (10.91)	5.80 (9.34)	2351	4	0.506 (0.308)	13.81 (2.72)	air cld	41 (5)	29.5 (100.0)
12th (2FH) Gear									
37.2 (27.7)	1860 (8.29)	7.49 (12.04)	2350	2	0.518 (0.315)	13.50 (2.66)	air cld	43 (6)	29.5 (100.0)
13th (3FL) Gear									
34.1 (25.4)	1320 (5.88)	9.67 (15.57)	2350	1	0.565 (0.343)	12.38 (2.44)	air cld	43 (6)	29.5 (100.0)

TRACTOR SOUND LEVEL WITHOUT CAB

	dB(A)
Maximum sound level - in 10th (1FH) gear	94.6
Bystander in 16th (4FH) gear	81.4

Location of Test: ISMA Via Milano 43, 24047 Treviglio BG Italy

Dates of Test: February, 1996

Manufacturer: S+L+H S.p.A. V.le F. Cassani 15, 24047 Treviglio BG Italy

FUEL OIL and TIME: Fuel No. 2 Diesel **Cetane No. NA Specific gravity converted to 60°/60° F** *(15°/15°C)* 0.839 **Fuel weight** 6.99 lbs/gal *(0.838 kg/l)* **Oil SAE** 30 **API service classification** SE/CD **Oil consumption for 10 hours** 0.57 lb *(260 gm)* **Transmission and hydraulic lubricant** AKROS Multi 95 fluid **Front axle lubricant** SAE 95 API GL-4

ENGINE: Make S+L+H Diesel **Type** three cylinder vertical **Serial No.** 1732 **Crankshaft** lengthwise **Rated rpm** 2350 **Bore and stroke** 4.134" × 4.547" *(105 mm × 115.5 mm)* **Compression ratio** 17.1 to 1 **Displacement** 183 cu in *(3000 ml)* **Starting system** 12 volt **Lubrication** pressure **Air cleaner** two paper elements **Oil filter** one full flow cartridge **Oil cooler** radiator for crankcase oil **Fuel filter** one paper element **Muffler** vertical **Cooling medium temperature control** air cooled

CHASSIS: Type front wheel assist **Serial No.** 001032 **Tread width** rear 51.8" *(1315 mm)* to 55.1" *(1410 mm)* front 52.8" *(1340 mm)* to 60.6" *(1540 mm)* **Wheel base** 80.9" *(1926 mm)* **Hydraulic control system** direct engine drive **Transmission** selective gear fixed ratio **Nominal travel speeds mph** *(km/h)* first 0.70 *(1.13)* second 0.86 *(1.39)* third 1.11 *(1.78)* fourth 1.37 *(2.21)* fifth 1.76 *(2.83)* sixth 2.17 *(3.50)* seventh 2.78 *(4.48)* eighth 3.45 *(5.56)* ninth 3.85 *(6.20)* tenth 4.77 *(7.68)* eleventh 6.10 *(9.82)* twelfth 7.58 *(12.20)* thirteenth 9.75 *(15.69)* fourteenth 12.08 *(19.44)* fifteenth 15.47 *(24.89)* sixteenth 19.01 *(30.60)* reverse 0.82 *(1.32)*, 1.30 *(2.10)*, 2.07 *(3.33)*, 3.27 *(5.26)*, 4.53 *(7.29)*, 7.20 *(11.59)*, 11.48 *(18.48)*, 18.12 *(29.17)* **Clutch** single dry disc operated by foot pedal **Brakes** wet multiple disc hydraulically operated by two foot pedals which can be locked together **Steering** hydrostatic **Power take-off** 540 rpm at 2077 engine rpm **Unladen tractor mass** 4885 lb *(2215 kg)*

REPAIRS AND ADJUSTMENTS: No repairs or adjustments

REMARKS: All test results were determined from observed data obtained in accordance with official OECD test procedures. The performance results on this summary were taken from OECD tests conducted under the Code I Standard Test Code procedure.

Figure 2.8. Continued.

CENTER OF GRAVITY

Horizontal distance forward from centerline of rear wheels	33.1 in *(841 mm)*
Vertical distance above roadway	30.5 in *(774 mm)*
Horizontal distance form center of rear wheel tread	0.2 in *(6 mm)* to the left

We, the undersigned, certify that this is a true summary of data from OECD Report No. **1610**, Nebraska Summary 217, September 23, 1998.

LEONARD L. BASHFORD
Director

M.F. KOCHER
R.D. GRISSO
G.J. HOFFMAN
Board of Tractor Test Engineers

TURNING ON A CONCRETE SURFACE

Turning radius—with brake applied right 134" *(3.40 mm)* left 140" *(3.56 mm)*
without brake right 150" *(3.80 mm)* left 159" *(4.03 mm)*
Turning space radius—with brake applied right 138" *(3.51 mm)* left 144" *(3.67 mm)*
without brake right 154" *(3.91 mm)* left 162 " *(4.14 mm)*

TIRES, BALLAST AND WEIGHT

		With Ballast	Without Ballast
Rear Tires	--No., size, ply & psi *(kPa)*	Two 14.9R28; 6;23 *(160)*	Two 14.9R28; 6;23 *(160)*
Ballast	--Liquid (total)	None	None
	--Cast iron (total)	340 lb *(155 kg)*	None
Front Tires	--No., size, ply & psi *(kPa)*	Two 11.2R20; 6;23 *(160)*	Two 11.2R20; 6;23 *(160)*
Ballast	--Liquid (total)	None	None
	--Cast Iron (total)	330 lb *(150 kg)*	None
Height of Drawbar		17.8 in *(451 mm)*	17.9 in *(455 mm)*
Static Weight with Operator	--Rear	3165 lb *(1435 kg)*	2845 lb *(1290 kg)*
	--Front	2555 lb *(1160 kg)*	2205 lb *(1000 kg)*
	--Total	5720 lb *(2595 kg)*	5050 lb *(2290 kg)*

THREE POINT HITCH PERFORMANCE (OECD Static Test)

CATEGORY: II
Quick Attach: None
Maximum Force Exerted Through Whole Range: 3830 lbs *(17.0 kN)*
i) Opening pressure of relief valve: NA
Sustained pressure with relief valve open: 2610 psi *(180 bar)*
ii) Pump delivery rate at minimum pressure
and rated engine speed: 12.5 GPM *(47.4 l/min)*
iii) Pump delivery rate at maximum
hydraulic power: 11.1 GPM *(42.0 l/min)*
Delivery pressure: 2180 psi *(150 bar)*
Power: 14.1 HP *(10.5 kW)*

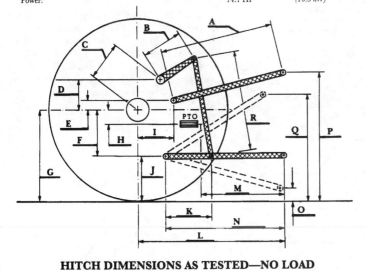

	inch	mm
A	23.3	*593*
B	9.8	*250*
C	14.2	*361*
D	12.4	*315*
E	13.2	*335*
F	6.1	*154*
G	25.2	*640*
*H	-1.2	*-30*
I	15.4	*390*
J	19.1	*486*
K	15.7	*400*
L	38.7	*983*
M	21.4	*543*
N	31.5	*800*
O	7.9	*200*
P	43.1	*1096*
Q	35.8	*910*
R	23.2	*590*

* PTO is above rear axle

HITCH DIMENSIONS AS TESTED—NO LOAD

Agricultural Research Division
Institute of Agriculture and Natural Resources
University of Nebraska—Lincoln
Darrell Nelson, Dean and Director

Fig. 2.9.

NEBRASKA OECD TRACTOR TEST 1747—SUMMARY 263
JOHN DEERE 8400T DIESEL
16 SPEED

Location of Test: Nebraska Tractor Test Laboratory, University of Nebraska, Lincoln, Nebraska 68583-0832

Dates of Test: April 8 - May 1, 1998

Manufacturer: John Deere Tractor Works, P.O. Box 270, Waterloo, Iowa 50704

FUEL OIL and TIME: Fuel No. 2 Diesel **Cetane No.** 50.6 **Specific gravity converted to 60°/60° F** *(15°/15°C)* 0.8471 **Fuel weight** 7.053 lbs/gal *(0.845 kg/l)* **Oil SAE** 15W-40 **API service classification** CD, CE, CF-4 **Transmission and hydraulic lubricant** John Deere Hy-Gard fluid **Total time engine was operated** 29.5 hours.

ENGINE: Make John Deere Diesel **Type** six cylinder vertical with turbocharger and air to air intercooler **Serial No.** *RG6081H040198* **Crankshaft** lengthwise **Rated engine speed** 2200 **Bore and stroke** (as specified) 4.56" × 5.06" *(115.8 mm × 128.5 mm)* **Compression ratio** 16.5 to 1 **Displacement** 496 cu in *(8132 ml)* **Starting system** 12 volt **Lubrication** pressure **Air cleaner** two paper elements and aspirator **Oil filter** one full flow cartridge **Oil cooler** engine coolant heat exchanger for crankcase oil, radiator for hydraulic and transmission oil **Fuel filter** one paper element and prestrainer **Fuel cooler** radiator for return fuel **Muffler** vertical **Cooling medium temperature control** two thermostats and variable speed fan

ENGINE OPERATING PARAMETERS: Fuel rate: 85.3-92.8 lb/h *(38.7-42.1 kg/h)* **High idle:** 2275-2325 rpm **Turbo boost** nominal 19.7-23.9 psi *(136-165 kPa)* as measured 22.8 psi *(157 kPa)*

CHASSIS: Type Tracklayer-rubber tracked **Serial No.** *RW8400T902165* **Tread width** 60.0" *(1524 mm)* to 88.0" *(2235 mm)* **Length of track on ground** 89.0" *(2260 mm)* **Hydraulic control system** direct engine drive **Transmission** selective gear fixed ratio with full range operator controlled powershift **Nominal travel speeds mph** *(km/h)* first 1.16 *(1.87)* second 1.49 *(2.39)* third 1.89 *(3.04)* fourth 2.41 *(3.88)* fifth 2.92 *(4.70)* sixth 3.30 *(5.31)* seventh 3.73 *(6.01)* eighth 4.21 *(6.78)* ninth 4.75 *(7.65)* tenth 5.36 *(8.63)* eleventh 6.07 *(9.77)* twelfth 6.85 *(11.02)* thirteenth 8.71 *(14.02)* fourteenth 11.13 *(17.91)* fifteenth 14.17 *(22.80)* sixteenth 18.10 *(29.13)* reverse 1.01 *(1.63)*, 2.55 *(4.10)*, 2.88 *(4.63)*, 5.53 *(8.90)* — 1600 engine rpm **Clutch** multiple wet disc hydraulically actuated by foot pedal **Brakes** wet multiple disc hydraulically actuated by foot pedal **Steering** electro-hydraulic differential steering controlled by steering wheel **Power take-off** 1000 rpm at 2180 engine rpm **Unladen tractor mass** 25000 lb *(11340 kg)*

POWER TAKE-OFF PERFORMANCE

Power HP (kW)	Crank shaft speed rpm	Gal/hr (l/h)	lb/hp.hr (kg/kW.h)	Hp.hr/gal (kW.h/l)	Mean Atmospheric Conditions
MAXIMUM POWER AND FUEL CONSUMPTION					
Rated Engine Speed—(PTO speed—1007 rpm)					
226.73 *(169.08)*	2200	13.00 *(49.21)*	0.404 *(0.246)*	17.44 *(3.44)*	
Maximum Power (2 hours)					
255.16 *(190.27)*	2000	13.60 *(51.47)*	0.376 *(0.229)*	18.77 *(3.70)*	
VARYING POWER AND FUEL CONSUMPTION					
226.73 *(169.08)*	2000	13.00 *(49.21)*	0.404 *(0.246)*	17.44 *(3.44)*	Air temperature
197.76 *(147.47)*	2258	11.87 *(44.92)*	0.423 *(0.257)*	16.66 *(3.28)*	75°F *(24°C)*
148.86 *(111.00)*	2269	9.40 *(35.58)*	0.445 *(0.271)*	15.84 *(3.12)*	Relative humidity
99.89 *(74.48)*	2280	6.98 *(26.41)*	0.493 *(0.300)*	14.32 *(2.82)*	43%
49.68 *(37.05)*	2290	4.81 *(18.19)*	0.682 *(0.415)*	10.34 *(2.04)*	Barometer
1.00 *(0.75)*	2299	2.81 *(10.63)*	19.752 *(12.015)*	0.36 *(0.07)*	28.65"Hg *(97.02 kPa)*

Maximum Torque 795 lb.-ft. *(1078 Nm)* at 1199 rpm
Maximum Torque Rise 47.1%
Torque rise at 1800 engine rpm 35%

DRAWBAR PERFORMANCE (Unballasted)
FUEL CONSUMPTION CHARACTERISTICS

Power Hp (kW)	Drawbar pull lbs (kN)	Speed mph (km/h)	Crank-shaft speed rpm	Slip %	Fuel Consumption lb/hp.hr (kg/kW.h)	Fuel Consumption Hp.hr/gal (kW.h/l)	Temp.°F (°C) cooling med	Temp.°F (°C) Air dry bulb	Barom. inch Hg (kPa)
Maximum Power—9th Gear									
195.75 *(145.97)*	15925 *(70.84)*	4.61 *(7.42)*	2208	3.28	0.468 *(0.284)*	15.09 *(2.97)*	188 *(87)*	73 *(23)*	28.75 *(97.36)*
75% of Pull at Maximum Power—9th Gear									
152.36 *(113.62)*	11926 *(53.05)*	4.79 *(7.71)*	2261	1.83	0.491 *(0.299)*	14.35 *(2.83)*	188 *(86)*	76 *(24)*	28.69 *(97.16)*
50% of Pull at Maximum Power—9th Gear									
103.06 *(76.85)*	7955 *(35.39)*	4.86 *(7.82)*	2272	0.97	0.563 *(0.342)*	12.54 *(2.47)*	184 *(84)*	76 *(24)*	28.68 *(97.12)*
75% of Pull at Reduced Engine Speed—11th Gear									
152.36 *(113.61)*	11933 *(53.08)*	4.79 *(7.71)*	1768	1.83	0.430 *(0.261)*	16.41 *(3.23)*	189 *(87)*	76 *(24)*	28.68 *(97.12)*
50% of Pull at Reduced Engine Speed—11th Gear									
103.05 *(76.85)*	7942 *(35.33)*	4.87 *(7.83)*	1782	0.89	0.471 *(0.287)*	14.97 *(2.95)*	187 *(86)*	76 *(24)*	28.68 *(97.12)*

Fig. 2.9. Continued.

DRAWBAR PERFORMANCE (Ballasted at 2000 RPM)
MAXIMUM POWER IN SELECTED GEARS

Power Hp (kW)	Drawbar pull lbs (kN)	Speed mph (km/h)	Crank-shaft speed rpm	Slip %	Fuel Consumption lb/hp.hr (kg/kW.h)	Hp.hr/gal (kW.h/l)	Temp.°F (°C) cooling med	Air dry bulb	Barom. inch Hg (kPa)
					3rd Gear				
122.57 (91.40)	27993 (124.52)	1.64 (2.64)	2212	13.60	0.546 (0.332)	12.92 (2.54)	183 (84)	57 (14)	28.70 (97.19)
					4th Gear				
138.77 (103.48)	25003 (111.22)	2.08 (3.35)	2074	8.56	0.494 (0.301)	14.27 (2.81)	186 (85)	62 (17)	28.70 (97.19)
					5th Gear				
163.86 (122.19)	24955 (111.01)	2.46 (3.96)	2037	9.06	0.470 (0.286)	14.99 (2.95)	183 (84)	67 (19)	28.70 (97.19)
					6th Gear				
181.26 (135.16)	24287 (108.03)	2.80 (4.50)	2046	8.92	0.467 (0.284)	15.10 (2.97)	187 (86)	70 (21)	28.70 (97.19)
					7th Gear				
205.38 (153.15)	23920 (106.40)	3.22 (5.18)	2087	9.19	0.469 (0.285)	15.03 (2.96)	185 (85)	72 (22)	28.70 (97.19)
					8th Gear				
219.68 (163.82)	22703 (100.99)	3.63 (5.84)	2005	5.60	0.435 (0.265)	16.21 (3.19)	188 (86)	61 (16)	28.76 (97.39)
					9th Gear				
224.21 (167.20)	20190 (89.81)	4.16 (6.70)	2002	3.85	0.426 (0.259)	16.57 (3.26)	186 (86)	59 (15)	28.80 (97.53)
					10th Gear				
224.72 (167.57)	17774 (79.06)	4.74 (7.63)	2001	2.95	0.426 (0.259)	16.57 (3.26)	188 (87)	59 (15)	28.80 (97.53)
					11th Gear				
223.31 (166.52)	15444 (68.70)	5.42 (8.73)	2006	2.26	0.428 (0.260)	16.47 (3.24)	190 (88)	62 (17)	28.79 (97.49)
					12th Gear				
221.96 (165.52)	13558 (60.31)	6.14 (9.88)	2003	1.72	0.429 (0.261)	16.43 (3.24)	193 (89)	62 (17)	28.79 (97.49)
					13th Gear				
219.69 (163.82)	10511 (46.76)	7.84 (12.61)	1998	1.10	0.433 (0.264)	16.28 (3.21)	193 (89)	61 (16)	28.77 (97.43)

DRAWBAR PERFORMANCE (Ballasted at 2200 RPM)
MAXIMUM POWER IN SELECTED GEARS

Power Hp (kW)	Drawbar pull lbs (kN)	Speed mph (km/h)	Crank-shaft speed rpm	Slip %	Fuel Consumption lb/hp.hr (kg/kW.h)	Hp.hr/gal (kW.h/l)	Temp.°F (°C) cooling med	Air dry bulb	Barom. inch Hg (kPa)
					3rd Gear				
118.91 (86.67)	27163 (120.82)	1.64 (2.64)	2226	14.26	0.560 (0.341)	12.60 (2.48)	183 (84)	57 (14)	28.70 (97.19)
					4th Gear				
134.47 (100.28)	22048 (98.07)	2.29 (3.68)	2198	5.31	0.501 (0.305)	14.07 (2.77)	185 (85)	58 (14)	28.70 (97.19)
					5th Gear				
157.31 (117.30)	21186 (94.24)	2.78 (4.48)	2201	4.88	0.476 (0.289)	14.83 (2.92)	186 (85)	66 (19)	28.70 (97.19)
					6th Gear				
174.50 (130.13)	20841 (92.70)	3.14 (5.05)	2202	5.03	0.471 (0.287)	14.97 (2.95)	184 (84)	68 (20)	28.70 (97.19)
					7th Gear				
196.04 (146.19)	20738 (92.24)	3.55 (5.71)	2202	5.31	0.467 (0.284)	15.11 (2.98)	185 (85)	72 (22)	28.70 (97.19)
					8th Gear				
201.48 (150.25)	18504 (82.31)	4.08 (6.57)	2201	3.18	0.453 (0.275)	15.58 (3.07)	187 (86)	62 (17)	28.76 (97.39)
					9th Gear				
201.43 (150.20)	16238 (72.23)	4.65 (7.49)	2202	2.34	0.454 (0.276)	15.55 (3.06)	188 (86)	58 (14)	28.80 (97.53)
					10th Gear				
200.31 (149.37)	14229 (63.29)	5.28 (8.50)	2205	1.96	0.456 (0.278)	15.45 (3.04)	188 (86)	60 (16)	28.80 (97.53)
					11th Gear				
199.83 (149.01)	12535 (55.76)	5.98 (9.62)	2196	1.57	0.455 (0.277)	15.50 (3.05)	188 (86)	61 (16)	28.80 (97.53)
					12th Gear				
197.91 (147.58)	10969 (48.79)	6.77 (10.89)	2197	1.18	0.462 (0.281)	15.26 (3.01)	188 (87)	62 (17)	28.78 (97.46)
					13th Gear				
193.43 (144.24)	8375 (32.75)	8.66 (13.94)	2203	0.94	0.470 (0.286)	14.99 (2.95)	187 (86)	61 (16)	28.77 (97.43)

Fig. 2.9. Continued.

DRAWBAR PERFORMANCE (Unballasted)
MAXIMUM POWER IN SELECTED GEARS

Power Hp (kW)	Drawbar pull lbs (kN)	Speed mph (km/h)	Crank-shaft speed rpm	Slip %	Fuel Consumption lb/hp.hr (kg/kW.h)	Hp.hr/gal (kW.h/l)	Temp.°F (°C) cool-ing med	Air dry bulb	Barom. inch Hg (kPa)
				3rd Gear					
110.25 (82.21)	25021 (111.30)	1.65 (2.66)	2254	14.54	0.577 (0.351)	12.23 (2.41)	182 (83)	62 (17)	28.78 (97.46)
				4th Gear					
131.27 (97.89)	22780 (101.33)	2.16 (3.48)	2173	9.19	0.517 (0.315)	13.64 (2.69)	185 (85)	64 (18)	28.77 (97.43)
				5th Gear					
154.88 (115.49)	22238 (98.92)	2.61 (4.20)	2156	8.72	0.491 (0.299)	14.35 (2.83)	184 (84)	71 (22)	28.77 (97.43)
				6th Gear					
172.89 (128.92)	22167 (98.60)	2.92 (4.71)	2143	8.86	0.485 (0.295)	14.54 (2.86)	187 (86)	72 (22)	28.76 (97.39)
				7th Gear					
195.78 (145.99)	22114 (98.37)	3.32 (5.34)	2155	9.06	0.479 (0.291)	14.72 (2.90)	183 (84)	72 (22)	28.76 (97.39)
				8th Gear					
206.36 (153.89)	22068 (98.16)	3.51 (5.64)	2023	9.45	0.466 (0.283)	15.15 (2.98)	188 (86)	75 (24)	28.72 (97.26)
				9th Gear					
215.06 (160.37)	19922 (88.62)	4.05 (6.52)	1999	6.26	0.446 (0.271)	15.80 (3.11)	189 (87)	74 (23)	28.74 (97.32)
				10th Gear					
216.31 (161.30)	17389 (77.35)	4.67 (7.51)	1997	4.17	0.444 (0.270)	15.88 (3.13)	191 (88)	75 (24)	28.73 (97.29)
				11th Gear					
215.97 (161.05)	15151 (67.39)	5.35 (8.60)	1999	2.97	0.443 (0.269)	15.92 (3.14)	195 (90)	75 (24)	28.73 (97.26)
				12th Gear					
215.78 (160.91)	13282 (59.08)	6.09 (9.81)	2004	2.29	0.446 (0.271)	15.82 (3.12)	194 (90)	75 (24)	28.71 (97.22)
				13th Gear					
214.03 (159.60)	10282 (45.73)	7.81 (12.56)	1999	1.44	0.447 (0.272)	15.79 (3.11)	192 (89)	75 (24)	28.70 (97.19)

REPAIRS AND ADJUSTMENTS: No repairs or adjustments

NOTE: The 8400T engine has an electronic control system which provides a vehicle protection system to avoid overloading the drive train. This system provides four different engine power levels. The engine produces 160 PTO Hp when the transmission is in gears 1 through 4 and the PTO is not engaged. The engine produces 180 PTO Hp when the transmission is in 5th gear and the PTO is not engaged. The engine produces 200 PTO Hp when the transmission is in 6th gear and the PTO is not engaged. The engine produces 225 PTO Hp in all other applications.

REMARKS: All test results were determined from observed data obtained in accordance with official OECD, SAE and Nebraska test procedures. For the maximum power tests, the fuel temperature at the injection pump return was maintained at 170°F (76°C). The performance results on this summary were taken from OECD tests conducted under the Code II Test Code procedure.

We, the undersigned, certify that this is a true and correct report of official Tractor Test No. **1747**, Summary 263, May 26, 1998.

LEONARD L. BASHFORD
Director

M. F. KOCHER
R. D. GRISSO
G. J. HOFFMAN
Board of Tractor Test Engineers

TRACTOR SOUND LEVEL WITH CAB	dB(A)
At 75% load in 9th Gear	76.8
Bystander in 16th gear	88.8

TRACKS, BALLAST AND WEIGHT	With Ballast	Without Ballast
Track Width	24.0 in (610 mm)	24.0 in (610 mm)
Ballast—Cast iron—Front (total)	2165 lb (982 kg)	None
Height of Drawbar	18.5 in (470 mm)	18.0 in (455 mm)
Static Weight with operator	27330 lb (12397 kg)	25165 lb (11415 kg)

Fig. 2.9. Continued.

43

Chapter 2
Power
Performance

THREE POINT HITCH PERFORMANCE (OECD Static Test)

CATEGORY: III
Quick Attach: yes
Maximum Force Exerted Through Whole Range: 15749 lbs   *(70.1 kN)*

i) Opening pressure of relief valve: NA
    Sustained pressure of the open relief valve: 2890 psi   *(199 bar)*

ii) Pump delivery rate at minimum pressure: 31.0 GPM   *(117.3 l/min)*

iii) Pump delivery rate at maximum
     hydraulic power: 29.7 GPM   *(112.4 l/min)*
     Delivery pressure: 2500 psi   *(172 bar)*
     Power: 43.3 HP   *(32.3 kW)*

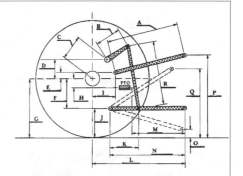

HITCH DIMENSIONS AS TESTED—NO LOAD

	inch	*mm*
A	28.9	*733*
B	19.5	*495*
C	22.9	*582*
D	22.2	*565*
E	10.2	*260*
F	11.0	*280*
G	33.6	*853*
H	3.2	*81*
I	15.6	*395*
J	22.6	*573*
K	28.3	*718*
L	48.5	*1231*
*L'	52.0	*1320*
M	25.5	*647*
N	41.6	*1056*
O	8.0	*203*
P	40.8	*1037*
Q	39.1	*993*
R	42.9	*1089*

*L' to end of Quick Attach

THREE POINT HITCH PERFORMANCE (SAE Static Test)

Observed Maximum Pressure psi *(bar)*	2890 *(199)*
Location	lift cylinder
Hydraulic oil Temperature 'F *('C)*	148 *(64)*
Location	hydraulic sump
Category	III
Quick Attach	yes

As per current SAE test procedures

Hitch point distance to ground level in. *(mm)*	8.0 *(203)*	16.1 *(408)*	24.1 *(613)*	32.1 *(814)*	40.0 *(1016)*
Lift force on frame lb.	15904	15964	16354	16348	15410
" " " " *(kN)*	*(70.7)*	*(71.0)*	*(72.8)*	*(72.7)*	*(68.6)*

As per current ASAE test procedures

Hitch point distance to ground level in. *(mm)*	8.0 *(203)*	16.1 *(408)*	24.1 *(613)*	32.1 *(814)*	40.0 *(1016)*
Lift force on frame lb.	17671	17634	18059	18053	16981
" " " " *(kN)*	*(78.6)*	*(78.4)*	*(80.3)*	*(80.3)*	*(75.5)*

JOHN DEERE 8400T DIESEL

Agricultural Research Division
Institute of Agriculture and Natural Resources
University of Nebraska–Lincoln
Darrell Nelson, Dean and Director

Fig. 2.10.

NEBRASKA OECD TRACTOR TEST 1702—SUMMARY 190
FORD 3930 8 x 8 DIESEL
8 SPEED
(CHASSIS SERIAL NUMBERS BE81400 AND HIGHER)

POWER TAKE-OFF PERFORMANCE

Power HP (kW)	Crank shaft speed rpm	Gal/hr (l/h)	lb/hp.hr (kg/kW.h)	Hp.hr/gal (kW.h/l)	Mean Atmospheric Conditions
MAXIMUM POWER AND FUEL CONSUMPTION					
Rated Engine Speed—(PTO speed—676 rpm)					
46.23 (34.47)	2201	3.16 (11.97)	0.481 (0.292)	14.61 (2.88)	
Standard Power Take-off speed (540 rpm)					
41.81 (31.18)	1759	2.68 (10.13)	0.450 (0.273)	15.62 (3.08)	
VARYING POWER AND FUEL CONSUMPTION					
46.23 (34.47)	2201	3.16 (11.97)	0.481 (0.292)	14.61 (2.88)	Air temperature
40.55 (30.24)	2266	2.82 (10.67)	0.488 (0.297)	14.38 (2.83)	75°F (24°C)
30.56 (22.79)	2292	2.31 (8.73)	0.530 (0.322)	13.25 (2.61)	Relative humidity
20.58 (15.34)	2313	1.79 (6.79)	0.612 (0.372)	11.47 (2.26)	62%
10.38 (7.74)	2333	1.41 (5.34)	0.954 (0.580)	7.36 (1.45)	Barometer
0.27 (0.20)	2347	0.98 (3.72)	25.131 (15.287)	0.28 (0.06)	28.87" Hg (97.77 kPa)

Maximum Torque 134 lb.-ft. *(182 Nm)* at 1249 rpm
Maximum Torque Rise 21.4%
Torque rise at 1755 engine rpm 13%

DRAWBAR PERFORMANCE
(UNBALLASTED—FRONT DRIVE ENGAGED)
FUEL CONSUMPTION CHARACTERISTICS

Power Hp (kW)	Drawbar pull lbs (kN)	Speed mph (km/h)	Crank shaft speed rpm	Slip %	Fuel Consumption lb/hp.hr (kg/kW.h)	Hp.hr/gal (kW.h/l)	Temp.°F (°C) cooling med	Air dry bulb	Barom. inch Hg (kPa)
Maximum Power—5th (H1) Gear									
37.19 (27.73)	3097 (13.77)	4.50 (7.25)	2206	8.49	0.592 (0.360)	11.86 (2.34)	191 (88)	65 (18)	28.96 (98.07)
75% of Pull at Maximum Power—5th (H1) Gear									
29.46 (21.97)	2321 (10.32)	4.76 (7.66)	2266	5.92	0.604 (0.367)	11.63 (2.29)	190 (88)	69 (21)	28.92 (97.93)
50% of Pull at Maximum Power—5th (H1) Gear									
20.22 (15.08)	1542 (6.86)	4.92 (7.91)	2294	4.24	0.699 (0.425)	10.05 (1.98)	188 (86)	70 (21)	28.91 (97.90)
75% of Pull at Reduced Engine Speed—6th (H2) Gear									
29.37 (21.90)	2317 (10.30)	4.75 (7.65)	1529	5.97	0.536 (0.326)	13.10 (2.58)	192 (89)	73 (23)	28.88 (97.80)
50% of Pull at Reduced Engine Speed—6th (H2) Gear									
20.26 (15.11)	1543 (6.86)	4.92 (7.93)	1549	3.87	0.561 (0.341)	12.52 (2.47)	186 (86)	74 (23)	28.87 (97.77)

Location of Test: Tractor Testing Laboratory, University of Nebraska, Lincoln, Nebraska 68583-0832

Dates of Test: October 2-27, 1995

Manufacturer: New Holland N.A., 500 Diller Avenue, New Holland, PA 17557

FUEL OIL and TIME: Fuel No. 2 Diesel **Cetane No.** 50.6 **Specific gravity converted to 60°/60° F** *(15°/15°C)* 0.8435 **Fuel weight** 7.023 lbs/gal *(0.842 kg/l)* **Oil SAE** 15W-40 **API service classification** CG-4,SH **To motor** 1.487 gal *(5.630 l)* **Drained from motor** 1.386 gal *(5.225 l)* **Transmission and final drive lubricant** Ford M2C 134-D fluid **Front axle lubricant** Ford M2C 134-D fluid **Total time engine was operated** 15.0 hours.

ENGINE: Make Ford New Holland Diesel **Type** three cylinder vertical **Serial No.** *BB537895* **Crankshaft** lengthwise **Rated rpm** 2200 **Bore and stroke** (as specified) 4.4" × 4.2" *(111.8 mm × 106.7 mm)* **Compression ratio** 16.3 to 1 **Displacement** 192 cu in *(3147 ml)* **Starting system** 12 volt **Lubrication** pressure **Air cleaner** two paper elements **Oil filter** one full flow cartridge **Oil cooler** radiator for transmission fluid **Fuel filter** one paper element and sediment bowl **Muffler** vertical **Cooling medium temperature control** one thermostat

ENGINE OPERATING PARAMETERS: Fuel rate: 21.6-23.4 lb/h *(9.8-10.6 kg/h)* **High idle:** 2325-2375 rpm

CHASSIS: Type front wheel assist **Serial No.** *BE03879* **Tread width** rear 59.6" *(1515 mm)* to 79.8" *(2026 mm)* front 55.0" *(1396 mm)* to 73.0" *(1855 mm)* **Wheel base** 84.1" *(2136 mm)* **Hydraulic control system** direct engine drive **Transmission** selective gear fixed ratio **Nominal travel speeds mph** *(km/h)* first 1.39 *(2.23)* second 2.05 *(3.30)* third 3.13 *(5.03)* fourth 4.59 *(7.38)* fifth 4.88 *(7.85)* sixth 7.23 *(11.63)* seventh 11.01 *(17.72)* eighth 16.17 *(26.03)*, reverse 1.38 *(2.22)*, 2.04 *(3.28)*, 3.11 *(5.00)*, 4.56 *(7.34)*, 4.85 *(7.81)*, 7.19 *(11.57)*, 10.95 *(17.63)*, 16.09 *(25.90)* **Clutch** single dry disc operated by foot pedal **Brakes** wet multiple disc operated by two foot pedals which can be locked together **Steering** hydrostatic **Power take-off** 540 rpm at 1756 engine rpm **Unladen tractor mass** 5446 lb *(2470 kg)*

REPAIRS AND ADJUSTMENTS: No repairs or adjustments

NOTE: The performance figures on this report apply to chassis serial numbers *BE81400* and higher.

Fig. 2.10. Continued.

45

Chapter 2
Power
Performance

DRAWBAR PERFORMANCE
(UNBALLASTED—FRONT DRIVE ENGAGED)
MAXIMUM POWER IN SELECTED GEARS

Power Hp (kW)	Drawbar pull lbs (kN)	Speed mph (km/h)	Crank-shaft speed rpm	Slip %	Fuel Consumption lb/hp.hr (kg/kW.h)	Fuel Consumption Hp.hr/gal (kW.h/l)	Temp.°F (°C) cooling med	Temp.°F (°C) Air dry bulb	Barom. inch Hg (kPa)
					3rd (L3) Gear				
32.74 (24.41)	4480 (19.93)	2.74 (4.41)	2248	14.71	0.648 (0.394)	10.83 (2.13)	189 (87)	58 (14)	28.96 (98.07)
					4th (L4) Gear				
35.13 (26.20)	3111 (13.84)	4.24 (6.82)	2205	8.55	0.630 (0.383)	11.14 (2.20)	192 (89)	61 (16)	28.96 (98.07)
					5th (H1) Gear				
37.19 (27.73)	3097 (13.77)	4.50 (7.25)	2206	8.49	0.592 (0.360)	11.86 (2.34)	191 (88)	65 (18)	28.96 (98.07)
					6th (H2) Gear				
36.73 (27.39)	1990 (8.85)	6.92 (11.14)	2202	5.09	0.601 (0.366)	11.69 (2.30)	191 (88)	68 (20)	28.96 (98.07)

DRAWBAR PERFORMANCE
(UNBALLASTED—FRONT DRIVE DISENGAGED)
FUEL CONSUMPTION CHARACTERISTICS

Power Hp (kW)	Drawbar pull lbs (kN)	Speed mph (km/h)	Crank-shaft speed rpm	Slip %	Fuel Consumption lb/hp.hr (kg/kW.h)	Fuel Consumption Hp.hr/gal (kW.h/l)	Temp.°F (°C) cooling med	Temp.°F (°C) Air dry bulb	Barom. inch Hg (kPa)
				Maximum Power—6th (H2) Gear					
36.57 (27.27)	2051 (9.12)	6.69 (10.76)	2205	7.80	0.601 (0.366)	11.68 (2.30)	192 (89)	69 (21)	28.95 (98.04)
				75% of Pull at Maximum Power—6th (H2) Gear					
28.79 (21.47)	1540 (6.85)	7.01 (11.28)	2262	5.61	0.649 (0.395)	10.82 (2.13)	191 (88)	75 (24)	28.85 (97.70)
				50% of Pull at Maximum Power—6th (H2) Gear					
19.74 (14.72)	1024 (4.55)	7.23 (11.64)	2286	3.82	0.751 (0.457)	9.36 (1.84)	189 (87)	75 (24)	28.85 (97.70)
				75% of Pull at Reduced Engine Speed—7th (H3) Gear					
28.78 (21.46)	1533 (6.82)	7.04 (11.34)	1482	5.38	0.567 (0.345)	12.38 (2.44)	194 (90)	75 (24)	28.85 (97.70)
				50% of Pull at Reduced Engine Speed—7th (H3) Gear					
19.73 (14.71)	1027 (4.57)	7.21 (11.60)	1498	3.94	0.586 (0.357)	11.98 (2.36)	188 (86)	75 (24)	28.85 (97.70)
				MAXIMUM POWER IN SELECTED GEARS					
				5th (H1) Gear					
34.75 (25.91)	3094 (13.76)	4.21 (6.78)	2232	14.84	0.627 (0.381)	11.20 (2.21)	190 (88)	67 (19)	28.96 (98.07)
				6th (H2) Gear					
36.57 (27.27)	2051 (9.12)	6.69 (10.76)	2205	7.80	0.601 (0.366)	11.68 (2.30)	192 (89)	69 (21)	28.95 (98.04)

REMARKS: All test results were determined from observed data obtained in accordance with official OECD, SAE and Nebraska test procedures. For the maximum power tests, the fuel temperature at the injection pump inlet was maintained at 136° F *(58°C)*. The performance figures on this summary were taken from a test conducted under the OECD Code II Restricted Standard Test Code procedure.

We, the undersigned, certify that this is a true and correct report of official Tractor Test No. **1702**, Summary 190, November 29, 1995.

LOUIS I. LEVITICUS
Engineer-in-Charge

L.L. BASHFORD
R.D. GRISSO
M.F. KOCHER
Board of Tractor Test Engineers

TRACTOR SOUND LEVEL WITHOUT CAB	Front Wheel Drive Disengaged dB(A)	Front Wheel Drive Engaged dB(A)
At 75% load in 5th (H1) Gear	98.0	98.5
Bystander in 8th (H4) Gear	88.5	—

TIRES, BALLAST AND WEIGHT

	Tested Without Ballast
Rear Tires—No., size, ply & psi *(kPa)*	Two 14.9-28; 6; 12 *(85)*
Front Tires—No., size, ply & psi *(kPa)*	Two 8.3-24; 6; 20 *(140)*
Height of Drawbar	16.5 in *(420 mm)*
Static Weight with Operator—Rear	3388 lb *(1537 kg)*
—Front	2222 lb *(1008 kg)*
—Total	5610 lb *(2545 kg)*

Fig. 2.10. Continued.

SUPPLEMENT TO NEBRASKA OECD TRACTOR TEST 1702—SUMMARY 190
NEW HOLLAND 3930 8x8 DIESEL
8 SPEED
CHASSIS SERIAL NUMBERS *091151B* AND HIGHER

POWER TAKE-OFF PERFORMANCE

Power HP (kW)	Crank shaft speed rpm	Gal/hr (l/h)	lb/hp.hr (kg/kW.h)	Hp.hr/gal (kW.h/l)	Mean Atmospheric Conditions
MAXIMUM POWER AND FUEL CONSUMPTION					
Rated Engine Speed—(PTO speed—676 rpm)					
52.54 (39.18)	2200	3.25 (12.31)	0.438 (0.267)	16.15 (3.18)	
Standard Power Take-off Speed (540 rpm)					
46.43 (34.62)	1757	2.71 (10.26)	0.414 (0.252)	17.13 (3.37)	
VARYING POWER AND FUEL CONSUMPTION					
52.54 (39.18)	2200	3.25 (12.31)	0.438 (0.267)	16.15 (3.18)	Air temperature
46.38 (34.59)	2291	3.09 (11.71)	0.472 (0.287)	15.00 (2.95)	83°F (28°C)
35.49 (26.46)	2320	2.63 (9.94)	0.524 (0.319)	13.51 (2.66)	Relative humidity
23.83 (17.77)	2344	2.08 (7.86)	0.617 (0.375)	11.48 (2.26)	61%
11.86 (8.84)	2363	1.57 (5.93)	0.936 (0.569)	7.57 (1.49)	Barometer
0.42 (0.31)	2368	1.10 (4.17)	18.757 (11.410)	0.38 (0.07)	28.93"Hg (97.96 kPa)

Maximum Torque 139 lb.-ft. *(188 Nm)* at 1699 rpm
Maximum Torque Rise 10.5%
Torque rise at 1757 rpm 10%

Location of Test: Nebraska Tractor Test Laboratory, University of Nebraska, Lincoln, Nebraska 68583-0832

Dates of Test: September 29, 1998

Manufacturer: New Holland N.A., 500 Diller Avenue, New Holland PA 17557

FUEL and OIL: Fuel No. 2 Diesel **Specific gravity converted to 60°/60°F *(15°/15°C)*** 0.8506 **Fuel weight** 7.082 lbs/gal *(0.849 kg/l)*

ENGINE: Make Ford New Holland Diesel **Type** three cylinder vertical with turbocharger **Serial No.** *332T/JD*785681* **Crankshaft** lengthwise **Rated engine speed** 2200 **Bore and stroke** 4.4" × 4.2" *(111.8 mm × 106.7 mm)* **Compression ratio** 17.5 to 1 **Displacement** 192 cu in *(3141 ml)*

CHASSIS: Type front wheel assist **Serial No.** *103947B*

NOTE: The performance figures on this summary apply to tractor chassis serial numbers *091151B* and higher.

We, the undersigned, certify that this is a true and correct supplement to official Tractor Test No. **1702**, Summary 190, October 19, 1998.

LEONARD L. BASHFORD
Director

M.F. KOCHER
R.D. GRISSO
G.J. HOFFMAN
Board of Tractor Test Engineers

Fig. 2.10. Continued.

THREE POINT HITCH PERFORMANCE (OECD Static Test)

CATEGORY: I
Quick Attach: none

Maximum Force Exerted Through Whole Range:	3056 lbs	*(13.6 kN)*
i) Opening pressure of relief valve:	NA	
Sustained pressure of the open relief valve:	2590 psi	*(178 bar)*
ii) Pump delivery rate at minimum pressure:	9.3 GPM	*(35.2 l/min)*
iii) Pump delivery rate at maximum hydraulic power:	8.1 GPM	*(30.7 l/min)*
Delivery pressure:	2250 psi	*(155 bar)*
Power:	10.6 HP	*(7.9 kW)*

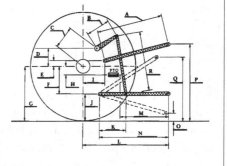

THREE POINT HITCH PERFORMANCE (SAE Static Test)

Observed Maximum Pressure psi. *(bar)*	2600.*(179)*
Location	remote outlet
Hydraulic oil temperature °F *(°C)*	169 *(76)*
Location	rear axle sump
Category	I
Quick attach	none

As per current SAE test procedures

Hitch point distance to ground level in. *(mm)*	8.3 *(211)*	13.0 *(330)*	17.7 *(450)*	22.4 *(569)*	27.2 *(691)*	32.1 *(815)*
Lift force on frame lb	3056	3353	3524	3542	3434	3245
Lift force on frame *(kN)*	*(13.6)*	*(14.9)*	*(15.7)*	*(15.8)*	*(15.3)*	*(14.4)*

As per current ASAE test procedures

Hitch point distance to ground level in. *(mm)*	8.3 *(211)*	13.0 *(330)*	17.7 *(450)*	22.4 *(569)*	27.2 *(691)*	32.1 *(815)*
Lift force on frame lb.	3271	3589	3772	3791	3676	3473
Lift force on frame *(kN)*	*(14.5)*	*(16.0)*	*(16.8)*	*(16.9)*	*(16.3)*	*(15.4)*

HITCH DIMENSIONS AS TESTED—NO LOAD

	inch	*mm*
A	28.3	*719*
B	10.0	*254*
C	12.9	*327*
D	10.1	*257*
E	7.5	*191*
F	8.0	*203*
G	24.0	*610*
H	4.7	*120*
I	8.7	*222*
J	16.0	*407*
K	18.9	*481*
L	36.7	*931*
M	20.9	*530*
N	34.0	*864*
O	8.0	*203*
P	34.1	*867*
Q	33.1	*841*
R	29.1	*740*

FORD 3930 8 X 8 DIESEL

Agricultural Research Division
Institute of Agriculture and Natural Resources
University of Nebraska–Lincoln
Darrell Nelson, Dean and Director

As an example of the use of Fig. 2.11, assume that a tractor was tested on a concrete surface and was found to have exerted 67.3 kW [90.3 HP] in overcoming a horizontal drawbar pull of 35.2 kN [7920 lb] at 6.89 km/hr [4.28 MPH] with 7.81% drive-wheel slip. The tractor's mass is 9978 kg [21,953 lb] of which 68.36 kN [15,367 lb] is the vertical static rear axle force [SRAF]. (Because the ratio DRAWBAR PULL/SRAF in Fig. 2.11 is a force ratio, one must multiply the SI system's drive axle weight reaction in kilograms by g to get kilonewtons.) The performance in the same gear on firm soil with a semimounted implement is desired.

1. Enter the upper left quadrant at 7.81% slip. Go horizontally left (dashed line) to the concrete curve, turn up, read 0.91 DRAWBAR POWER/AXLE POWER ratio. Axle power = 67.3 [90.3]/0.91 or 73.9 kW [99 HP].

2. Determine the no-load forward speed:

6.89 km/hr [4.28 MPH]/(1 − 0.0781) = 7.47 km/hr [4.64 MPH]

3. Determine SRAF/AXLE POWER ratio:

68.36/73.9 = 925 N/kW [15,367/99 = 155 lb/HP]

4. Enter lower right quadrant at the no-load speed (see 2 above) and move to the right (solid line). Turn up at the 925 [155] SRAF/AXLE POWER line. Terminate at the curve, S, in the firm soil area, upper right quadrant.

5. Move horizontally left from the terminal point in 4 above through the slip axis (9%) to the curve, S, in the firm soil area, upper left quadrant. Drop vertically to read a DRAWBAR PULL/SRAF value of 0.44. The drawbar pull is thus 0.44 × 68.36 [15,367] = 30 kN [6761 lb].

6. From the turning point in 5 above, continue horizontally left (solid line) to the firm soil curve, turn upward to read 0.78 = DRAWBAR POWER/ AXLE POWER. The drawbar power is 0.78 × 73.9 [99] = 57.6 kW [77.2 HP].

7. From the terminal in 4 above, move down, parallel to lines having constant actual speed (dashed line), to the horizontal axis. Then drop vertically to the 925

[155] SRAF/AXLE POWER turning line. Go horizontally left and read 6.8 km/hr [4.22 MPH] actual forward speed. This tractor could be expected to exert a maximum pull of 30 kN [6761 lb] on a semimounted implement at 6.8 km/hr [4.22 MPH] with a 9% slip on firm soil.

Tire equipment is a very important variable in estimating tractor performance. Fig. 2.11 predicts the results for the smallest commercial (most economical) tires that can carry the mass load indicated. Table 2.1 lists the rated load capacities of agricultural drive tires.

Oversize and/or dual tires on drive wheels would be expected to give greater performance than that predicted by Fig. 2.11. The effect of dual tires on drawbar performance can be deduced by comparing the results of Nebraska tests with the predicted values of Fig. 2.11. The departure of actual test values from predicted values is not great when engine power is the constraint on performance. Using data from Test 1175, a test of a dual-drive-wheel tractor conducted in 1975,

SRAF = 60.9 kN [13,700 lb] R = 19.5/104.8 = 0.186

A table of performance in each gear can be developed. (The R curves must be extrapolated.)

Gear	DB PULL/ SRAF from test	Speed km/hr	[MPH]	Predicted Slip, %	Actual Slip, %	Error %
2	1.09	3.35	[2.08]	22	14.8	21
3	0.93	6.07	[3.77]	14	9.16	13
4	0.79	7.26	[4.51]	10	7.08	9
6	0.64	9.00	[5.59]	7	5.62	6
7	0.64	9.16	[5.61]	7	5.54	6
8	0.54	10.81	[6.72]	5.5	4.75	4.5

The predictor chart overestimates the slip and thus underestimates the efficiency for converting engine power to drawbar power for dual drive wheels. The error is not great (less than 10%) for normal field speeds. The chart underestimates substantially the performance of the dual-wheeled tractor at slow-speed, heavy-pull drawbar loadings.

Similarly, the prediction accuracy of Fig. 2.11 can be tested by using the 5th gear, MFWD-engaged data from Fig. 2.10. SRAF = 2545 kg [5610 lb]. (With MFWD engaged, all the tractor weight is on drive wheels.) R = 420/2136 [16.5/84.1] = 20; slip = 8.49%. Speed is 7.25 km/hr [4.5 MPH]. Drawbar power is

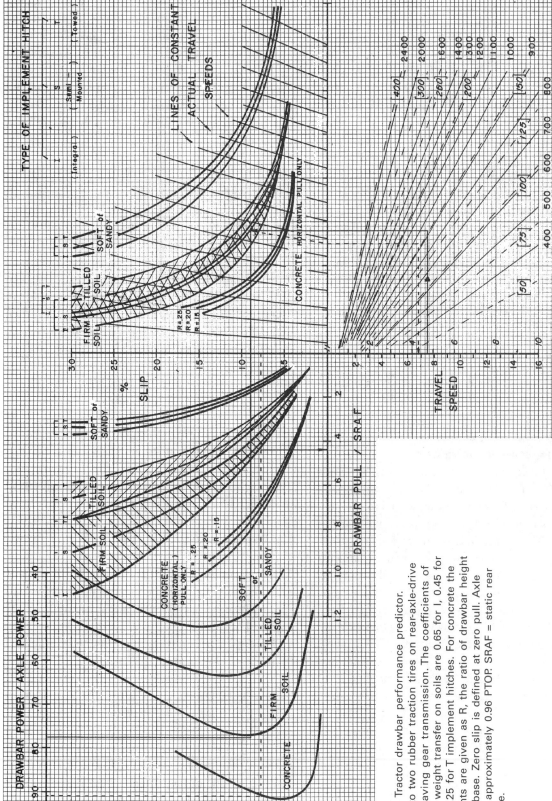

Fig. 2.11. Tractor drawbar performance predictor. Limited to two rubber traction tires on rear-axle-drive tractor having gear transmission. The coefficients of dynamic weight transfer on soils are 0.65 for I, 0.45 for S, and 0.25 for T implement hitches. For concrete the coefficients are given as R, the ratio of drawbar height to wheelbase. Zero slip is defined at zero pull. Axle power = approximately 0.96 PTOP. SRAF = static rear axle force.

27.73 kW [37.19 HP]. Derived value: Drawbar Pull/SRAF = 13.77/2496 kN [3097/5610 lb] = 0.55

Enter chart, upper left-hand quadrant, at 0.55. Go up to R = 0.20 then left to concrete and go up to get Drawbar Power/Axle Power = 0.905. Axle power is 27.73/0.905 = 30.64 kW [37.19/.905= 41.1HP]. Return to intersection of R = 20 with 0.55 Drawbar Pull/SRAF. Go right through slip = 5.5% into upper, right quadrant and turn vertically down at R = 20. Intersect the derived value, SRAF/Axle Power = 2545 kg × 9.807/30.64 kW = 815 [5610lb/41.1 HP = 136], turn left to read no-load speed 7.5 km/hr [4.66 MPH]. Loaded speed can be calculated as 0.945 × 7.5 = 7.08 km/hr [0.945 × 4.66 = 4.40 MPH]; or loaded speed can be read from the chart by dropping down from the upper right quadrant along a line parallel to Lines of Constant Actual Travel Speeds to read 7.08 km/hr [4.4 MPH]. Comparisons are:

Item	Tested	Predicted by chart
Slip %	8.49	5.5
Loaded speed		
km/hr	7.25	7.08
MPH	4.50	4.40

The error is not great since the chart is most effective for concrete surfaces.

It may be concluded that at normal field speeds and on a good tractive surface, the performance of 4-wheel-drive and dual-tire tractors is not much more efficient than the rear-axle-drive tractor with single-drive tires. The greater difference comes at low speeds and heavy pulls. These differences can be expected to be even more pronounced in soft soil conditions where flotation is important.

Fig. 2.11 is effective in predicting both maximum and part-load performances of a tractor. The example solution shows how OECD test data can be converted to maximum expected field performance. But typically, tractors are loaded less than maximum:
1. To be able to accelerate the load from a standstill
2. To provide for some reserve to handle the normal variations in soil and crop energy requirements
3. Inability to always match an implement operation to the full power capability of the tractor.

The machinery manager must keep in mind that some *load factor*, a ratio of actual load to maximum load, should be used in actual field situations.

As an example of operations at a load factor, consider the situation in which it is known that an integral chisel plow requires 10.7 kN [2405 lb] of pull in firm soil at a desired speed of 6.8 km/hr [4.23 MPH]. One wishes to predict the power and the no-load speed (transmission gear) required for a tractor used with radial tires and ballast, SRAF = 3643 kg [8030 lb], and a PTOP = 64.59 kW [86.62 HP]. Compute

DBP = 10.7 × 6.8/3.6 = 20.2 kW
  = [2405 × 4.23/375 = 27.1 HP]

DB PULL/SRAF = 10,700/(3643 × 9.807) = 0.30
  = [2405/8030 = 0.30]

Enter the upper left quadrant of Fig. 2.11 at 0.30 and go vertically to the line, I, in the firm soil area. Move horizontally to the left to the firm soil line and then turn up to read 0.735 DRAWBAR POWER/AXLE POWER. Axle power then is

DBP/0.735 = 20.2/0.735 [27.1/0.735]
  = 27.48 kW [36.87 HP]

PTOP = axle power/0.96 = 27.48/0.96 [36.87/0.96]
  = 28.63 kW [38.4 HP]

The load factor is the ratio of average actual PTOP to the maximum available PTOP:

28.63/64.59 [38.4/86.62] = 0.44

The % slip is found to be 5.5 from the horizontal line extended to the right. The desired no-load forward speed would be 6.8/(1–0.06) [4.23/(1–0.06)] or 7.23 km/hr [4.50 MPH].

The actual forward speed might have been estimated from Fig. 2.11. Compute

SRAF/AXLE POWER = 3643×9.807/27.48 = 1300
  = [8030/36.87 = 218]

Enter the lower right quadrant at 7.23 km/hr [4.50 MPH]. Go horizontally to the right and turn up at the 1300 [218] line to the firm soil line, I, in the upper quadrant. Follow a line parallel to the lines of constant actual speed back down to the axis. Then drop vertically to the 1300 [218] line. Go left and read 6.8 km/hr [4.23 MPH] on the speed axis.

Fig. 2.11 is useful for predicting the effect of changes on tractor performance. A criterion commonly used for drawbar performance is *tractive efficiency*, the ratio of drawbar power to axle power. This ratio is really a measure of tire efficiency and is the axis located along the top of the upper left quadrant in Fig. 2.11. As an example of the use of this criterion, consider the removal of 2000 kg [4400 lb] of ballast from the rear axle of the tractor in the previous example problem. Such a weight reduction seems to be poor management since wheel slip increases to 13.0% slip (DRAWBAR PULL/SRAF = 0.66). But the tractive efficiency increases from 0.735 to 0.750. More of the axle power is converted to useful work because the reduced weight has decreased the tractor's rolling resistance.

Examination of the upper left quadrant of Fig. 2.11 shows the percent slip values at which tractive efficiency is maximum for all four surfaces: concrete, 4–8%; firm soil, 8–10%, tilled soil, 11–13%; soft soil, 14–16%. Ballasting the tractor according to expected drawbar pulls is the only way to place the operation within these desirable slip values and realize the benefits of maximum tractive efficiency. Note that the penalty, in terms of reduced tractive efficiency, is greater for overballasting than for underballasting. As far as the machinery manager is concerned, ballasting tractors becomes a trade-off between the costs of ballast and extra fuel consumption and the costs of tire wear and labor for removal of unneeded ballast for light drawbar pulls.

Ballasting can cause excessive soil compaction. Especially susceptible to compaction are wet soils with a high clay content. A moderate amount of soil compaction is used in planting seeds to insure good soil-to-seed contact for moisture transfer and rapid seed germination. Excessive compaction will cause reduced rainfall infiltration, poor soil aeration, decreased root penetration, denitrification because of poor internal soil drainage, and increased draft of tillage implements in subsequent years. Surface compaction can be reduced by expanding the surface area of contact with multiple wheels, bigger tires, and with track-laying tractors. Some research shows that deep compaction is a function of axle load only. The effects of deep compaction can last for years on some soils with little chance of rectification.

The farm manager will be concerned as to yield reductions associated with soil compaction. The soil compaction phenomena is highly related to the soil type and to the amount of seasonal rainfall. For organic and sandy soils in years of frequent rainfall, yield reductions may be minimal or even enhanced with wheel-traffic compaction. In a six-year study, John Siemens, Agricultural Engineering Department, University of Illinois, found compacted soils increased corn yields in two of the six years, but on the average, yields were reduced about 3% for both corn and soybeans. Machinery managers should be aware of the potential damage to the productivity of soils from ballasted tractors and from transports carrying heavy loads over farm fields.

The total load (including ballast, filled mounted chemical tanks, and attached implements) carried by the tractor tires is limited to the values shown in Table 2.1. These data are extracted from the 1999 Yearbook published by the Tire and Rim Association, Copley, Ohio. Drive wheel tire load limits for bias ply (tire cords molded in at an angle to the wheel radii), radial ply (tire cords placed parallel to the radii), and metric (manufactured to metric dimensions) tires are shown as are those for steering wheel tires.

Distinctive markings identify each tire type. A tire may have any of the following code markings molded into the sidewall:

16.9　L　R　38　SL

where: 16.9　is the nominal section width in inches,
　　　L　indicates a lower section height than standard,
　　　R　signifies radial construction,
　　　38　is the nominal rim diameter in inches,
　　　SL　means the tire is limited to agricultural use.

Other markings may include:
　　　R1　identifies tire with a regular tread,
　　　R2　identifies tire with a deep tread for rice and cane fields,
　　　R3　a shallow tread tire,
　　　R4　an intermediate tread tire for industrial use.

The strength of nonmetric tires is indicated by the ply rating. Greater strength is indicated by the higher ply number. Higher ply numbers allow for greater inflation pressures which allow the tire to carry a heavier load.

TABLE 2.1. Tire Static Load Limits, kg[lb], at Various Inflation Pressures, kPa [psi].

Diagonal (bias) ply drive-wheel tires. Speeds limited to 40 km/hr [25 MPH]

Pressures	80 [12]	110 [16]	140 [20]	150 [22]	170 [24]	180 [26]	190 [28]	220 [32]	260 [38]
12.4-24	710[1570]	850[1870](4)	975[2150]	1030[2270]	1090[2400](6)	1120[2470]	1180[2600]	1285[2830](8)	
14.9-24	*1000[2200]	1180[2600]	1360[3000](6)	1450[3200]	1500[3300]	1600[3520](6)	1650[3640]		2000[4400]
17.5L-24	1180[2600]	1400[3080](6)	1600[3520]	1650[3640](8)	1750[3860]	1850[4080]	1900[4180](10)		
12.4-28	775[1710]	900[1980](4)	1030[2270]	1090[2400]	1150[2540](6)				
14.9-28	*1060[2340]	1285[2830]	1450[3200]	1550[3420](6)	1600[3520]	1700[3740](8)	1800[3960]	1950[4300](10)	
18.4-28	*1550[3420]	1850[4080](6)	2120[4680]	2240[4940]	2360[5200]	2430[5360](10)	2575[5680]	2800[6150](12)	
14.9-30	*1120[2470]	1320[2910]	1500[3300]	1600[3520]	1650[3640]	1750[3860](8)	1800[3960]	1950[4300](10)	
18.4-30	*1600[3520]	1900[4180]	2180[4800]	2300[5080]	2430[5360]	2500[5520](10)			
18.4-34	*1700[3740]	2000[4400](6)	2300[5080](8)						
14.9-38	*1250[2760]	1500[3300]	1700[3740](6)						
18.4-38	*1800[3960]	2120[4680](6)	2430[5360](8)	2575[5680](8)	2725[6000]	2900[6400](10)	3000[6600]	3250[7150](12)	
18.4-42	*1900[4180]	2240[4940]	2575[5680](8)	2725[6000]	2900[6400]	3000[6600](10)			
20.8-42	*2300[5080]	2725[6000]	3150[6950]	3350[7400](10)					

Bold figures in parentheses denote ply rating for which the loads and inflations are maximum.

* marked values are for duals only.

When used as duals, tire loads must be reduced to 0.88 of table loads, 0.82 if triples.

For transport service and operations which do not require sustained high tire torque, the table loads above may be increased 33% for speeds up to 16 km/hr[10 MPH]; 22% up to 24 km/hr[15 MPH]; 11% up to 32 km/hr [20 MPH]; none for 40 km/hr[25 MPH]

Diagonal (bias) steering-wheel tires. Speeds limited to 40 km/hr [25 MPH]

Pressures	170[24]	190[28]	220[32]	250[36]	280[40]	300[44]	Maximum
7.50-16 SL	450[990]	500[1100]	545[1200]	600[1320]	630[1390]	670[1480](6)	775[1710]@390 kPa[56 psi] (8)
11-16 SL	850[1870]	950[2090]	1030[2270](6)	1120[2470]	1180[2600](8)	1250[2760]	1550[3420]@410 kPa[60 psi] (12)
11L-16 SL	670[1480]	750[1650]	800[1760]	875[1930]	950[2090]	1000[2200](10)	1250[2760]@440 kPa[64 psi] (12)
11-24 SL	1090[2400]	1215[2680]	1320[2910]	1450[3200]	1550[3420](8)	1650[3640]	1750[3860]@333 kPa[48 psi] (10)

For above tires used in cycling loading service at speeds below 8 km/hr[5 MPH], loads may be increased 67%

For transport speeds, table values may be increased 50% up to 16 km/hr[10 MPH]; 28% up to 24 km/hr[15 MPH]; 11% up to 32 km/hr[20 MPH]

TABLE 2.1. Continued

Radial ply drive-wheel tires. Speeds limited to 48 km/hr [30 MPH]

Pressures	40[6]	60[8]	70[10]	80[12]	100[14]	110[16]	120[18]*	160[24]**	210[30]***
13.6R24	630[1390]	750[1650]	850[1870]	950[2090]	1030[2270]	1120[2470]	1180[2600]	1450[3200]	1600[3520]
14.9R24	750[1650]	875[1930]	1000[2200]	1120[2470]	1215[2680]	1320[2910]	1400[3080]	1550[3420]	1700[3740]
16.9R24	900[1980]	1060[2340]	1215[2680]	1360[3000]	1600[3520]	1500[3300]	1700[3740]	2120[4680]	2300[5080]
14.9R28	800[1760]	950[2090]	1060[2340]	1180[2600]	1400[3080]	1320[2910]	1500[3300]	1800[3960]	2060[4540]
16.9R28	975[2150]	1150[2540]	1320[2910]	1450[3200]	1700[3740]	1600[3520]	1850[4080]	2240[4940]	2500[5520]
14.9R30	825[1820]	975[2150]	1120[2470]	1215[2680]	1360[3000]	1450[3200]	1550[3420]	1850[4080]	2120[4680]
16.9R30	1000[2200]	1180[2600]	1360[3000]	1500[3300]	1650[3640]	1750[3860]	1900[4180]	2300[5080]	2575[5680]
16.9R38	1120[2470]	1320[2910]	1500[3300]	1700[3740]	1850[4080]	2000[4400]	2120[4680]	2575[5680]	2900[6400]
18.4R38	1360[3000]	1600[3520]	1800[3960]	2000[4400]	2180[4800]	2360[5200]	2575[5680]	3000[6600]	3450[7600]
20.8R38	1650[3640]	1950[4300]	2180[4800]	2430[5360]	2650[5840]	2900[6400]	3075[6800]	3650[8050]	4125[9100]
18.4R42	1400[3080]	1700[3740]	1900[4180]	2120[4680]	2300[5080]	2500[5520]	2725[6000]	3150[6950]	3650[8050]
20.8R42	1700[3740]	2060[4540]	2300[5080]	2575[5680]	2800[6150]	3075[6800]	3250[7150]	3875[8550]	4375[9650]

Bold numbers denote maximum loads for the number of star(*) markings molded on the tire, referred to as 1-star, 2-star, 3-star.
For transport service above 32 km/hr[20 MPH], inflation pressure should be 60 kPa[8 psi] or above .
When used as duals, tire loads must be reduced to 0.88 of table loads, 0.82 if triples.
For transport service and operations that do not require sustained high torque, the table loads may be increased 34% for speeds up to 16 km/hr[10 MPH]; 11% up to 24 km/hr[15 MPH]; 7% up to 32 km/hr[20 MPH]; none for over 32 km/hr[25 MPH]

Radial ply metric drive-wheel tires. Speeds limited to 48 km/hr [30 MPH]

Pressures	40[6]	60[9]	80[12]	100[15]	120[17]	160[23]	Maximum
480/70R30	1180[2600]	1400[3080]	1650[3640]	1900[4180]	2120[4680]134B	2575[5680]141B	3550[7850]152B@320 kPa[46 psi]
380/85R34	1030[2270]	1250[2760]	1450[3200]	1650[3640]	1850[3860]127B	2300[5080]137B	2650[5840]142B@240 kPa[35 psi]
520/70R38	1500[3300]	1800[3960]	2120[4680]	2360[5200]	2725[6000]143B	3350[7400]150B	

All the notes above for radial tires apply also to metric radials also except the bold numbers are maximum for the Load Index and Speed Symbol shown. The above loads may be increased 70% (with a 40 kPa[6 psi] increase in pressure) for speeds under 8 km/hr[5 MPH] for operations without sustained high torque.
These tires may be used for cyclic loading service.

An example metric tire marking is:

380/85 R 34

where: 380   is the nominal section width in mm,
       85   is the aspect ratio (100 × section height/section width).
       R   indicates radial; B for bias; D for diagonal construction,
      34   is the nominal rim diameter in inches.

Metric tires also have designations for speed and load limits.

Speed Symbol	Limiting Speed	
A2	10 km/hr	[5 MPH]
A3	15 km/hr	[10 MPH]
A5	25 km/hr	[15 MPH]
A6	30 km/hr	[20 MPH]
A8	40 km/hr	[25 MPH]
B	50 km/hr	[30 MPH]

The strength of a metric tire is reported by the Load Index. This index is a numerical code associated with the maximum load that a tire can carry at the speed indicated by its Speed Symbol under specified conditions. The Service Designation includes both codes and is marked for example as 143 B.

The machinery manager can use Table 2.1 to insure long tire life by honoring the load limits for specific inflation pressures when adding ballast to tractor tires. Notice in Table 2.1 that the allowable load capacities are reduced with lower inflation pressures. The loads with metric units (kg) are in mass units and must be multiplied by g to get a force. The notes indicate modifications in the use of this table.

Excessive torque can cause a drive-wheel tire to slip on the rim and lose pressure. The maximum pull at the tire-ground surface contact point should not exceed the rated static load excluding momentary and occasional peak loads.

Table 2.2 lists the load limits for drive wheels having cyclic loading such as self-propelled combines, cotton pickers, sprayers and any other self-driven implement that gains or loses weight as it operates. Ballast is usually not needed for these tires nor is tire torque a problem. Soil compaction may be a serious problem however as these machines may have to operate when the soil is most susceptible to compaction. The machinery manager can refer to Table 2.2 and select the lowest tire pressure possible

TABLE 2.2. Tire Static Load Limits, kg[lb], at Various Inflation Pressures, kPa[psi]. Diagonal (bias) ply drive wheel tires used in cycling loading field service. Speeds limited to 8 km/hr[5 MPH].

Tire size	Ply rating	Max. loads	Infl. pressure
13.6-28	6	2400 [5300]	190 [28]
	8	2810 [6150]	250 [36]
	10	3180 [7000]	320 [46]
16.9-34	6	3360 [7400]	170 [24]
	8	3960 [8800]	220 [32]
	10	4420 [9700]	250 [36]
18.4-34	8	4300 [9500]	180 [26]
20.8-34	8	4960 [10900]	170 [24]
	10	5620 [12300]	190 [28]
11.2-3	8	41780 [3900]	170 [24]
13.6-3	8 6	2810 [6150]	190 [28]
15.5-3	8 6	2990 [6600]	180 [26]
16.9-3	8 6	3560 [7800]	170 [24]
	8	4180 [9250]	220 [32]
18.4-38	8	4540 [10000]	180 [26]
	10	5420 [12000]	230 [34]
	12	6080 [13400]	290 [42]
20.8-38	8	5240 [11500]	170 [24]
	10	5900 [13000]	190 [28]
	14	7250 [16000]	290 [42]
20.8-42	10	6250 [13800]	190 [28]

Note: When used as duals, tire loads must be reduced to 0.88 of table loads.

for the loads imposed and the increased tire deflection will spread the existing load over a greater area and limit the compaction. The note for Table 2.2 cautions that the load limits for a tire as one of a dual pair is less than that for a single tire as occasionally substantial overload may shift onto only one tire of the pair.

Fuel Consumption

Estimates of fuel consumption are of interest to farm machinery managers. Table 2.3 estimates fuel consumption for full and part loads. The data were compiled from Nebraska Tractor Test data before 1980 for gasoline and LP gas fuels. The data for diesels were reported in 1984 by A. Khalilian, D. Batchelder, K. Self, and J. Summers (Agricultural Engineering Department, Oklahoma State University). These data are for new tractors prepared for official testing. Aged tractors might not equal these efficiencies. The coefficient of variation for this data was 9.1%, which means two-thirds of all the data would be contained in a range ± 9.1% about the mean.

The use of Table 2.2 is illustrated by the following:

TABLE 2.3. Fuel Efficiency, kW·hr/L [HP·hr/gal]

Loading, % max.	Gasoline	LP Gas	Diesel Nat. aspirated	Diesel Turbo	Diesel Turbo and cooled
100	2.17 [11.01]	1.78 [9.06]	2.90 [14.72]	3.07 [15.58]	3.09 [15.68]
80	1.96 [9.95]	1.68 [8.55]	2.84 [14.41]	2.82 [14.31]	2.86 [14.52]
60	1.63 [8.30]	1.47 [7.50]	2.60 [13.19]	2.55 [12.94]	2.59 [13.15]
40	1.28 [6.45]	1.17 [5.95]	2.13 [10.81]	2.10 [10.66]	2.15 [10.91]
20	0.83 [4.20]	0.83 [4.20]	1.38 [7.00]	1.36 [6.90]	1.42 [7.21]

Note: Full-open governor control.

Estimate the fuel consumption of a 100 kW [134 HP], turbocharged, intercooled diesel tractor operating at half load.

One must interpolate between 40% load, 2.15 [10.91], and 60% load, 2.59 [13.15]. The calculations are:

$$21.1 \, \text{L/hr} = 50 \, \text{kW} \times \frac{1 \, \text{L}}{2.37 \, \text{kW} \cdot \text{hr}}$$

$$\left[5.157 \, \text{gal/hr} = 67 \, \text{HP} \times \frac{1 \, \text{gal}}{12.03 \, \text{HP} \cdot \text{hr}} \right]$$

Engine Oil Consumption

Engine oil consumption is an item of rather small importance but it should not be overlooked. Consumption is defined as the total volume of new oil placed in the engine in a given time period. The term consumption includes both the amount of oil used up in an engine and the amount of oil drained from the engine. Table 2.4 has been developed from Nebraska Tractor Test data and manufacturers' recommended oil change periods. The recommended oil change periods vary widely.

Power Requirements

Efficient power performance includes the selection of implements that neither overload nor fail to use adequately the power available from a tractor or a self-propelled engine. The previous sections show how tractor drawbar performance can be predicted. This section lists data and discusses procedures for predicting the power requirements of common field machine operations.

Field machine power requirements consist of functional requirements and rolling resistance requirements. The functional requirements are those that relate directly to the processing of soils, seeds, chemicals, or crops. Rolling resistance (RR) is usually an undesirable parasitic power requirement arising from the necessity for moving heavy machinery over soft field surfaces. Transport wagons have only rolling resistance power requirements. Tillage implements have only functional resistance if their weight is carried on their tooling and is necessary for tool penetration. A land roller or packer is a unique machine in that its rolling resistance is also its functional resistance. Most other implements have combinations of rolling and functional resistances. Trailed harvesting machines provide a functional load

TABLE 2.4. Average Crankcase Oil Consumption (assumes 150-hr change interval)

Maximum PTOP kW [HP]	Gasoline Sump capacity L [gal]	Gasoline Consumption L/hr [gal/hr]	Diesel Sump capacity L [gal]	Diesel Consumption L/hr [gal/hr]	LP Gas Sump capacity L [gal]	LP Gas Consumption L/hr [gal/hr]
0-15 [0-20]	2.6 [0.7]	0.017 [0.005]	5.3 [1.4]	0.035 [0.009]		
15-30 [20-40]	5.7 [1.5]	0.038 [0.010]	5.9 [1.6]	0.039 [0.011]		
30-45 [40-60]	6.2 [1.6]	0.041 [0.011]	7.0 [1.8]	0.047 [0.012]		
45-60 [60-80]	7.5 [2.0]	0.050 [0.013]	9.0 [2.4]	0.060 [0.016]	7.6 [2.0]	0.05 [0.013]
60-75 [80-100]	8.4 [2.2]	0.056 [0.015]	11.4 [3.0]	0.076 [0.020]	9.1 [2.4]	0.06 [0.016]
75-100 [100-134]	11.4 [3.0]	0.076 [0.020]	15.0 [4.0]	0.100 [0.027]	9.8 [2.6]	0.07 [0.017]
100-150 [134-200]			16.6 [4.4]	0.111 [0.029]		
150+ [200+]			20.2 [5.3]	0.135 [0.035]		

for the tractor's PTO drive and a rolling resistance load for its drawbar. Ground-driven machines load both their functional and rolling resistances through the tractor drawbar.

Functional requirements depend on soil and crop conditions, which are highly variable. Listed here are the relative drawbar pulls for moldboard plows in different soil types.

Sandy soil	1.0	Heavy clay	3.3
Sandy loam	1.6	Heavy clay sod	3.6
Silt loam	2.0	Moist gumbo	5.5
Clay loam	2.3	Dry adobe	7.8
Clay	2.8		

In addition to soil type, tillage draft varies with soil moisture. The draft of plows is increased if the soil is either too wet or too dry. Minimum draft seems to occur in the moisture range that produces a friable consistency of the soil. Root development, organic matter content, and depth of operation provide additional variability. Forward speed causes significant variation in plow draft.

The functional power requirements of most harvesting implements are sensitive to variable crop conditions. High crop yields can lead to high machine feed rates or throughputs and will cause a substantial increase in the power required to operate the harvester.

Variations in crop conditions and uneven feed rates can cause significant variation in power requirements. Merlin Hansen (Deere and Co.) reports the following ratios of operational torque peaks to average torque of unity for several PTO-driven machines: forage harvesters, 1.8; corn pickers, 1.2; combines, 2.0; rectangular balers, 5.1.

The rolling resistance power requirements of implements can be predicted. Fig. 2.12 from American Society of Agricultural Engineers (ASAE) Data D230.3 (1977) shows the expected coefficients of rolling resistance for conventional pneumatic tires used in agriculture that have a tire section width equal to about half the rim diameter such as 7.50–16 or 15.5–38. (Refer to Tables 2.1, 2.2 and 13.5, for tire sizes.) The outside diameter of tires is approximately 90% rim diameter plus twice their section width. The outside diameter of a 10.50–16 tire would thus be

$$(267 + 267 + 406) \times 0.9 = 846 \text{ mm}$$
$$[(10.5 + 10.5 + 16) \times 0.9 = 33 \text{ in.}]$$

Using this tire as an example, one would enter the left portion of Fig. 2.12 at 0.84 m [33 in.], go vertically (dashed line) to a surface condition of interest (tilled, settled soil line), and then go horizontally left to read 0.14 at 110 kPa [16 psi] inflation pressure; or go to the right graph for a

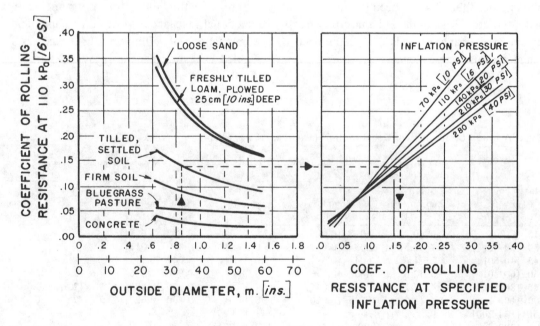

Fig. 2.12. Effect of tire diameter and inflation pressure on coefficient of rolling resistance.

different pressure, 210 kPa [30 psi], and drop down to read a different coefficient of 0.16. At rated inflation pressures, extra wide tires could have as much as 50% less rolling resistance on sand and 50% more on concrete.

Mounted and semimounted implements may have rolling resistance too. If some of the implement weight is carried on the tractor while operating, that weight times the coefficient of rolling resistance (RR) for the rear tractor tires is really implement rolling resistance. Such rolling resistance need not be included when computing drawbar pull for use in Fig. 2.11.

Power requirements can be quoted in many different ways depending on the characteristics of the operation and the machine. Quite often a force instead of a power requirement is reported to remove the effect of variations in forward speed. The variation due to different sizes of implements is removed by reporting draft per unit of effective width, per machine unit, or per row. Consequently, both speed and size variables are eliminated when technical work or energy requirements are quoted.

The unit of energy in the SI system is the joule, defined as one newton of force exerted through one metre of distance (J = N·m). In agriculture, the product of power and time per area has utility and is the energy unit used. A manager can estimate power requirements to obtain a certain capacity, or conversely, given the power available, estimate the capacity. The common unit for field work is kW·hr/ha [HP·hr/a]. For harvesters where yield affects power so greatly, an energy unit of kW·hr/t [HP·hr/T] is more appropriate. Rolling and functional resistances are combined for most tillage and ground-driven machine reports, while functional requirements only are reported for most PTO- and self-propelled machines.

Table 2.5 lists power, draft, and energy requirements for many field machines. Compiled from many sources and including estimates, it attempts to report ranges in which 90% of all actual operations would be included. These values are not expected to be very descriptive for any individual operation. A machinery manager can make determinations for a specific operation through use of an inexpensive pull meter (Fig. 2.13).

An example can illustrate how the power requirements for a field operation can be computed: Estimate the total power required for a 4-m [13-ft] light tandem disk harrow pulled behind a 4-m [13-ft], 500 kg [1100 lb] stalk chopper equipped with two 6.50-16 tires inflated to 170 kPa [24 psi]. Operating speed is 8 km/hr [5 MPH].

Fig. 2.13. Hydraulic pull meter.

Stalk chopper: From Table 2.5, the average power is 11 kW/m [4.5 HP/ft].

PTOP = 4 × 11 = 44 kW [13 × 4.5 = 58.5 HP]

From Fig. 2.12, the coefficient of RR at 170 kPa [24 psi] and for a tire diameter of 66 cm [0.9 x (16 + 6.5 + 6.5) = 26 in.] is 0.18.

$$DBP = \frac{0.18 \times 500 \times (9.807/1000) \times 8}{3.6} = 1.96 \, kW$$

$$\left[\frac{0.18 \times 1100 \times 5}{375} = 2.64 \, HP \right]$$

Disk harrow: From Table 2.4, the average force is 2.05 kN/m [140 lb/ft]. Total draft is 8.2 kN [1820 lb].

DBP = 8.2 × 8/3.6 = 18.22 kW [1820 × 5/375 = 24.3 HP]

Total power is 64.18 kW [85.44 HP].

Table 2.5. Field Machinery Power Requirements at 4.8 km/hr [3 MPH].

Machine	Draft, Force per Unit Width kN/m [lb/ft]		Energy or Work kW·hr/ha [HP·hr/a]	
Tillage Implements				
Plow, moldboard or disk				
(18-cm [7-in.] depth)				
light soils	3.2-6.3	[220-430]	8.7-17.5	[4.7-9.5]
medium soils	5.3-9.5	[350-650]	14.6-25.8	[7.9-14]
heavy soils	8.5-16.6	[580-1140]	22.1-46.1	[12-25]
Vertical disk plow				
(8-13 cm [3-5 in.] deep)	2.6-5.8	[180-400]	7.4-16.2	[4.0-8.8]
Lister, hard ground				
(1 m [38 in.] spacing)	5.8-14.6/unit	[400-1000/unit]	4.18-12.5	[2.6-6.8]
Subsoiler (40-cm [16-in.] deep)				
2-m [6-ft] spacing)				
light soils	16.0-26.3/unit	[1100-1800/unit]	7.2-12.0	[3.9-6.5]
medium soils	23.3-36.5/unit	[1600-2500/unit]	10.1-15.7	[5.5-8.5]
Land plane	4.4-11.7	[300-800]	12.2-31.3	[6.6-17]
Chisel plow				
(18-23 cm [7-9 in.] deep)	2.9-13.1	[200-900]	8.1-36.9	[4.4-20]
Field cultivator				
(8-13 cm [3-5 in.] deep)	0.9-4.4	[60-300]	2.4-12.0	[1.3-6.5]
Blade cultivator				
(2 m [6.5 ft] blade)				
7.5 cm [3 in.] deep	4.0-4.7	[275-320]	6.9-13	[6-7]
Disk harrow				
single gang	0.7-1.5	[50-100]	2.0-4.0	[1.1-2.2]
light tandem	1.5-2.6	[100-180]	4.0-7.4	[2.2-4.0]
heavy tandem	80-150% of weight		7.4-12.9	[4.0-7.0]
Spike-tooth harrow	0.3-0.9	[20-60]	0.7-2.4	[0.4-1.3]
Spring-tooth harrow	1.0-4.4	[70-300]	2.1-12.2	[1.7-6.6]
Rod weeder	0.5-1.8	[35-120]	1.5-4.8	[0.8-2.6]
Land roller or packer	0.3-0.9	[18-60]	0.7-2.4	[0.4-1.3]
Rotary tiller[a]				
(8-10 cm [3-4 in.] forward slice)	12.2-24.5 kW/m	[5-10 HP/ft]	25.8-51.6	[14-28]
Rotary harrow	10.0-15.0 kW/m	[4-6 HP/ft]	20.0-30.0	[11-17]
Seeders				
Row planter (1-m [38 in.] spacing)	0.45-0.80/row	[100-180/row]	1.1-2.4	[0.6-1.3]
fertilizer attachs.	1.1-1.8/row	[250-400/row]	3.1-5.2	[1.7-2.8]
Grain drill	1.0-1.6	[30-100]	1.1-3.9	[0.6-2.1]
No-till drill	2.3-3.5	[160-240]	6.4-9.7	[3.5-5.3]
Air seeder (draft only)	3.5-3.9	[240-270]	9.7-10.8	[5.3-6.0]
Chemical Applicators				
NH₃ applicator	5.1-7.3	[350-500]	4.4-6.5	[2.4-3.5]
Sprayer (10-80 cm[32 in.])[b]	0.2 kW	[0.3 HP]	0.02-.04	[.01-.02]
Broadcast distributor	0.7-2 kW	[1-3 HP]	0.2-0.4	[0.1-0.2]
Field distributor	0.3-1.2	[20-80]	0.9-3.1	[0.5-1.7]
Cultivation				
Row cultivators				
shallow	0.6-1.2	[40-80]	1.6-3.3	[0.9-1.8]
deep (8 cm [3 in.])	0.9-1.8	[60-120]	2.4-4.8	[1.3-2.6]
Rotary hoe	0.4-0.9	[30-60]	1.3-2.4	[0.7-1.3]
	PTO Power kW/m [HP/ft]		Energy or Work kW·hr/ha [HP·hr/a]	
Grain Harvesting				
Windrower (swather)[b]	1.4-1.9	[0.6-0.8]	2.9-4.0	[1.6-2.2]
Combines				
small grain	3.6-12.0	[1.5-4.9]	7.5-25	[4.1-13]
corn (0.76 m [30 in.] row)	7-11/row	[10-15/row]	20-30	[11-16.5]
Corn picker[b]	1.5-3.7/row	[2-5/row]	4.4-8.8	[2.4-4.8]

Table 2.5. Continued.

	PTO Power kW/m [HP/ft]		Energy or Work kW·hr/ha [HP·hr/a]	
Forage Harvesting				
Mower, cutterbar[b]	1.0-1.2	[0.4-0.5]	2-2.5	[1.1-1.4]
Mower, rotary disc[b]	3.5-6.5	[1.4-2.7]	7.3-13.5	[3.9-7.4]
Conditioner, cutterbar[b]	3.7-4.9	[1.5-2.0]	7.7-10.2	[4.1-5.5]
Conditioner, rotary disc[b]	5.6-10.4	[2.2-4.2]	11.6-21.7	[6.1-11.6]
Side-delivery rake[b]	0.5-0.8	[0.2-0.3]	1.0-1.7	[0.6-0.8]
Stalk chopper[b]	4.9-17.1	[2.0-7.0]	6.5-20	[3.5-11]
Baler, rectangular[b]				
hay			1.0-1.5	[1.2-2.7]
straw			1.4-1.5	[1.7-2.7]
Baler, round[b]			1.6-1.9	[1.9-2.3]
Forage harvester, flail[b]			0.9-2.0	[1.1-2.5]
Forage harvester, shear bar[b]				
green forage			0.8-2.0	[1.0-2.5]
wilted forage			1.2-4.1	[1.5-5.0]
hay, straw			1.6-4.1	[2.0-5.0]
corn ensilage			1.2-3.3	[1.5-4.0]
Tub grinder			3.0-9.1	[3.7-11.1]

	PTO Power kW/row [HP/row]		Energy or Work kW·hr/ha [HP·hr/a]	
Special Crop Harvesters				
Cotton picker, spindle				
(single pick,1-m [38-in.] rows)	7.5-11.2	[10-15]	15.6-23.3	[8.5-12.6]
Cotton stripper[b]				
(1-m [38-in.] rows)	1.5-2.2	[2-3]	3.1-4.6	[1.7-2.5]
Beet topper[b]				
(0.56-m [22-in.] rows)	3.7-5.2	[5-7]	7.7-10.8	[4.2-5.9]
Beet lifter				
(0.56-m [22-in.] rows)	1.5-3.0	[2-4]	13-26	[7.1-14.1]
draft force	2-4 kN/row	[450-900 lb/row]		
Potato harvester				
(1-m [38-in.] rows)	0.7-1.5	[1-2]	14.4-23.5	[7.8-12.8]
draft force	2.2-3.5 kN/row	[500-800 lb/row]		

[a]Has negative rolling resistance.

[b]Does not include implement or wagon rolling resistance.

Part-Load Operation

C. J. Ricketts and J. A. Weber (University of Illinois) found a rather uniform distribution of the time an "average" tractor spends at each power loading. The data in Table 2.6 are based on their work with a 35 kW [48 HP] tractor used in actual on-farm work.

It would appear that the power performance of tractors at part load is a very important consideration for good power and machinery management. The effects on fuel efficiency and wear are of primary interest. Ricketts and Weber indicated the fuel efficiency of their gasoline tractor with a diagram similar to that in Fig. 2.14. Lines of constant fuel efficiency are superimposed on the horsepower-engine speed graph. The dotted lines indicate the performance when the governor control or hand-operated speed control lever is set at fractions of the full-open position. Potential fuel economy is demonstrated by considering a partial

TABLE 2.6. Distribution of Tractor Power Loading

% Maximum Power	% Total Time
Over 80	16.8
80-60	23.9
60-40	22.6
40-20	17.5
Less than 20	19.2

load of 40% on the tractor. If the governor control is set at full-open the fuel efficiency is about 12%. By adjusting the transmission gear ratio it is possible to reduce the governor control setting to one-fourth open and still maintain the same forward speed and power output. In doing so the fuel efficiency will improve to 17%.

In a similar study, Samuel G. Huber and Benson J. Lamp (Ohio State University) found that although there are differences in individual tractors, considerable fuel savings can be realized by reducing the governor

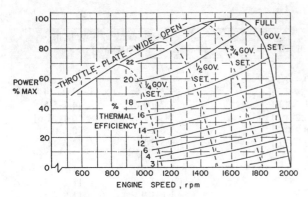

Fig. 2.14. Fuel efficiency for partial governor settings.

control setting at part loads. Fig. 2.15 and 2.16 compare the efficiency for full-open setting to the optimum setting for gasoline and diesel tractors. Note in Fig. 2.16 that the greatest fuel efficiency occurs at less than maximum power for this particular diesel tractor.

The effect of increased fuel efficiency with a

reduced governor control setting is demonstrated in Fig. 2.9. The average fuel efficiencies of the 2000 rpm drawbar tests are over 4% greater than those for the 2200 rpm tests.

One should be cautious about improving part-load power performance by reducing the governor control setting. While the fuel efficiency is increased, the torque load on the tractor is greater and the tractor may be overloaded. *Overload* describes the condition when a tractor has received demands for torque beyond that amount available at maximum horsepower and consistent with the governor control setting. Overloading causes extra heavy forces on crankshaft and connecting rod bearings, pistons, and transmission gear teeth and bearings. Depending on the reserve built into these components by the designer and on the effectiveness of lubrication, overloading can lead to early failure. Certainly more tractors are damaged by overloading than by operating at high engine speeds.

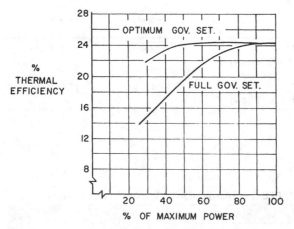

Fig. 2.15. Potential fuel efficiency for gasoline tractor with reduced governor settings.

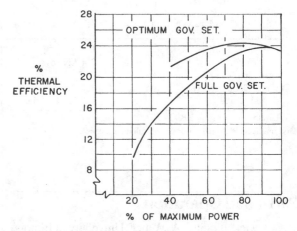

Fig. 2.16. Potential fuel efficiency for diesel tractor with reduced governor settings.

Practice Problems

2.1. a. What power is required to lift a mounted plow 60 cm [23.6 in.] in 3 s? The plow's mass is 300 kg [660 lb].

   b. Compute the DBP of a tractor pulling a 9-kN [2025-lb] load at 8 km/hr [5 MPH].

   c. What power is a tractor producing if a dynamometer test shows a scale reading of 900 N [202 lb] at 1000 rpm? The scale arm length is 0.3 m [1 ft].

   d. During a sharp turn the power steering system of a tractor requires a hydraulic fluid flow of 1.25 L/s [19.8 gal/min] at a pressure of 14,000 kPa [2030 psi]. What is the power requirement?

   e. A tractor cab heater fan draws 6 A at 12 V. Compute EP.

2.2. A forage blower elevates silage into a silo 20 m [65.6 ft] high at the rate of 500 kg [1100 lb]/min. The power efficiency of the blower is 8%. What PTO torque is required from the tractor at 1000 rpm?

2.3. Predict maximum pull, drive-wheel slip, and maximum forward speed at maximum tractive efficiency for a 75-kW [100 HP] PTO, 10,000-kg [22,000-lb] 2-wheel drive tractor on tilled soil while pulling

   a. Towed or trailed implement

   b. Fully mounted or integral implement. Use the static weight distribution assumed by Fig 2.11.

2.4. Use a load factor of 0.8 and select the PTOP necessary to pull a 6–40 cm [6–16 in.] semimounted moldboard plow having an average draft of 7 kN/m [480 lb/ft] at 8 km/hr [5 MPH] on firm soil. Predict the total tractor mass, including ballast, to produce maximum tractive efficiency.

2.5. Use Fig. 2.10 to compare the predicted slip and travel speed for the tractor in Fig. 2.10, front drive disengaged, to that for the test data for maximum power in sixth gear.

2.6. Calculate the various fuel efficiencies for a tractor delivering 75 kW [100 HP] while using $0.26/L [$1.00/gal] fuel at the rate of 23 L/hr [6.076 gal/hr]. The fuel density is 0.84 kg/L [7 lb/gal] and its energy value is 36,000 kJ/L [129,170 Btu/gal].

   a. kW·hr/L [HP·hr/gal]

   b. kg/(kW·hr) [lb/(HP·hr)]

   c. ¢/(kW·hr) [¢/(HP·hr)]

   d. % thermal efficiency

2.7. A 6-row planter with 1-m [39.4-in.] row spacing is converted to 75-cm [29.5-in.] spacing. Use average draft figures for planters with attachments and find the increase in energy per hectare [acre] that this change would cause.

2.8. Find the total implement power requirements for a 2-row mounted cotton stripper on a tractor with 7.50–16 tires in front and 14.9-38 tires on the rear. Assume the 1-t [1.1-T] implement's weight is distributed one-third on the front wheels and two-thirds on the rear wheels. Stripping speed is 5 km/hr [3.125 MPH]. Use inflation pressures of 170 kPa [24 psi] front, 110 kPa [16 psi] rear.

2.9. An 800-kg [1760-lb] wagon is pulled behind the tractor in 2.8 to catch the yield. Its tires are sized to fit 16-in. rims and are to be the smallest possible to carry a 3000-kg [6600-lb] load in the wagon. Predict the wagon's rolling resistance and power requirement when full.

2.10. A 3-t [3.3-T] tractor is to pull a 2-t [2.2-T], 2-row (1-m [38-in.] spacing), 2-wheeled forage harvester at 4 km/hr [2.5 MPH] to produce corn silage in a field yielding 24.7 t/ha [11 T/a]. A 1-t [1.1-T] 4-wheeled wagon will carry 4 t [4.4 T] when filled. Tire equipment is 9.00–24 for the harvester and 9.00–16 for the wagon. Assume 170 kPa [24 psi] inflation pressures. The tractors rear axle carries 75% of the static weight.

a. Use average power and energy requirements for firm soil and predict the PTOP for the tractor when the wagon is full.

b. Are there any tires overloaded during this operation? See Table 13.5.

OPERATOR PERFORMANCE **3**

The third component of economic performance of a machine system is operator performance. A manager of equipment may be quite knowledgeable about machine and power performance; but unless the machine operator's performance also is high the total system performance may be low.

A manager must consider the type, amount, and value of operator labor required when planning for mechanized agricultural production. In addition, the manager now is required by law to provide a safe environment and safety education as related to equipment operation.

The operation of farm machinery is not physically strenuous but is fatiguing because of the need for continual alertness. Fig. 3.1 shows most of the many controls and instruments that must receive the attention of a combine operator. The major controls are steering (1); engine governor control (2); combine drive clutch (3); one-hand hydrostatic propulsion speed control including header lift, reel lift, and automatic header height control switches (4); transmission gear shift (5); and propulsion brakes (6). Controls out of view are a header reversing clutch, an unloading auger clutch, and a swing-out control for the unloading auger. Also not shown are the cab environmental controls for heating, air conditioning, and lights.

Gages (A) include coolant temperature, engine oil pressure, battery charge, and fuel level. An hour meter (H) indicates accumulated operating time. As a help for attracting the operator's attention to the gages, a bank of warning lights (W) indicates any out-of-normal conditions for oil pressures, coolant level and temperature, battery charge, air filter resistance, brakes applied, and rear wheel drive engaged.

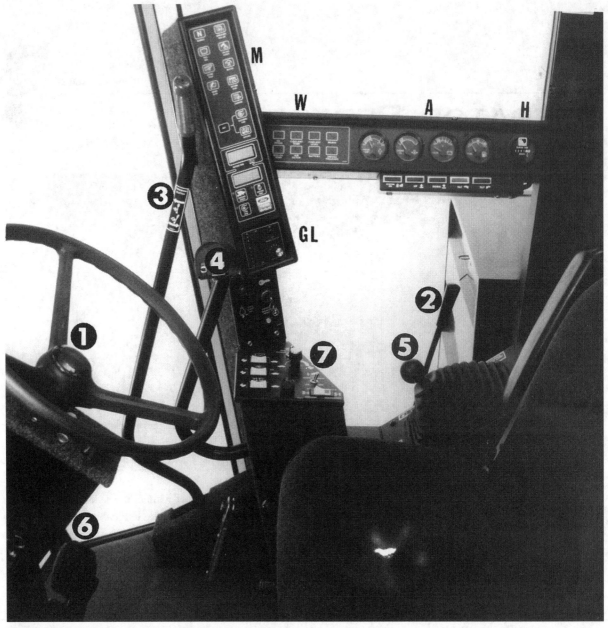

Fig. 3.1. Controls and instruments in a modern combine cab.

A speed control console (7) allows for the adjustment of combine cylinder, header drive, beater, and reel rpm. A grain loss monitor (GL) indicates the operating efficiency of the combine. A monitoring instrument panel (M) provides digital readings of ground speed; indicates engine, cylinder, and fan rpm; includes warning lights for the non-operation of the straw spreader or chopper, rear beater, tailings return elevator, clean grain elevator, cleaning fan and cylinder; shows the position of the unloading auger and

stone trap; and indicates an overfull grain bin. A skilled operator will understand and make use of all the controls and instrumentation provided by the manufacturer.

The need for alertness increases with the size and complexity of machines. Small, simple machines may require only steering activities from an operator. Large, complex machines require only a little more attention to steering but much more activity in monitoring the machine's operation. In cultivating with a

2-row machine it is rather easy to observe the effectiveness of the 10–12 shovels or sweeps; but with a 12-row cultivator there will be 60 to 72 different points to watch, many of which are far from the operator and difficult to see. In some critical operations such as seeding, where malfunction can mean unseeded areas producing no income, the costs of inattention can be substantial.

Operator Skill

The individualism of people is an important factor in labor performance. K. Von Bargen (University of Nebraska) found that differences in ability, motivation, alertness, and training of an operator can have significant effects on operator performance. In cases under similar field environments, one operator of a 4.27-m [14-ft] windrower was able to do 2 ha/day [5a/day] extra because among other things he could consistently average 15 cm [6 in.] greater width of cut than could the other operators. Fig. 3.2 shows a windrower operator undercutting about 15 cm [6 in.]. Fig. 5.7 shows an overlap of 10% for a wide tillage implement.

Von Bargen analyzed field times for several operator-machine combinations. He defined the potential capacity of the combinations as based on operating time plus the time required for turns, for travel between fields, and for delays due to necessary machine support activi-

ties. (This base is not directly comparable to field efficiency.) Time for the operator; for machine maintenance, service, and repair; and for undesirable delays were defined as idle or lost time.

Fig. 3.3, 3.4, and 3.5 show the variations in operator-machine performance. The points on the curves are to be read as percentage of total time in which the machine was operating at the given percentage of potential capacity. The areas under the curves represent total time and are equal to each other. The data for the operators are as follows:

Operator A. Age 29, experienced with balers, a former farm operator.

Operator E. Age 25, one season's experience with SP balers, general farm background, very adept mechanically.

Operator F. Age 17, no previous windrower experience, general farm background.

Operator H. Age 19, one season's experience in pull-type bale wagon operations, general farm background.

Operator I. Age 17, no previous experience with bale wagons, general farm background.

Operator J. Age 18, no windrowing experience, general farm background.

Operator K. Age 18, custom baling experience, general farm background.

Fig. 3.2. A 93% effective cut, E_w = 0.93.

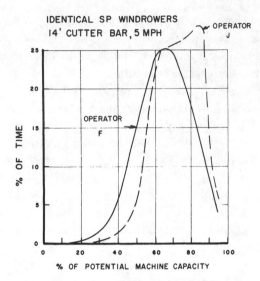

Fig. 3.3. Operator-machine performance, SP windrowers.

Good performance was defined as operating at 65% of potential capacity or higher. In Fig. 3.3, Operator J achieved this performance 69% of the time, while Operator F could attain this performance only 48% of the time as determined by comparing the areas under the curves. In Fig. 3.4 all three baler operators had their peak percent of time above 65%; but Operator E spent 66% of the total time above 65, while Operator K spent only 48%.

These curves can be misleading. For the bale wagons (Fig. 3.5), Operator H was classified as the superior operator even though Operator I spent a greater amount of time above the 65% mark. Operator H was able to load bales so fast that he wasted a considerable amount of time waiting for more bales from the baler, another illustration of the need for a systems approach to machinery management analyses. In general, this analysis tends to penalize operators who drove at faster field speeds. High field speeds reduce theoretical field or operating time relative to

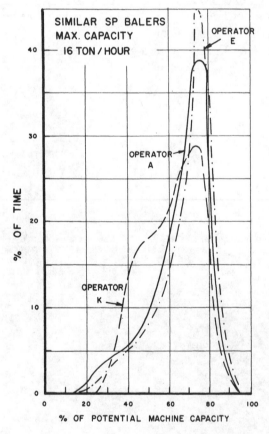

Fig. 3.4. Operator-machine performance, SP baler.

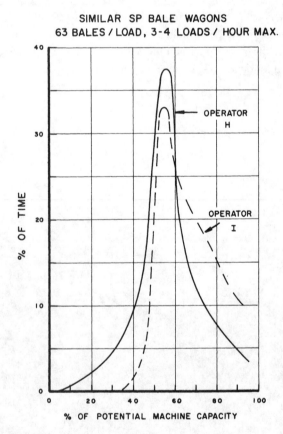

Fig. 3.5. Operator-machine performance, SP bale wagons.

total time. As with field efficiencies in Chapter 1, time efficiencies look poor even though the actual accomplishments in acres per hour or bales per hour may be greater. The determination of an optimum field speed is one of the real challenges for the machinery manager and the machine operator.

These studies do not permit definite conclusions about the type of labor required for high-performance machines. There is no real indication that either operator age or experience gave increased performance. The factors of motivation and mental alertness are probably of greater significance but are difficult for a farm machinery manager to determine before hiring an operator.

Operator Aids for Control

Several automatic mechanical-hydraulic systems have been developed to relieve the operator of constant monitoring and control functions. Fig. 3.6 shows a mechanical sensor that acts when a trailing, row-crop harvester head is off the row enough so that the plant stalks strike the sensor. Hydraulic pressure is then triggered to actuate a cylinder between the header and the hinged tongue which automatically realigns the header on the row. With this assistance, the tractor driver is relieved of the need for precise steering.

Another example of mechanical machine control is illustrated in Fig. 3.7. The delivery spout of the forage harvester is controlled by the position of the wagon tongue relative to the harvester. When the harvester makes a turn, the spout is turned by a hydraulic motor to point to the wagon without the attention of the operator. Additionally, the motor oscillates the spout back and forth during straight-ahead operation to distribute the forage evenly in the wagon.

A combine header height system, as shown in Fig. 10.9, is an effective automatic control for harvesting low-growing, podded crops such as soybeans. The fingers gage the height of the cutterbar and through a hydraulic system raises or lowers the head to maintain a desired height above the soil surface.

There are mechanical aids that allow the steering function to be done automatically. Mechanical sensors follow a furrow, a ridge, a windrow, or a plant row and signal the steering system to make corrections. The operator disengages the system for field turns but is relived of steering for more attention to machine monitoring and control.

Instrumentation

Farm machines can have considerable instrumentation to assist the operator in making appropriate decisions and then taking proper actions. Use of instrumentation can make field operations more efficient and can be conserving of both the operator and

Fig. 3.6. Guidance control for a row crop head.

Fig. 3.7. Mechanical control for a delivery spout.

the machine. Machinery managers and operators should understand these instrumentation systems for optimum use of the information developed, for maintenance purposes, and for determining replacement needs.

The instrumentation systems on farm machines can be grouped into three levels of technology. The first level is one of display only, sometimes called *status indicators*. Gages are mounted on an instrument panel to be monitored by the operator. Typically in the past, such gages have been analog in nature in that the relative position of a pointer sweeping across a semicircular face indicates the amount of the reading. More recently, digital instruments have replaced most analog gages. Digital instruments display a single measurement number at any instant. These digital displays are frequently larger, lighted, and provide data that are easily read. Examples of this level of instrumentation are the common engine sensing systems for coolant temperature, oil pressure, ammeter, voltmeter, engine speed, air cleaner resistance, and fuel levels. Examples of implement operation display systems are seed counters for planting machines, spray delivery, speed indicators for combine cylinders and cotton picker drums, combine cylinder-concave clearance and stone trap door position, and shape of a round bale being formed. The operator must monitor these displays and take corrective action when conditions warrant. Some systems produce a sound signal or a flashing light when an operating condition is outside an acceptable range.

A second level of instrumentation technology provides the operator with increased control over the machine functions. This level has been increasingly employed in farm machinery with the advances in solid-state electronics. Such circuitry typically converts a sensed condition into an electronic signal, displays a value to the operator, and then sends the operators electronic response to a hydraulic actuator which controls the machine adjustment. Such adjustments can be made instantly without stopping the forward progress of the machine. Variations in soil fertility and condition can occur over a very short distance of field travel. Quick adjustments to application rates of fertilizer, seed, and pesticides allows this instrumentation level to meet the demanding needs of modern precision agriculture.

Fig. 3.8 illustrates a second level monitor display for a baler. Touch panels provide control of bale density and the unloading of bales into an accumulator (refer to Fig. 11.20). The digital display indicates the

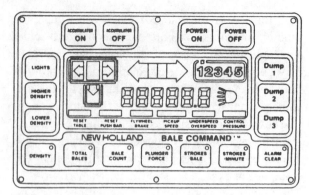

Fig. 3.8. Baler monitor in a tractor cab.

bale density, the bale count and bales per time, the plunger strokes per minute and per bale, the location of the bales on the accumulator, and the forces on the plunger. Indicators advise the tractor operator to drive to the left or right over the windrow to produce a more uniform bale cross section and which of the 5 knotters missed a tie. Warning lights and a horn indicate failures of density-control hydraulic pressure, over- and underspeeds, pick-up plugs, stoppages, flywheel brake activation and nonready accumulator status.

An important second-level instrument is the radar-like sensor of Fig. 3.9. This sensor measures the machine's actual ground speed as opposed to speed indicators based on wheel revolution. These data can be processed into a display of forward speed and drive-wheel slip. These data are also used to display the amount of area processed. The operator enters the effective width of the implement into the memory of the microprocessor, which combines time and the distance traveled to display an instantaneous area performance rate. When combined with the data from a chemical or seeding system, a current application rate

Fig. 3.9. Sensor for measuring true ground speed.

is computed and displayed to the operator. At this level the operator is a judgment link between the sensed display and the control that can modify the machine function.

The third level of instrumentation technology senses conditions and makes adjustments to machine operations without operator involvement. Such systems can be engaged or disengaged at the discretion of the operator who is always able to take over command from such automatic systems. Global positioning systems (GPS) can be utilized to identify the position of the machine in the field, and record it in memory. Fig. 3.10 shows the necessary signal travel from the satellite to the machine. Fig. 3.11 shows the instrumentation needed to use GPS. In addition to position location this equipment monitors grain moisture and yield and area covered for combine harvesting. The equipment is portable to seeders and fertilizer applicators to provide application rate control. The parallel swathing indicator assists the operator in steering machines so that overlap and skipped strips are avoided. This feature is particularly valuable for night time operations where it is difficult to see the swath edge even if lighted. An additional correctional signal added to GPS (Differential Global Positioning System or DGPS) allows location accuracy to about one-fourth of a metre. This location can be automatically compared with a map of the field attributes stored in memory. An appropriate implement adjustment signal is then created and an adjustment made for that specific sector of the field in which the implement is operating. Yield maps are developed during harvest using GPS to relate the instantaneous yield to location in the field. These data can be recorded on cards for later reading into a computer which prints the map. Soil fertility maps are developed from a grid sampling of the field.

An eventual technology level may avoid the maps by having instantaneous sensing of soil conditions and weed and pest infestations with consequent implement

Fig. 3.11. GPS antenna, display console, and indicator for parallel swaths.

Fig. 3.10. GPS signal travel.

response. Electronic machine vision is a potential for guiding tractors along rows, identifying weeds for spraying controlled amounts, etc. GPS has the potential for guiding implements around fields which may make farm machinery field operations completely driver free. As with all technologies, adoption will depend on the economic benefits to be gained.

Amount of Labor

The amount of farm labor required for production depends on the amount of mechanization and upon the size of the implements. Table 3.1 reports the amount of labor used on Illinois grain farms by size of farm. As farm size increases, the efficiency of labor use increases. Such efficiencies are possible for large farms since high-capacity equipment under the control of a single operator can be economically employed. The area per worker for large farms is about 300 ha [740 a]. This labor efficiency may be near the ultimate if these farms are using the highest-capacity machines available commercially. Even larger farms having multiple machines would not be expected to have greater labor efficiency since the additional machines require additional operators. Until larger machines become available, the 300 ha [740 a] area per worker may be the effective amount that a corn and soybean farm worker can operate. This productivity is about 2700% of the 1890 value of 27.5 a [11 ha] per worker reported by *The Illinois Agriculturist* (14[3], 1909).

Value of Labor

The value of labor for machine operations can be determined in several ways, each of which may have merit.

One obvious evaluation method is to portion out the cost of actual hired labor according to the hours of time spent operating the equipment. Such a method determines the value of farm labor by having it match off-farm labor rates. This is a very realistic way for evaluating hired labor, but it is not too pertinent a criterion for the manager-operator's wage.

Meeting labor competition from other farm enterprises is one way of determining the cost of the manager-operator's time. Perhaps $8 can be earned for each hour spent in a livestock enterprise, which can be expanded to compete for time with the field operations. If so, the time operating machinery is worth $8 per hour. Needless to say, a rather detailed economic analysis of each farm enterprise is needed before such data are available.

Another method is to consider labor for operating machines not as an expense but as an investment with an opportunity for profit. As suggested at the beginning of Part I, Economic Performance, machine operations are undertaken only when they increase the value of a product. If the value is increased beyond the power and machinery costs, the difference may be assigned as a wage to the manager-operator.

The simplest analysis suggests that the manager-operator's return be divided into two categories: A return for labor at an arbitrarily fixed rate and a return for management of the profits left over.

Safety

The safety of farm machinery operators is a major concern for machinery managers. Despite precautions, agriculture does suffer many accidents. The National Safety Council (NSC) reported that in 1990 the agriculture industry had an accident death rate of 42 per 100,000 workers. This rate was exceeded only by the 43 for mining and quarrying industries. Other high-risk industries were construction with a rate of 33 and transportation and public utilities at 22. The death rate for all industries combined was only 9. The agriculture rate does not include persons under 14 years of age. But NSC data report that 10% of all accidental

TABLE 3.1. Grain Farm Area per Farm Worker, Northern and Central Illinois

Category	% of total	Avg. Farm Size		All Labor (months/yr)	Hired Labor (months/yr)	Area/Worker	
		ha	[a]			ha	[a]
Small	8.5	105	[259]	11.9	0.5	106	[262]
Medium small	56.1	223	[550]	13.2	1.2	203	[502]
Medium large	24.0	374	[923]	16.8	3.9	267	[660]
Large	11.4	610	[1507]	24.7	8.4	296	[731]
All farms	100.0	296	[732]	15.4	2.7	231	[571]

Source: 1990 Summary of Illinois Farm Business Records, Circular 1316, Cooperative Extension Service, College of Agriculture, University of Illinois, Urbana.

deaths in agriculture are for persons under age 14. The disabling injury rate for farm workers doing farm work was 1520 per 100,000. One must conclude that agriculture is a high-risk industry. Some good news is that the death rate has fallen 20% in the past 10 years. It should be noted that the farm worker is one and a half times more at risk in highway travel than in farm work.

Farm powered-equipment accidents are the major source of agricultural injuries and deaths. Listed below here are the percentages of each cause of farm accident.

Machinery	48.7%
Drowning	9.4%
Struck by object	9.1%
Falls	7.0%
Firearms	5.5%
Animals	3.4%
Electric current	3.1%
Suffocation	2.8%
Burns	2.0%
Lightning	1.4%
Poison	1.0%
Other	6.5%

The 48% value for machinery hasn't changed appreciably for the past 10 years. Table 3.2 reports the agencies involved in accidents and the relative injury. The top five listings are machines that are the responsibility of the farm machinery manager.

In previous studies, farm machinery accidents were related to specific machines as follows:

Tractors	27%
Wagons	16%
Elevators	11%
Combines	7%
Corn pickers	3%

Mowers	3%
Balers	2%
Other	31%

Other data collected by the NSC show that most accidents occur in June and July. Persons of ages 25 to 44 are at greatest risk. Accidents appeared to peak at 10 a.m. and 2 p.m.

Table 3.3 reports an older study by NSC and relates the number of machinery injuries to the activity of the operator and the use of the machine (not tractor). Repairing and replacing components of a stopped machine produce the greatest number of injuries in the individual categories.

Tractor accidents causing fatalities amounted to 9.9 deaths/100,000 tractors in 1990. Of these deaths, 52% were from overturning the tractor, 33% from being overrun, and 3% from being caught in the PTO.

Federal legislation has been enacted in the United States to promote safety at work. Several regulations relating to farm equipment operation have been published by the Occupational Safety and Health Administration (OSHA), U.S. Department of Labor. These regulations apply whenever an employee is operat-

TABLE 3.2. Farm Accidents by Agency (% of total)

Agency	All Injuries	Permanently Impaired	Deaths
Machinery	17.6	47.5	13.3
Tractors	7.9	7.5	26.7
Trucks	4.1	3.7	3.3
Other vehicles	10.2	7.5	13.3
Power tools	4.7	10.0	0.0
Subtotal	(44.5)	(76.2)	(56.6)
Animals	16.9	7.5	0.0
Hand tools	7.5	2.5	0.0
All other	31.1	13.8	43.4

Source: Occupational Injuries in Agriculture: A 35-State Summary, NSC, 1988.

TABLE 3.3. Distribution of Injuries by Activity of Victim for Each Use of Machine (%)

Activity	All Injuries	Use of Machine					
		Stopped, Not Running	Stopped, Running	Harvesting	Loading	In Transit	Other
Operating	15.4	6.6	23.5	22.4	3.3	23.1	18.5
Repairing	14.4	32.8	7.4	5.2	6.7	7.7	6.7
Adjusting	11.5	14.8	14.7	6.9	10.0	7.7	9.2
Cleaning	9.0	9.8	13.3	17.3	0.0	0.0	5.1
Riding	4.9	1.6	2.9	8.6	6.7	15.4	5.9
Feeding material	4.6	0.0	4.4	5.1	13.3	0.0	7.6
Bystander	3.4	0.8	4.4	1.7	3.3	15.4	5.0
Other	36.8	33.6	29.4	32.8	56.7	30.7	42.0
Total	100.0	100.0	100.0	100.0	100.0	100.0	100.0
% Injuries	100.0	29.8	16.6	14.1	7.3	3.2	29.0

ing equipment. Compliance is not mandatory if only family members operate the equipment. The equipment operation areas judged to be most hazardous have received initial attention and regulation. Additional regulations may be expected in the future.

Highlights of the present regulations pertaining to farm field equipment and operations are listed here.

Tractors

A. All tractors with 15-kW [20-HP] engines must be equipped with rollover protective structures (ROPS) and seat belts. Figure 2.10 shows a ROPS designed to meet OSHA standards. Enclosed tractor cabs must meet the same standards. Low-profile tractors and tractors mounted with equipment incompatible with ROPS are exempt, but only while the tractor is being used in operations where vertical clearance is insufficient or where mounted equipment occupies the ROPS space. After such operations are completed, the ROPS must be reinstalled.

B. The employer must instruct the employee at the time of the initial assignment to the tractor and annually thereafter as follows:

1. Securely fasten your seat belt if the tractor has ROPS.
2. Where possible, avoid operating the tractor near ditches, embankments, and holes.
   on rough, slick, or muddy surfaces.
3. Stay off slopes too steep for safe operation.
4. Watch where you are going, especially at row ends, on roads, and around trees.
5. Do not permit others to ride.
6. Operate the tractor smoothly with no jerky turns, stops, or starts.
7. Hitch only to the drawbar and hitch points recommended by the tractor manufacturer.
8. When the tractor is stopped, set brakes securely and use park lock if available.

Implements

A. All farm field and farmstead equipment, regardless of the date of manufacture, must be provided with guarding of all PTO drives. The shielding must be sturdy enough to withstand the effects of a 113-kg [250-lb] person stepping on the installed shield. When the PTO is disconnected and removed, the protruding stub shaft from the tractor shall be guarded.

All new farm machines must have guards or shielding over moving parts. These include rotating shafts; ground-drive power trains; and the pinch or nip points that occur as belts feed onto pulleys, chains feed onto sprockets and idlers, and gears and other mechanisms mesh. Even the functional parts of machines (cutter bars, reels, snapping rolls, etc.) must be shielded as much as possible and yet permit their intended function. Warning signs and motion signals are required where shields may have to be removed to get at moving parts for service and maintenance reasons.

B. At the time of the initial assignment and at least annually thereafter, the employer must instruct the employee in the safe operation and servicing of the equipment including at least the following:

1. Keep all guards in place when the machine is in operation.
2. Permit no riders on farm field equipment other than persons required for instruction or assistance in machine operations.
3. Stop engine, disconnect the power source, and wait for all machine movement to stop before servicing, adjusting, cleaning, or unclogging the equipment except where the machine must be running to be properly serviced or maintained, in which case the employer shall instruct employees as to all steps and procedures necessary to safely service or maintain the equipment.
4. Make sure everyone is clear of the machinery before starting the engine, engaging power, or operating the machine.
5. Lock out electrical power before performing maintenance or service on farmstead equipment.

Noise

OSHA has set noise limits for the environment surrounding workers. The basic permissible noise intensity is 90 decibels on the A scale of a sound level meter, 90 dB(A), at slow response for an eight-hour day. The slow response setting averages out intermittent high-level noises such as hammering. Higher levels of noise are permitted for shorter periods (Table 3.4). Exposure to various sound levels are accommodated by formula. The ratios of the actual exposure time to permissible exposure time at each sound level must sum to less than one. In effect, this formula requires that part of the day must include exposures to less than 90 dB(A) if there is any expo-

Table 3.4. Permissible OSHA Noise Exposures

Level, dB(A)	hr/day
90	8
92	6
95	4
97	3
100	2
102	1 1/2
105	1
110	1/2
115	1/4
140	0

sure to levels above 90 dB(A). This scale is logarithmic. An increase of 10 dB(A) will approximately double the loudness to the human ear.

The machinery manager may be able to reduce the noise of farm machinery to acceptable levels; but if not, ear plugs or ear muffs must be provided and must be worn.

Anhydrous Ammonia Application

The handling of anhydrous ammonia has been considered hazardous by OSHA. Workers are required to wear impervious gloves and goggles or a face shield for protection in the case of an accidental release of ammonia. A water supply of at least 19 L [5 gal] must be available on the tank-transporting vehicle, which must have a drawbar equipped with a safety chain and which must not whip or swerve dangerously from side to side when being towed. The tank must be securely fastened to the vehicle.

For fertilizer applicator tanks of 1000 L or 1 m³ [250 gal] capacity or less, tanks may be filled by venting to open air provided the bleeder valve orifice does not exceed 1.11 cm [7/16 in.] in diameter.

Other regulations cover the design of tanks, hoses, and valves. The responsible machinery manager will rent or purchase only equipment that meets OSHA standards and will see that this equipment is maintained.

Emblem

The slow-moving-vehicle (SMV, 40 km/hr [25 MPH]) emblem is required for all farm equipment driven on roads (Fig. 5.4). The emblem must be located in a vertical position 0.6 to 1.8 m [2 to 6 ft] above the ground and in a central position on the rear of the vehicle.

Various administrative responsibilities are required by OSHA. If ten or more workers are employed on a farm or a ranch, OSHA requires records of job-related deaths, injuries, and illnesses. A supplemental record of each incident and an annual summary are also required. All farmers with at least one employee are required to post a copy of the OSHA poster, which explains employee rights and responsibilities. A farmer must report to OSHA within 48 hr the details of a farm accident resulting in one or more deaths or the hospitalization of five or more employees. Farmers must also cooperate with surveys conducted by the Bureau of Labor Statistics. Farms and other businesses are subject to inspection by OSHA compliance officers if an employee files a complaint, if a major accident occurs, or if the farm is randomly selected for inspection.

Census statistics show that hired employees make up less than 27% of the U.S. farm labor force; thus, many farms need not comply with OSHA standards. A safety-conscious farmer, however, will comply with OSHA standards since in all probability it is family members that will be protected. Additionally, a farmer must provide for those occasions when temporary labor will operate the machines.

II

COSTS

Most management decisions for farm machinery involve an accurate knowledge of costs. Maintaining an accurate record of costs is an indispensable part of the machinery manager's job. The cost of maintaining and operating machinery is a deductible business expense for income tax purposes, which is sometimes the sole reason for keeping good cost records.

But such records aren't being used to their maximum extent unless these records are also used to aid in decision-making processes that are the core of good machinery management. Disappointingly, the use of cost data has to be made by estimation only since the actual cost of use of a machine is never known until the machine has been sold or worn out and junked. One can't postpone decisions until then to determine the cost of operations. The ideal situation is some running account wherein the manager can estimate closely the cost per unit of production of a particular machine for any operation and at any moment.

Representative machinery costs have been a popular subject for agricultural experiment stations. Publications on costs are available for most field machine operations in many sections of the nation. The applicability of such costs suffers from rapid changes in prices and technology and costs are often of historical interest in only a few years. In no way can it be said that there are standard costs for field operations as there are in some manufacturing industries.

The determination of the field machinery cost of operations depends on so many factors that each farm's machinery system must be treated as a special case. Significant differences in machine use, price levels, energy requirements, fuel costs, and labor costs suggest that the farm machinery manager will have to develop individual standard costs and use the average costs obtained by others only for comparison purposes.

The ultimate goal of the machinery manager is to maximize enterprise profits by getting the greatest output from machines at a minimum cost. Note again that it is the cost of use of a machinery system that is to be minimized, not the cost of use of individual implements. A long-range management viewpoint must also be maintained; that is, when the economy of a machine is being evaluated, one must keep in mind that the present machine is only one of a succession of machines whose costs are to be minimized. All machines are eventually dropped from use because the farming practice for which they are needed is dropped or they are replaced with a successor machine. The time of replacement of a machine is perhaps the most important decision-making activity of the machinery manager.

This section describes the elements that make up the cost of a machine. Depreciation is described in detail and income tax provisions relating to depreciation are examined. Guidelines for determining the time of replacement are presented at the end of the chapter.

Chapter 4

COST DETERMINATION

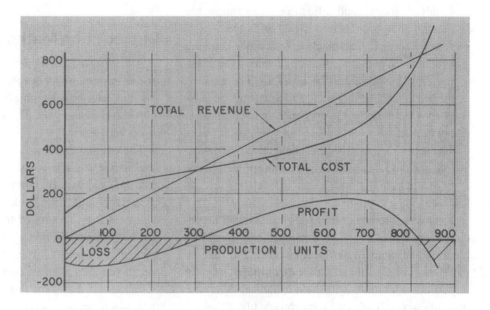

As this sample graph illustrates, costs influence profit substantially. For farmers who don't have control over product prices, the revenue curve is straight and pretty well fixed; thus, profit depends wholly on the position of the cost curve. If the cost curve can be lowered, profit will begin at a lower production and reach higher values.

The size of a profitable enterprise is also defined by the graph. If the cost curve shown reflects the cost curve for machinery, then the size of an operation can be limited by machinery costs. A detailed knowledge of machinery costs is essential to a profitable farm business.

Machinery costs are divided into two categories, *fixed costs* and *variable costs*. Variable costs increase proportionally with the amount of operational use given the machine, while fixed costs are independent of use.

It is not always clear which category some of the specific costs belong in. The costs of interest on the machinery investment, taxes, housing, and insurance are dependent on calendar-year time and are clearly independent of use. The costs of fuel, lubrication, daily service and maintenance, power, and labor are clearly costs associated with use. The two remaining cost items, depreciation and the cost of repairing, seem to be functions of both use and time.

Depreciation

Depreciation, often the largest cost of farm machinery, measures the amount by which the value of a machine decreases with the passage of time whether it is used or not. The value declines because of several factors.

1. The parts of the machine become worn with use and cannot perform effectively. These parts are the economically irreparable mechanisms in a machine; for example, the basic frame may be worn or distorted. The field capacity of the machine may decrease and/ or its material performance may suffer.
2. The expense of operating the machine at its original performance increases as more power, labor, and repair costs for the same unit of output are required. Repair and adjustment can renew the machine but at an increased rate with age.
3. A new, more efficient machine or practice becomes available. When this situation develops the existing machine is said to be *obsolete*. The existing machine may be functionally adequate but because of new technology it is uneconomic to continue to operate it.
4. The size of the enterprise is changed and the existing machine's capacity is not appropriate for the new situation.

It may be inferred from this discussion that depreciation is more likely a function of time for machines having small annual use. Obsolescence and rust are likely to end the life of these machines. Only for large farm operations will the mechanical deterioration due to use likely end a machine's life.

Machine Life

Three concepts of machine life concern the machinery manager: physical life, accounting life, and economic life. The *physical life* (also called service life) is terminated when a machine can't be repaired because of an irreplaceable or irreparable part failure. Most equipment manufacturers stock repair parts for years after a model line has been discontinued and the physical life of a farm machine can be very long.

An *accounting life* is the predicted life in hours of use for a machine based on the surveyed use of existing machines and upon the design life used by the manufacturer (see Table 4.5). The predicted life in years can be estimated by dividing the wear-out life by the hours of annual use.

The *economic life* of a machine is a more pertinent measure of the time period for which depreciation should be estimated. Economic life is defined as the length of time from purchase of a machine to that point where it is more economical to replace it with a second machine than to continue with the first. At this time a machine may still have considerable service life but be uneconomic because of high rate-of-repair costs, technological obsolescence, or a change in the farm enterprise. The machine may still be retained in some lesser role as part of the machine system or it may be sold to someone for whom it will have continued economic life. In such an event the price paid by the second owner will determine the depreciation cost to the first owner.

Depreciation Methods

An *estimated value* method may be the most realistic determination of depreciation. At the end of each year the value of a machine is compared with its value at the start of the year. The difference is the amount of depreciation.

The estimated values of aged farm machines are established at farm sales, at specialized machinery auctions, and by farm equipment dealers through use of "guides" and "blue books." These publications are compiled from reports of trading practices by dealers themselves and are used as a guide for quoting trade-in allowances for aged equipment. The values are quoted on an "as is" basis.

Table 4.1 shows average yearly depreciation for farm machines as a percentage of the purchase price. The mower conditioner, cotton and forage harvester data is from published "as is" values in 1991. The remaining data were developed by T. L. Cross, University of Tennessee, and G. M. Perry, Oregon State University, 1994. Variations as much as 10% about these values can be expected, depending on the popularity of specific models. Inflation effects during the 20-year period are included. It is apparent that the greatest depreciation occurs during the first year of life. The depreciation in succeeding years declines until a constant 1% occurs. The accumulated depreciation is the sum of the individual year values. For example, the accumulated depreciation for a plow at the end of year 10 is 72% of the purchase price.

Other reports state depreciation on usage and not age. A 1999 Midwest study showed tractors depreciating at 1.35% of the purchase price, P, per 100 hours of use. Self-propelled combines depreciated at 2.54% P/100 hr.

The depreciation method used for farm equipment is of concern to the federal government, as the cost of depreciation is a deductible item for income tax purposes. The Internal Revenue Service (IRS) suggests two specific methods of calculating depreciation.

1. *Straight-Line Method.* The annual depreciation charge using Eq. 4.1, is

$$D = (P - S)/L \qquad (4.1)$$

where P = purchase price
  S = salvage or selling price
  L = time between buying and selling, yr.

This method is the simplest, as it charges an easily calculated, constant amount each year.

2. *Declining-Balance Method.* A uniform rate is applied each year to the remaining value (includes salvage value) of the machine at the beginning of the year. The depreciation amount is different for each year of the machine's life. Eq. 4.2 expresses the relationships by formulas.

$$D = V_n - V_{n+1}$$

$$V_n = P\left(1 - \frac{x}{L}\right)^n \qquad V_{n+1} = P\left(1 - \frac{x}{L}\right)^{n+1} \qquad (4.2)$$

where D = amount of depreciation charged for year n+1

  n = number representing age of the machine in years at *beginning* of year in question

  V = remaining value at any time

  x = ratio of depreciation rate used to that of straight-line method (x may have any value between 1 and 1.5.

Table 4.2 lists depreciations as percentages of P for the year in question if one would rather not solve

TABLE 4.1. Estimated Value Annual Depreciation (% of purchase price)

Machine	1	2	3	4	5	6	7	8	9	10	11	12	13	14	15	16	17	18	19	20
Tractors	30	6	6	5	4	4	3	3	3	2	2	2	1	1	1	1	1	1	1	1
Combines	30	12	8	6	5	4	4	3	3	2	2	2	2	2	2	1	1	1	1	1
Cotton, forage harvesters	54	5	4	3	3	3	3	3	3	2	2	2	1	1	1	1	1	1	1	1
Balers	32	7	4	4	3	3	3	2	2	1	1	1	1	1	1	1	1	1	1	1
Mowers	44	5	3	2	2	2	2	2	2	1	1	1	1	1	1	1	1	1	1	1
Mower-conditioners	50	5	5	4	4	3	3	3	3	2	2	1	1	1	1	1	1	1	1	1
Plows	59	2	2	2	2	1	1	1	1	1	1	1	1	1	1	1	1	1	1	1
Disk harrows	30	8	5	5	3	3	3	3	2	2	2	2	2	2	1	1	1	1	1	1
Planters	25	5	4	4	3	3	2	2	2	2	2	2	1	1	1	1	1	1	1	1
Manure spreaders	41	6	6	5	5	3	3	3	2	2	2	2	2	1	1	1	1	1	1	1

TABLE 4.2. Remaining Values of Machines Expressed as a Percentage of Purchase Price for Each Year of Life (10-yr life and 10% salvage value assumed for depreciation methods)

Method	0	1	2	3	4	5	6	7	8	9	10
Straight-line	100	91	82	73	64	55	46	37	28	19	10
Sinking-fund[a]	100	94	88	80	72	64	55	45	34	23	10
Double-declining-balance	100	80	64	51	41	33	26	21	17	13	10
Trade-in values											
Tractors	100	64	61	57	54	51	49	46	44	42	40
Major implements	100	59	54	50	46	43	39	37	34	31	29
Minor implements	100	50	45	40	36	32	28	25	22	20	18

[a] Sinking fund is computed with 8% interest compounded annually.

Eq. 4.2. In no case is the machine allowed to depreciate beyond the salvage value.

Other depreciation methods are allowed by the IRS as long as they are consistent and do not permit depreciation to accumulate faster than for the double-declining-balance method for the first two-thirds of the machine's life.

A method of calculating depreciation used by engineering economists is the *sinking-fund method*. This method considers the problem of depreciation as one of establishing a fund that will draw compound interest. Uniform annual payments to this fund are of such a size that by the end of the life of the machine the funds and their interest have accumulated to an amount that will purchase another equivalent machine. By formulas, the values for the sinking-fund annual payment (SFP) and the value at the end of year n are

$$SFP = (P - S)\frac{i}{(1+i)^L - 1}$$

$$V_n = (P - S)\left[\frac{(1+i)^L - (1+i)^n}{(1+i)^L - 1}\right] + S$$

(4.3)

Examination of the depreciation and trade-in value curves in Fig. 4.1 indicates that this method of evaluating depreciation is the most remote from actual trade-in values, as the rate of depreciation is the greatest near the end of the machine's life. Such a depreciation schedule occurs because the time value of money is taken into account.

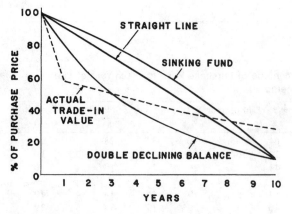

Fig. 4.1. Remaining value for three methods of depreciation compared with actual trade-in values.

The time value of money can be summarized by saying that a present dollar is worth more than a future dollar, or that delayed payments are most profitable if the money can be used to earn interest in the meantime. To illustrate, consider the worth of a present dollar invested at 10% interest compounded annually (Table 4.3). If $5000 must be spent for new machinery in 10 years, one need deposit only $1930.50 = $5000/2.594 to have the amount needed.

The sinking-fund method is primarily advantageous for use with a planned replacement interval policy. It is not a very flexible program if the chance for early obsolescence is great. Very possibly the sinking-fund method most closely approximates the actual depreciation of equipment with a slow, early depreciation rate and a fast, final rate near the end of the machine's life.

Fig. 4.1 and Table 4.2 compare the various methods of depreciation with the actual trade-in values. As all depreciation methods accomplish the same thing by the tenth year, the primary difference is in the *rate* of depreciation at specific times in the machine's life.

Inflation

In times of substantial monetary inflation, a machinery manager must include the effects of inflation on machinery planning. Inflation causes increased prices for goods and services in future years. Decisions involving a time span of more than 1 year are made using values expressed in *constant dollars*, dollars from which the effect of inflation is deducted. The effect of inflation, the *inflation factor*, is equal to $(1 + i_i)^n$, where i is the inflation rate and n is the number of years under consideration. Table 4.3 shows the inflation factors used to obtain constant dollar values. The price of a $1000 machine would be expected to rise to $1000 × 1.967 ($1.07^{10}$) = $1967 in 10 years; 10 years from now the price of the machine in constant dollars as of today would be $1967/1.967 or $1000. The price of a machine today based on constant dollars of 5 years ago is $1000/1.403 or $713.

Table 4.3 also shows the error that can occur in planning when inflation is ignored. A 10% interest rate might seem to be an adequate return for an investor; but when the 7% inflation rate is compounded annually and deducted from the 10% return (Table 4.3), the real return on investment is obtained. Interest rates asked by investors include any expectation of inflation. If they did not, investors could be paid

TABLE 4.3. Value of Money (compounded)

Time	Values at 10% Interest	Inflation 7% Annually	Real Rate of Return
Today	1.000	1.000	1.000
1 yr	1.100	1.070	1.028
2 yr	1.210	1.145	1.057
3 yr	1.331	1.225	1.086
5 yr	1.611	1.403	1.148
10 yr	2.594	1.967	1.318
20 yr	6.727	3.870	1.737

Rapid Depreciation

Depreciation has for a long time been recognized by the IRS as a cost of doing business and as such is deducted from the gross income of a business when calculating income tax. Rapid depreciation of equipment is a management accounting practice permitted by the IRS. Some machinery managers use rapid depreciation schedules to recover the investment early in a machine's life before obsolescence, accident, or wear-out ends the machine's usefulness and also to avoid income tax. Other machinery managers prefer slow depreciation schedules that spread the reduction-in-tax benefits over the actual life of the machine.

In 1998 the IRS in Section 179 permitted a farm equipment buyer to treat part or all of the purchase price as an expense rather than as a capital expenditure that must be depreciated over several years. This total expense deduction may not exceed $18,500 per taxpayer (including wife) per year. In effect this policy can provide a rapid and complete depreciation in the very first year if the price of the machine is $18,500 or less. The IRS raised the limit to $19,000 for 1999, $20,000 for 2000, and will raise it to $24,000 for 2001 and 2002, and $25,000 for 2003 and afterward.

Several limitations exist for using Section 179. The election to use this cost deduction must be made in the first year the machine is placed in service, and the deduction is allowed only to the extent that it is less than the tax bill. The allowable deduction is reduced one dollar for every dollar the price is over $200,000; that is, only $14,500 can be deducted (in 1998) for a $204,000 investment. The costs not allowed by the limits in the first year can be carried over to the second year. In no case can the deduction exceed $200,000 plus the allowable Section 179 deduction. The deduction is allowed only if the machine is used more than 50% for farming. Passenger cars, vans, and pickup trucks used 100% for farming were limited to a first year deduction of $2960 in 1994. Any deductions must reflect the proportion the machine or vehicle is used in the business.

Any Section 179 deduction must reduce the basis of that machine when depreciation is computed; that is, if a $10,000 deduction is taken on the purchase of a $100,000 combine, depreciation methods must start with the initial value (basis) of $90,000. Such a procedure models the "actual-trade-in-value" plot in Fig. 4.1.

The Modified Accelerated Cost Recovery System (MACRS) developed by the IRS is a way of securing

back an amount worth less than the original loan. The real interest rate, i_r, is a function of the nominal interest rate, i_n, and the rate of inflation, i_i, as shown in

$$i_r = (i_n - i_i)/(1 + i_i) \qquad (4.4)$$

In Table 4.3, $(0.10 - 0.07)/1.07 = 0.028$ or 2.8%, the real return on the investment when the increase in value due to inflation is discounted.

Any time machinery planning involves a time span of several years, the effects of inflation must be discounted. The inflation factor, $(1 + i_i)^n$, allows the calculation of both constant dollar values and the real rate of return of an investment.

Fig. 4.2 shows farm equipment prices in the years after 1973. These price ratios and the expression for the inflation factor, $(1 + i_i)^n$, permit the determination of inflation rates of 5% for 1965-73, 18% for 1974-76, 9% for 1977-78, 10% for 1979-82, 3% for 1983-86, 4% for 1987-89, and 3% for 1990-91. Inflation has varied from 2 to 3% in the years following.

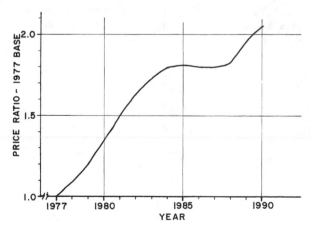

Fig. 4.2. Average inflation in machinery prices and R&M parts.

rapid depreciation for equipment placed in service after 1986. Autos and trucks used in farming are counted as 5-year property, farm machinery and fence as 7-year property, grain bins and other single purpose structures 10-year property, and farm buildings are 20-year property. Allowable depreciation must be spaced over the years designated in the property class. Either straight-line or declining-balance methods of depreciation may be used.

The most rapid depreciation allowed is the 1.5 declining-balance rate for equipment placed in service after 1988 (2.0 for equipment before 1989). Any time this yearly depreciation amount falls below that for the straight-line method applied to the beginning-of-the-year value (basis) for that tax year, the straight-line amount is used. Straight-line depreciation over the whole life is acceptable but does not produce accelerated depreciation. Once the straight-line method is elected, that method must be followed during the rest of the life of the equipment.

Since 1981 the amount to be depreciated is the amount "at risk," which is effectively the purchase price. In effect, the selling or salvage value is assumed to be zero. If the machine is eventually sold for more than its depreciated value, this gain is recognized as ordinary income by the IRS. The gain is computed to be the smaller of (1) the total depreciation taken, or (2) the positive amount remaining when the depreciated value is subtracted from the selling price. If a used machine is traded in for a new machine at a dealership, the transaction is a nontaxable exchange, but the basis or undepreciated value of the new machine is the basis for the old machine added to the cash difference paid.

A convention allowing depreciation to start at midyear has been adopted by the IRS. This convention treats all machinery placed in service, or disposed of, in a tax year as if the event occurred at midyear. A half-year depreciation is allowed even though the machine was purchased in December. A midquarter convention is required if 40% or more of the total year's purchases occurred during the last 3 months.

Table 4.4 shows the IRS allowable depreciation for 5- and 7-year properties and for both types of depreciation when using the midyear convention. The declining-balance depreciations are converted over to straight-line depreciation when permissible.

Machinery managers are not required to take rapid depreciation. The IRS allows an Alternate Depreciation System (ADS), which provides for the straight-line method and the half-year convention applied to the purchase price (plus trade-in basis, if any). Property lives are 5 years for autos and light duty trucks; 10 years for farm equipment, grain bins, and fences; and 25 years for farm structures that are not single purpose.

These descriptions of depreciation practices are abstracted from the *IRS Farmer's Tax Guide*. These guidelines are offered for information only. A tax specialist should be consulted when filing a tax return.

Renting and Leasing

The possibility of renting or leasing equipment as a means of reducing operating costs is currently of interest. Payments for renting or leasing are wholly deductible from income tax as a business expense.

Renting should be distinguished from leasing. *Renting* is defined as use of equipment for small periods of time. The renting period may be terminated at any time by either party. The term *leasing* is reserved for contracts of perhaps several year's duration. The contract cannot be canceled by either party without recovery of damages.

Typical rent charges for farm machinery are expressed as percentages of purchase price for definite time periods as shown on the next page.

TABLE 4.4. IRS-Permitted Rapid Depreciation (% of the unadjusted basis)

Year	1.5 Declining Balance		2.0 Declining Balance		Straight Line	
	5 yr	7 yr	5 yr	7 yr	5 yr	7 yr
1	15.00	10.71	20.00	14.29	10.0	7.14
2	25.50	19.13	32.00	24.49	20.0	14.29
3	17.85	15.03	11.52	12.49	20.0	14.28
5	16.66	12.25	11.52	8.93	20.0	14.29
6	8.33	12.25	5.76	8.92	10.0	14.28
7	¼	12.25	¼	8.93	¼	14.29
8	¼	6.13	¼	4.46	¼	7.14

1 day	1%
1 week	5%
1 month	15%
2 months	25%
3 months	35%

As shown later, the cost of renting farm machinery for 1 month nearly equals the fixed cost percentage used in the approximate cost analysis. Even if the lessor furnishes the necessary repair, cost of renting will surpass the cost of owning if the machine's operating season is greater than 1 month. Renting or leasing is especially advantageous to the machinery manager if there is much chance of early obsolescence.

Because several variables are involved in leasing, no absolute conclusions can be drawn as to the economy of leasing equipment. As with renting, the lease payments are deductible in the eyes of IRS; but the agreement must be a true lease and not a conditional sales contract wherein the machine is sold to the lessee for a nominal sum at the end of the leasing period.

In an appraisal of agricultural leases, W. H. M. Morris (Agricultural Economics Department, Purdue University) reached the following conclusions.

1. A lease has the objective of providing a method for obtaining the use of a piece of equipment without paying cash.
2. The risk of obsolescence falls on the lessee.
3. The possibility of borrowing money on a chattel mortgage at a relatively low rate and the liberalized income tax situation for equipment purchases typically provide a less costly alternative to a farmer than a lease on the terms usually available.
4. A lease can be a least-cost alternative for a farmer in the 20-30% income tax bracket if the return on the farmer's operating capital is 12% or more.
5. A lease is more attractive to a farmer if the machine can be purchased for a nominal sum at the end of the lease period. However, such a lease might be considered a contract sale if investigated by IRS.

Interest on Investment

The interest on investment in a farm machine is included in operational cost estimates. Even if the investment money is not actually borrowed, a charge is made since that money cannot be used for some other interest-paying enterprise. Nominal interest rates include expected inflation. The investment in a machine is greatest during the first year, as depreciation charges reclaim a portion of the investment each year.

Interest charges are usually computed when operating costs are being determined and may be calculated so that the result will be constant or an equal yearly charge throughout the life of the machine. When the straight-line method of depreciation is used, a constant annual interest charge is determined by calculating the average investment in that machine over its full life. The average investment is equal to one-half the sum of the initial value and the trade-in value. The annual interest charge will then be the product of the interest rate and the average investment.

A *capital recovery factor* (CRF) can be used to combine the total depreciation and interest charges into a series of equal annual payments at compound interest. These payments plus the interest on the undepreciated amount, S, can be used to estimate the capital consumption (CC) of farm equipment.

$$CC = (P - S)\ CRF + S\,i \qquad (4.5)$$

$$\text{where CRF} = \frac{i\,(1+i)^L}{(1+i)^L - 1}$$

$$i = \text{ annual interest rate, decimal}$$

Using typical values, L = 10 years, i = 0.10 compounded annually, S = 0.1 P, and the expected capital consumption per year would be 15.6% P. (See Table 4.7).

Property and Sales Tax

Farm machinery may be taxed at the same rate as other farm property. Wide variations exist from state to state in the amount of sales tax and in the methods of determining the assessed valuations. The property tax rate can vary widely within a state. Assuming a 4% sales tax, a property tax rate of $4/$100 evaluation, and an assessment equal to 50% value, the annual cost of taxes would be about 1.5% of the purchase price when spread over a 10-year life.

Insurance

While it is not a universal practice to insure farm machinery against loss by fire or windstorm, the insurance charge is justifiable. One who does not insure is in effect carrying the risk alone.

The rates for farm equipment are variable and fall

between $0.30/$100 and $0.60/$100 coverage per year. Most companies will insure up to two-thirds of the value of the equipment. Assuming the average investment to be approximately one-half the original price and a rate of $0.50/$100 valuation, the annual charge for insurance would be 0.25% of the original price.

Shelter

Machinery shelter has not been shown to increase machinery life, but it can increase a machine's resale value. In a survey of dealer opinions, Neil Meador (University of Missouri) found that sheltered machinery was valued from 10-23% more at trade-in time than machinery stored outdoors. The higher percentage was for the more complex machines, such as combines used in moist, warm climates. The lower value was for tillage implements. Many farmers think that machinery shelter provides indeterminate values such as a more attractive appearance of the farmstead, easier maintenance, and easier repair.

Most farmers do provide storage for their field equipment. In a central Illinois survey, 41% of the farms had all their equipment stored in enclosed buildings and 43% had heated working space for making repairs. About 20% of the disk harrows and field cultivators were stored outside without even a protective roof.

A charge must be made against machines for shelter or paint or other deterioration prevention practices. An enclosed building dedicated to machinery storage and having a 20-year life may be expected to generate an annual charge of 12% of its purchase price. Dividing this charge by the value of the machinery stored will produce an annual storage cost for the machine of 0.5-1% of its purchase price.

Many farm machines are inexpensively sheltered in structures such as empty grain storage or unused animal shelters. The cost of shelter under these conditions is likely to be no more than 0.2% of the purchase price of the machine.

Variable Costs

Costs that vary with use, variable costs, are added to the fixed costs to arrive at the annual cost for a machine's operation. Operator labor, engine fuel and oil, and scheduled maintenance costs are associated directly with the amount of machine use. The cost of repair is not related directly to use; wear costs can be expected to occur with use, but part failures occur in a random fashion.

Variable costs may be greater than fixed costs for some high-use machines. G. E. Fairbanks, G. H. Larson, and D. Chung (Agricultural Engineering Department, Kansas State University) found the variable costs for tractors to be about 64% of all costs, with the greatest single cost being that for operator labor (Fig. 4.3). The greatest single cost of self-propelled combines was depreciation, and variable costs were only 38% of all costs.

The actual estimation of variable costs is usually based on hours of use. The cost for operator labor is the labor rate in $/hr multiplied by the expected hours of use for the machine. The fuel and oil consumption is measured or estimated from Tables 2.3 and 2.4 and multiplied by their respective prices. Any fuel tax refunds are usually considered when determining the price of gasoline. The remaining costs, repair and maintenance, are not so easily estimated. Maintenance costs include labor, lubricants, and filters used to protect the machine. While the maintenance activity is very important, its costs are relatively minor and are often included with repair as a repair and maintenance (R&M) cost.

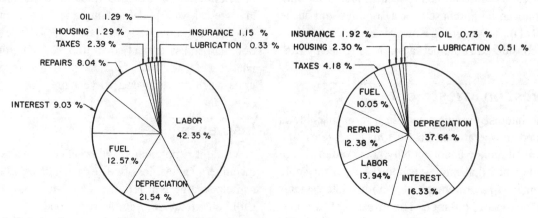

Fig. 4.3. Relative importance of all cost items for diesel tractors (left) and SP combines (right).

Repair

Repair costs are the expenditures for parts and labor for (1) installing replacement parts after a part failure and (2) reconditioning renewable parts as a result of wear. The anticipated annual cost of repair for any one machine is highly uncertain. Plowshares, cutter bars, etc., will wear with use and will need periodic replacing at different intervals for different conditions. Many of the gears, chains, belts, and bearings, seem to fail at random. Only repair records kept on many typical machines used in actual farm operations can indicate the average or expected repair costs.

Costs incurred for replacing worn parts might be expected to approach a predictable, constant level related directly to the amount of machine use; but in fact wear is uneven because of differences in crops, soils, weather, degree of machine maintenance, adjustments, load factors, design and factory errors, etc. New machines experience about as many breakdowns as do old ones (Fig. 1.21).

Appendix E lists 1990 repair and maintenance costs for grain farms on an area-of-work basis. But the average value is not very helpful for predicting costs if the data are highly variable. The standard deviation for these data is about equivalent to the mean, which implies that 16% of the machines had a rate double that of the average while many others had zero repair.

Repair rates that reduce some of the variation can be computed. A constant repair rate of $/hr of use is popular for both tractors and implements, but it does not include any differences in machine size or accumulated usage nor account for inflation effects from one year to the next. Of particular utility because of its simplicity is the rate expressed as $/area. This rate is a good predictor for many types of implements for which repair and maintenance costs seem to be about constant per area even though the sizes of the implements vary. Inflation effects, however, make such dollar rates quickly obsolete.

A rate expressed as a percentage of a machine's list price divided by the amount of use is an effective predictor of R&M costs. Such a unit reduces variations due to machine size since large machines will have higher prices. Inflation effects are minimized if the inflation during the data gathering period represents current inflation. The machinery manager should know that the list price, and not the actual price paid, must serve as the reference price.

An R&M unit of %P/100 hr gives numerical value

that are easy to use for estimating machine costs. In a presentation before the ASAE, C. A. Rotz, USDA, East Lansing, Michigan, and W. Bowers, Oklahoma State University, reviewed previous data for R&M and proposed new values for 1991 machines (Table 4.5). These values represent R&M costs averaged over the life of the machine and are not expected to predict the R&M cost for any specific year. The total R&M cost over the life of the machine is based on

TABLE 4.5. Repair and Maintenance Costs, Percentage of List Price

Machine	Wear-out life, hr	R&M, Lifetime	R&M/ 100 hr
Tractors			
2-wheel drive	12,000	100	0.83
4-wheel drive	16,000	80	0.50
Tillage			
Moldboard plow	2,000	100	5.00
Heavy-duty disk	2,000	60	3.00
Tandem disk harrow	2,000	60	3.00
Chisel plow	2,000	75	3.75
Field cultivator	2,000	70	3.50
Spring-tooth harrow	2,000	70	3.50
Roller harrow	2,000	40	2.00
Rotary hoe	2,000	60	3.00
Row crop cultivator	2,000	80	4.00
Rotary tiller	1,500	80	5.33
Seeders			
Row crop planter	1,500	75	5.00
Grain drill	1,500	75	5.00
Harvesters			
Combine	2,000	60	3.00
Combine, self-propelled	3,000	40	1.33
Forage harvester	2,500	65	2.60
Forage harvester, SP	4,000	40	1.25
Sugar beet harvester	1,500	70	6.67
Potato harvester	2,500	70	2.80
Cotton picker, stripper, SP	3,000	80	2.67
Hay Machines			
Mower, cutterbar	2,000	150	7.50
Mower, rotary	2,000	175	8.75
Mower-conditioner	2,500	80	3.20
Mower-conditioner, rotary	2,500	100	4.00
Windrower, self-propelled	3,000	55	1.83
Side-delivery rake	2,500	60	2.40
Rectangular baler	2,000	80	4.00
Large, rectangular baler	3,000	75	2.50
Large, round baler	1,500	90	6.00
Miscellaneous			
Fertilizer spreader	1,200	80	6.67
Boom sprayer	1,500	70	4.67
Orchard sprayer	2,000	60	3.00
Bean puller/windrower	2,000	60	3.00
Beet topper/stalk chopper	1,200	35	2.92
Forage blower	1,500	45	3.00
Forage wagon	2,000	50	2.50
Grain wagon	3,000	80	2.67

the presumed wear-out life as shown in column 1. This wear-out life is the same life as the accounting life discussed previously.

As an example of the use of %R&M/100, suppose a $7000 field cultivator was used 80 hr per year. The R&M cost would be estimated as

$7000 ×.035/100 = $2.45 per hr, or $196 per yr.

A more detailed study shows that R&M costs are not constant over a machine's life but increase from a very low value during the first year to some relatively constant value after a period of use. Relationships between the rate and the hours of accumulated use were obtained from a 1967 Midwest survey of the repair costs of over 1100 tractors and 5000 field implements. Fig. 4.4 indicates the relationship for tractors. Little expense was reported during the first season, probably because the warranty period was in effect. The repair rate rose rapidly to a peak near the 2000-hr point of accumulated use. This point may have been the time of the first engine overhaul. A second but lesser peak occurred at 5000 hr of use, and then the rate settled to a lower and essentially constant rate.

Grain drill and baler R&M costs also show cyclic peaks (Fig. 4.5). Apparently some major repair occurs after definite intervals of use. Disk harrows and field cultivators, however, exhibited a steadily increasing R&M rate with use. Eventually one would expect these implements also to arrive at a constant repair rate related to the replacement of disk blades and shovels or sweeps.

The other machines included in this study (SP and PTO combines, corn pickers, moldboard plows, row planters and cultivators, mowers, and forage harvest-

ers) had R&M rates that seemed completely insensitive to the amount of use. The randomness of failure seemed to outweigh any predictable effects due to wear.

Another meaningful way to consider R&M costs is to report the accumulated costs for accumulated use. The point plots of average values found for each increment of use can be fitted statistically with smooth curves having a mathematical relationship of the form.

$$Y = (c1)X + (c2)X^2 + (c3)X^3 \qquad (4.6)$$

where   Y = ratio of accumulated R&M costs to list price (constant dollars)
X = units of accumulated use/1000, and
c1, c2, c3 = constants for specific machines

In a study from 1965 to 1973 of 45 Illinois corn and soybean farms averaging 215 ha [531 a] of cropland, continuous R&M histories were obtained for typical machines used for corn and soybean row crop culture. This study included only machines purchased new. The range in accumulated usage was substantially less than for the 1967 study. The more recent study covers the period of use from new up to about the point where the R&M rate becomes relatively constant. The R&M costs over this early period of a farm machine's life are significant for those who would replace machines substantially before their wear-out life.

Fig. 4.6 shows the plot, the fitted curve, and the standard deviation of Y vs. X for 79 diesel tractors averaging about 75 kW [100 HP] in size. The data points are the mean values for use increments of 200 accumulated hours. The plotted curve has a very high

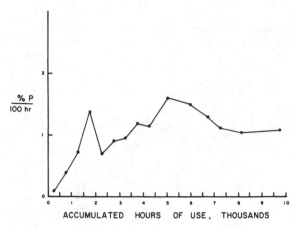

Fig. 4.4. R&M rates for tractors.

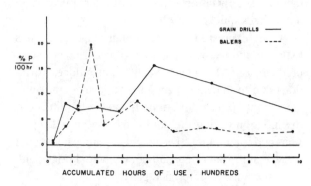

Fig. 4.5. R&M rates for balers and grain drills.

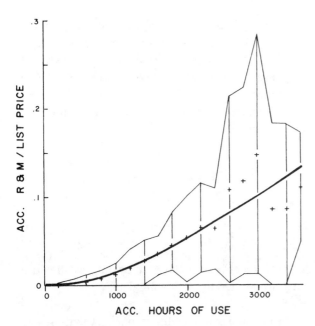

Fig. 4.6. Accumulated R&M costs for diesel tractors.

place equivalent limits on both large and small machines.

Approximate Annual Costs

Estimating future costs for farm machine use is important to machinery managers. With good estimations the costs of production can be predicted, alternatives such as custom work can be evaluated, and size selection and replacement decisions can be made. The actual costs are determined at the end of each year from cost records.

Approximate costs are determined by summing expected fixed and variable costs. A machine life must be estimated to predict CC (Eq. 4.5). Capital consumption is a fixed cost that is unaffected by subsequent monetary inflation if the purchase price is considered as an investment financed over the length of life of the machine at a competitive interest rate at the time of purchase. Charges for shelter, insurance, and taxes are not so free from inflation but can be assumed as fixed costs (0.025 P) with only minor error. The variable costs per year must be added to the fixed costs to produce a complete annual cost estimate. The variable costs are most sensitive to inflation.

Eq. 4.7 expresses the approximate annual costs for a field machine in constant dollars. The two parts of the equation represent the fixed and the variable costs, respectively. Note that the term cA/(Swe) is the hours of expected use per year.

correlation with the data points and shows a linear trend at the higher levels of accumulated use. This trend is consistent with the 1967 study, which suggests a constant R&M rate after 3000 hours. The vertical lines define a standard deviation on each side of the mean and indicate highly variable data.

Table 4.6 lists the values of the coefficients (c1, c2, c3) for Eq. 4.6. The last column shows the limit in accumulated use for which Eq. 4.6 applies. These limits are expressed in units of area per unit width to

TABLE 4.6. Coefficients for Eq. 4.6 (constant dollars)

Machine	c1	c2	c3	Acc. Use Limit ha/m [a/ft]
Tillage				
chisel plows and field cultivators	0.0091 [0.0037]	0.0381 [0.005 2]	-0.0024 [-0.000 16]	360 [272]
disk harrows	-0.0017 [-0.0007]	0.0169 [0.002 8]	-0.0027 [-0.000 18]	600 [458]
moldboard plows	0.0180 [0.0073]	0.4408 [0.072 2]	-0.1452 [-0.009 62]	422 [318]
rotary hoe	-0.0198 [-0.0080]	0.1425 [0.023 3]	-0.0526 [-0.003 49]	210 [158]
row cultivator	-0.0032 [-0.0013]	0.0488 [0.008 0]	-0.0249 [-0.001 65]	385 [290]
Planting				
row planters	0.0082 [0.0033]	0.0003 [0.000 05]	0.1054 [0.006 99]	385 [290]
Harvesting				
combines, SP	-0.0035 [-0.0014]	0.0571 [0.009 4]	0.0020 [0.000 13]	460 [348]
corn pickers	-0.0148 [-0.0060]	0.3917 [0.064 2]	0.3626 [0.024 03]	250 [190]
Tractors				
diesel	-0.0042	0.019 5	-0.002 23	4000 hr
gasoline	-0.0134	0.040 3	-0.004 39	3000 hr
LP gas	-0.0003	0.009 2	0.002 26	3000 hr
Miscellaneous				
stalk chopper	0.0398 [0.0161]	0.0751 [0.012 3]	-0.0178 [-0.001 18]	230 [174]

$$AC = \frac{(FC\%)P}{100} + \frac{cA}{Swe}[(R \& M)P + L + O + F + T] \quad (4.7)$$

where   AC = annual cost of operating machine, $/yr
   FC% = annual fixed cost percentage
   P = purchase price of machine
   c = constant, 10 [8.25]
   A = annual area use in ha [a]
   S = forward speed, km/hr [MPH]
   w = effective width of action of machine, m [ft]
   e = field efficiency, decimal
   R&M = repair and maintenance costs, decimal of purchase price per hour, Table 4.5
   L = labor rate, $/hr
   O = oil cost, $/hr
   F = fuel cost, $/hr
   T = cost of tractor use by machine, $/hr (T = 0 if self-propelled)

The fixed cost percentage (FC% in Eq. 4.7) is the sum of capital consumption (CC from Eq. 4.5) and the fixed cost percentages for taxes, insurance, and shelter; which is about (2.5%) P. Values of CC are given in Table 4.7 for various years of life and at nominal interest rates with salvage or resale values equal to (10%) P in constant dollars.

With a 12% interest rate, for example, a machine that will only last 5 years and be worth 10% P at that time will have a capital consumption of 26% P for each of those 5 years. For other resale values, interest rates, and years of life, Eq. 4.5 must be solved.

The variable costs must be computed for each year if actual dollars, rather than constant dollars at age of purchase, are desired. The variable costs are bracketed in Eq. 4.7 and must be multiplied by the inflation factor $(1 + i_i)^n$ for year n after the purchase of the machine.

A typical approximate cost calculation might assume a 10-year life, a 12%-interest rate, an inflation rate of 3%, a machine purchase price of $10,000, and a resale value of (10%) P in constant dollars at the time of purchase. The approximate annual cost in actual dollars for 80 hours of use in year 5 for a $25/hr operating cost would be

$$\frac{(17+2.5)\times10,000}{100} + 80\times25\times1.03^5 = \$4,269$$

Cost Records

Cost records of machinery operations should be kept to provide the following data:

1. Deductible expenses for income tax purposes
2. Cost of production data
3. Information on which to base equipment replacement decisions

Records in categories 1 and 2 need not be identical. Record 1 is a report to the Internal Revenue Service in which one seeks to avoid (not evade) taxes. Record 2 is used to analyze the actual costs of production. The rapid depreciation of equipment is desirable for 1 but artificial for 2. Growing a ton of corn with new equipment should not cost more than doing so with older, fully depreciated equipment; but if actual trade-in value depreciation is used (Table 4.2), that will be the case. The true equipment cost of production cannot be determined until the average costs over the economic life of the equipment are established by their replacement.

Record 3 accounts realistic and actual equipment expenses each year (see Fig. 4.7 and 4.8). Note that in a running account record of this nature the common denominator is the remaining value at the beginning of each year and not the purchase price as used in the approximate cost percentage method of calculating costs. Remaining value is determined at the end of each year (which need not be January 1 to January 1) as the difference between the remaining value at the beginning of the year and the depreciation charged for that year. Any method of depreciation can be used with this record.

The percentage charges for interest on investment, shelter, taxes, and insurance (ISTI) are not numerically comparable with the charges developed earlier in the section on approximate costs. The earlier

TABLE 4.7. Capital Consumption, CC, Percentage of Purchase Price (assumes S = 0.10 P)

Interest Rate, %	Years of Life									
	1	2	3	4	5	6	7	8	9	10
4	94	48	33	25	21	18	15	14	13	11
6	96	50	34	27	22	19	17	15	14	13
8	98	51	36	28	23	20	18	16	15	14
10	100	53	37	29	25	22	19	18	17	16
12	102	54	39	31	26	23	21	19	18	17
14	104	56	40	32	28	25	22	21	20	19
16	106	58	42	34	29	26	24	22	21	20
18	108	59	43	35	31	28	25	24	23	22
20	110	61	45	37	32	29	27	25	24	23

MACHINE _Chisel plow_ MFG _____ MODEL _____ SERIAL NO. _____ PURCHASE PRICE _$3250_

DATE PURCHASED _Aug.30 '85_ AGE WHEN PURCHASED _New_

	1	2	3	4	5	6	7	8	9	10
END OF YEAR	1985	1986	1987	1988	1989	1990	1991			
REMAINING VALUE, RV	3,000	2,700	2,400	2,100	1,900	1,700	1500			
INFL.FCTR.,$(1+i_i)^n$	1.03	$(1.03)^2$	$(1.03)^2 1.04$	$(1.03)^2 1.04^2$	$(1.03)^2 1.04^3$	$(1.03)^2 1.04^3 1.07$				
DEFL. VAL., $RV/(1+i_i)^n$	2,913	2,545	2,175	1,830	1,592	1,331				
ANNUAL COST, $/Yr										
(a) Depreciation, apparent	250	300	300	300	200	200				
(b) Depreciation, constant $	337	368	370	345	238	261				
(c) $ISTI_i(i+.055)$ of RV $(i=10\%)$	155	419	372	326	295	264				
(d) Repair & Maintenance	87	112	220	135	207	242				
(e) $(c+d)/(1+i_i)^n$	235	501	537	402	421	396				
(f) Ha or (Ac) per yr.	500	550	550	620	580	437				
(g) Hours per yr. _6 acres/hr_	83	92	92	103	97	73				
(h) Unit cost$(a+c+d)/f$ or g _$/acre_	0.98	1.51	1.62	1.23	1.21	1.62				
REPLACEMENT ANALYSIS (in constant $)										
(i) Acc. cost, (sum b+e)	572	1,441	2,348	3,095	3,754	4,411				
(j) Acc. use, (sum f or g) _acres_	500	1,050	1,600	2,220	2,800	3,237				
(k) Ave. acc. cost, i/j _$/ac_	1.14	1.37	1.47	1.39	1.34	1.36				
OPERATING COST, $/yr	_old tractor →_	_new tractor →_								
(ℓ) Tractor fixed costs	3,015	2,985	4,713	4,323	4,145	2,649				
(m) Fuel & Oil	750	790	770	826	808	690				
(n) Labor _$6/hr_	498	552	552	618	582	438				
(o) Sum ℓ,m,n/(f or g) _$/acre_	8.53	7.87	10.97	9.30	9.54	8.64				
TOTAL COST PER USE										
(p) h+o _$/acre_	9.51	9.38	12.59	10.53	10.75	10.26				

Fig. 4.7. Implement cost record.

charges were based on an average value over the life of the machine, while in the running account record they are based on remaining value. There is some question as to whether shelter charges can properly be based on remaining value, but the error if any would be small. The sum of the four percentages, I + S + T + I, may be estimated as i + 0.055 of the remaining value, RV, where i is the interest rate to be charged and the 0.055 estimates the cost of shelter, taxes, and insurance.

Since there must be one record form for each machine, the back of the form is a convenient place to record individual repair costs. These items, which include parts, maintenance costs, and repair and service labor costs, are summed each year and entered on line d.

The complete costs for an implement (Fig. 4.7) cannot be fully determined until the tractor record (Fig. 4.8) has been completed. The tractor's role in a machinery system is to provide service to implements; thus its costs must be recharged to the implements. Quite often this charge is the greatest single cost of operating an implement as shown in Fig. 4.7, where a new tractor (Fig. 4.8) was purchased in 1987 to power the plow. A common method of apportioning the tractor costs is to divide them according to the time spent with each implement. For example, the fixed and R&M costs of the tractor in 1987 were $19,516 (10,000 + 9300 + 216), and the tractor was used a total of 381 hours (Fig. 4.8). The cost per hour is shown in (h) to be $51.22. This value is multiplied by the hours of use for the chisel plow, 92 (g, Fig. 4.7), to get the $4713 value in (ℓ) for 1987. The machinery manager *must* record the hour meter readings at the start and finish of each machine operation to accurately assign tractor fixed costs to implement operations.

Fuel and oil costs should be determined from actual pump records. Lacking good records, estimated consumptions may be determined from Tables 2.3 and 2.4. It is helpful to charge the cost of the tractor's fuel

MACHINE _Tractor_ MFG_____ MODEL____ SERIAL NO._____ PURCHASE PRICE $ _70,000_

DATE PURCHASED _March '87_ AGE WHEN PURCHASED ____New____

	1	2	3	4	5	6	7	8	9	10
END OF YEAR	1987	1988	1989	1990	1991					
REMAINING VALUE, RV	60,000	51,000	43,000	36,000	30,000					
INFL. FCTR., $(1+i_i)^n$	1.04	(1.04)(1.07)	$1.04(1.07)^2$	$1.04(1.07)^2 / 1.08$						
DEFL. VAL., $RV/(1+i_i)^n$	57,692	45,830	36,113	27,995						
ANNUAL COST, $/yr										
(a) Depreciation, apparent	10,000	9,000	8000	7000						
(b) Depreciation, constant $	12,308	11,862	9,717	8118						
(c) ISTI,(i+.055)of RV (i=10%)	9300	7905	6,665	5580						
(d) Repair & Maintenance	216	305	290	482						
(e) $(c+d)/(1+i_i)^n$	9,150	7,378	5,841	4,714						
(f) Ha or Ac per yr										
(g) Hours per year	381	410	350	360						
(h) Unit cost (a+c+d)/f or g $/hr	51.22	41.98	42.73	36.28						
REPLACEMENT ANALYSIS (IN CONSTANT $)										
(i) Acc. cost, (sum b+e)	21,458	40,698	56,256	69,088						
(j) Acc. use,(sum f or g) hrs	381	791	1141	1501						
(k) Ave. acc. cost, i/j $/hr	56.32	51.45	49.30	46.03						
OPERATING COST, $/yr										
(ℓ) Tractor fixed costs										
(m) Fuel & Oil										
(n) Labor										
(o) Sum ℓ,m,n/(f or g)										
TOTAL COST PER USE										
(p) h+o										

Fig. 4.8. Tractor cost record.

and oil consumption directly to the implement. When the cost record form is used for tractors, (ℓ), (m), (n), and (o) can be omitted as these calculations are seldom needed.

The labor charge is determined from the hourly rate for one worker. Operation of two implements in tandem behind one tractor means that each implement will assume one-half the cost of labor. In such a situation the two implements should share equally the tractor's fixed costs but should divide the fuel and oil consumption according to their respective power requirements.

The total costs of operation for an implement can be displayed and analyzed with running cost accounting sheets such as those in Fig. 4.7 and 4.8. The unit costs, line h, are the costs per acre of annual use, f, for depreciation, interest on the value of the machine for a year, shelter, insurance, taxes, and repair and maintenance. The operating costs that accumulate with time can be presented as costs per unit area. For 1987 in Fig. 4.7, the sum of tractor fixed costs, fuel and oil costs, and labor costs is $6035. When divided by the annual acres of use, 550, a cost of $10.97 per acre is obtained and reported on line o. This amount added to the $1.62 on line h produces a total cost of $12.59/acre on line p. This cost for chisel plowing serves the farm manager when making production decisions.

These cost sheets are versatile. They can be used equally well for new or used equipment. If a tractor is replaced, the effects are incorporated without problems in Fig. 4.7. A self-propelled implement's costs are recorded effectively on such a sheet but would not use line ℓ, tractor fixed costs. The effects of inflation can be traced on this sheet but do not enter into determination of annual costs.

Replacement

Since mechanization of agriculture is nearly complete in the United States, the purchase of new machines results from a need to replace older, inadequate machines. This replacement decision is one of the most important decisions a machinery manager must make. Replacement of a machine is indicated when

1. Accidents have damaged the implement beyond repair,
2. Field capacity of the machine is inadequate because of an increase in the scope of the operation,
3. A new machine or farm practice makes the old machine obsolete,
4. Performance of a new machine is significantly superior, or
5. Anticipated costs for operating the old machine exceed those for a replacement machine.

Estimations of yearly costs are adequate for determining crop production costs and for deciding if machine ownership is profitable; but the *time of replacement* decision depends on the *accumulated* costs over a period of years. Fig. 4.9 compares yearly costs and accumulated costs during the life of a machine. The first year's costs are high because of the very real marketplace depreciation obtained from the estimated value method. The yearly costs drop to their lowest value and then begin to rise if the annual repair costs increase with age. The accumulated cost curve drops more gradually and levels out at the point where it crosses the yearly cost curve. The *optimum replacement time* is at the crossing point since the accumulated cost curve is at a minimum and is expected to rise after that point. The costs in Fig. 4.9 need to include only depreciation, ISTI, and repair as all other costs are assumed to be independent of the time of replacement.

Machine repair rates really determine the time of replacement. Fig. 4.10 traces the accumulated cost-time histories for machines having repair rates constant with use and for those having rates increasing with use. When repairs (and maintenance) are exactly proportional to use, as in the case for replacement sweeps and shovels for tillage implements, the implement should never be replaced for economic reasons since its accumulated cost per area gets lower each year of life. But when the repair rate increases with age or use, the accumulated cost per area does reach a definite minimum point.

The smooth curves in Fig. 4.9 and 4.10 might represent the averages for many machines, but the cost-time history curve for a specific machine is likely to be erratic. Since replacement decisions are usually made on individual machines, the good machinery manager will need to keep a cost record for each machine.

The cost records of Fig. 4.7 and 4.8 can indicate to a machinery manager the optimum time for replacement of a machine. As the replacement decision involves a time span of several years, it is necessary to calculate costs in constant dollars to avoid distortions caused by inflation. The rate of inflation must be known for each year of the time period to convert the

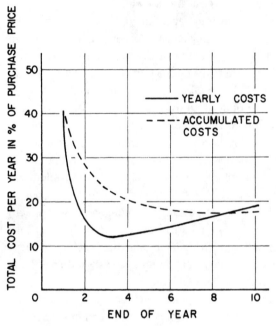

Fig. 4.9. Time of replacement: when yearly costs equal accumulated costs.

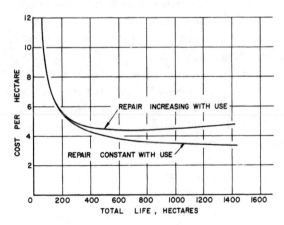

Fig. 4.10. Effect of repair costs on accumulated costs.

remaining values and annual ISTI and R&M expenditures to constant dollars as of the beginning of the record. For example, over the first 4 years of ownership the inflation rate in Fig. 4.7 was 3% compounded annually for years 1 and 2 and 4% for years 3 and 4. The inflation factor for 1988 is consequently $1.03 \times 1.03 \times 1.04 \times 1.04$, or 1.1475. The remaining value, RV, of the chisel plow in 1988 is 2100/1.1475, or $1830, in January 1985 dollars. Depreciation in constant dollars is determined from the deflated remaining values to be 2175 - 1830, or $345. The constant dollar costs for ISTI and R&M are shown in line e: (326 + 135)/1.1475, or $402.

The accumulated data are shown in the bracketed Replacement Analysis section. The accumulated values for cost and use are shown in lines i and j, respectively. The average accumulated cost, line k, is the indicator for replacement. As long as the values in line k decline each year, the machine should be retained. In 1990, the total accumulated costs since 1985 were $4411. Dividing by the accumulated 3237 acres of use produces an average cost of $1.36 for each acre worked since the chisel plow was purchased. It appears that the average accumulated cost reached its lowest point in 1989, and it will be uneconomic to use this implement any longer. Refer to Problem 4.8 to estimate whether this implement should be repaired or replaced. If the average accumulated cost per acre is expected to increase above $1.34 in 1991 because of an anticipated major repair, the machinery manager would know that the machine should have been replaced and not repaired.

An optimum replacement time can be predicted analytically if valid expressions for accumulated costs and obsolescence can be determined. An average expected use per year is required to relate the expected R&M costs to the annual fixed costs. Replacement time is the point at which accumulated costs divided by accumulated use is minimum.

While any depreciation method can be used to predict the optimum replacement time, the most realistic method depends on marketplace remaining value (Table 4.2). An expression that approximates current depreciation for either tractors or implements as a function of time and price can be developed from remaining values as

$$\text{acc. D} = (0.205 + 0.053n - 0.000\ 64n^2)P$$

where acc. D = accumulated depreciation through year
$\qquad$ n of use
$\qquad$ P = purchase price of machine

Other fixed costs (interest, shelter, taxes, insurance) can be accumulated by summing the percentage of remaining values over the years. Since taxes, insurance, and shelter charges are about a constant for both the existing and replacement machines, only the accumulated interest charges on the remaining value need be considered. The equation of a curve fitted to the points and multiplied by the rate gives

$$\text{acc. ISTI} = (0.23 + 0.8n - 0.022n^2)IP$$

where acc. ISTI = accumulated ISTI costs through year n
$\qquad$ I = annual charge for ISTI as a decimal of the purchase price (annual interest rate)

The accumulated R&M costs are described by multiplying Eq. 4.6 by P and using values of coefficients from

$$\text{acc. R \& M} = \left[(c1)\text{ha}/100 + (c2)(\text{na}/100)^2 + (c3)(\text{na}/100)^3 \right]P$$

where acc. R&M = accumulated repair and maintenance costs through year n
$\qquad$ a = average use, units/yr
$\qquad$ c1, c2, c3 = constants for specific machines, Table 4.6
$\qquad$ P = new list price (may be different from purchase price)

The units for a, c1, c2, and c3 must be consistent.

Obsolescence should be considered in determining the proper time of replacement. Fairbanks, Larson, and Chung (Kansas State University) have proposed that the cost of obsolescence increases at a constant rate each year. An obsolescence factor, ObF, can be defined as the rate of drop in machine value with time as desirable new features are added to new models. Introduction of the self-tying baler caused a large drop in value of the hand-tie models. In stony fields, automatic resetting trip beams on moldboard plows make other trip beams obsolete. Introduction of hydraulics, independent PTO, traction-assist systems, shift-on-the-go transmissions, and other important features caused a drop in the value of tractors not so equipped. The summation of these drops in value over a period of time leads to an estimate of the rate and cost of obsolescence.

An obsolescence factor can be estimated. If a $4000 machine is worthless because of obsolescence after 10 years of its 20-year life, the cost of obsolescence would be $2000/10 or $200/yr assuming straight-line depreciation. An obsolescence factor can be defined for this machine as 200/4000 or 0.05 of the list price per year. An accumulated obsolescence cost can be expressed as

$$\text{acc. Ob} = (P/2)(ObF)n^2$$

where acc. Ob = accumulated obsolescence costs
ObF = obsolescence factor, percentage of list price per year

Optimum replacement time is during that year when accumulated costs per unit of use is minimum. Dividing the total accumulated costs by accumulated use produces a cost per unit of use that can be minimized with the methods shown in Appendix B. The optimum replacement year, n, is found by a trial-and-error solution of

$$n^3 + (y/x)n^2 - z/x = 0 \qquad (4.8)$$

where $x = 2(c3)a^3/1000^3$

$$y = \frac{OBF}{2} + \frac{(c2)a^2}{1000^2} - 0.000,64 - 0.022I$$

$$z = 0.205 + 0.23I$$

As an example of the use of Eq. 4.8, the optimum replacement year for a corn planter used 200 ha/yr with a 10% ISTI charge (i = 0.10), a 1% obsolescence factor (ObF = 0.01), c1 = 0.0082, c2 = 0.0003, and c3 = 0.1054 is

$$x = 2 \times 0.1054 \times \left(\frac{200}{1000}\right)^3 = 0.001,686,4$$

$$Y = \frac{0.01}{2} + 0.0003\left(\frac{200}{1000}\right)^2 - 0.000,64 - 0.0022$$

$$z = 0.205 + 0.023 = 0.228.$$

Then

$$n^3 + 1.288\, n^2 - 135.2 = 0.$$

Constructing a table of values:

n	n^2	n^3	$1.288\, n^2$	$(n^3 + 1.288\, n^2 - 135.2)$
1	1	1	1.288	-132.912
2	4	8	5.152	-122.048
3	9	27	11.592	-96.608
4	16	64	20.608	-50.592
5	25	125	32.200	22.000

The value of the left side of Eq. 4.8 changes from negative to positive during the fourth year (n = 4+). The proper value for n to solve Eq. 4.8 is thus 4+ yr; therefore the optimum time for replacement is after the fourth year.

The machinery manager should be aware that general conclusions about optimum replacement times are highly dependent on the used equipment market. The accumulated depreciation charge based on marketplace value can change drastically with changes in used equipment prices, which in turn depend on the demand for used equipment. Only a broad mix of machine use and farmer purchasing and replacement policies permits the development of an optimum replacement policy.

Practice Problems

For these problems, use the following prices unless otherwise stated: labor, $8.00/hr; fuel, $0.30/L [$1.14/gal]; oil, $1/L [$3.75/gal].

4.1. a. Use the proper equation for each depreciation method and find the depreciation during the fifth year of life for a minor implement that was purchased new for $2000 and has an estimated life of 10 years. The salvage value is expected to be $100. (Do not use Table 4.2.)

  1. Straight-line

  2. Double-declining-balance

  3. Sinking-fund with i = 8% (0.08)

  b. What is the depreciation during the fifth year by the estimated value method? Refer to Table 4.2.

4.2. Using a fixed cost percentage, find the fixed costs per hour for a $60,000 tractor that is used 400 hr/yr. Assume a 12% interest rate and a 10 year life.

4.3. A 2-m [6.56-ft] rotary mower is priced at $3600 and is to be used for 100 ha [247 a] of mowing annually. Determine the approximate annual operation cost per hectare [acre]. The average forward speed is 8 km/hr [5 MPH] and the expected field efficiency is 0.80. An $18,000, 40-kW [53.6-HP] tractor, used 422 hr/yr for other work, will power the mower. The tractor uses 10 L/hr [2.65 gal/hr] of diesel fuel. The crankcase oil is changed every 200 hours.

  a. Determine the approximate annual operation cost per hectare [acre].

  b. How does the answer for (a) compare with the cost estimates shown in Appendix C?

  c. In what general ways may a machinery manager reduce operation costs?

4.4. Use typical approximate costs and determine the break-even use for a 5-m [16.4-ft], $89,000, self-propelled combine working in soybeans planted in 70-cm [28-in.] rows when compared with medium use costs listed in Appendix C. Assume the average field speed to be 4.5 km/hr [2.79 MPH] and a field efficiency of 0.70. Use an average energy requirement of 10 kW·hr/ha [5.5 HP· hr/a] (Table 2.5) to predict fuel consumption of the 82-kW [110-HP], turbo-diesel engine. The combine weighs 11,500 kg [25,300 lbs] with the yield collection tank half-full. Assume a 0.10 rolling resistance coefficient. [Hint: A break-even determination is made by equating the expression for the cost of one alternative to that for another alternative. The resulting equation is then solved for the variable of interest.]

4.5 A farmer is considering the purchase of a $7000 bale thrower for a rectangular PTO baler. The purchase would save the $8.00/hr hired labor cost for handloading the bales on a trailing wagon. The wagon unloading costs are considered to be similar for either system. Find the area of alfalfa harvested annually at which the employment of the bale thrower breaks even with the employment of extra labor. Assume a 10-year life for the thrower and an extra fuel consumption of 2 L/hr [0.53 gal/hr]. Use the R&M cost of balers for the thrower. The baler capacity is 10 t/hr [11 T/hr] without the thrower and 9 t/hr [9.9 T/hr] with the thrower. Yields of the four cuttings are 5, 3.7, 3.7, and 2.5 t/ha [2.2, 1.65, 1.65, and 1.1 T/a].

4.6. A farmer has been renting a fertilizer spreader for $8/ha [$3.25/a]. It has a field capacity of 4.0 ha/hr [10 a/hr]. He is considering buying a $6000 spreader with a capacity of 6.0 ha/hr [15 a/hr]. A $40,000 tractor, used 300 hr/yr exclusive of spreading fertilizer, will power the spreaders for either alternative. Fuel, oil, and R&M are estimated to be $2.50/ha [$1.00/a] for either alternative. Find the break-even area for these alternatives if the labor charge is $5.00/hr. (Hint: A cubic equation can be solved rather quickly by trial and error with a calculator. Find the answer to the nearest whole unit of area.)

4.7   A 4-year-old moldboard plow is purchased for $2500 and is expected to be depreciated out in the next six years. The list price new was $5000 and the expected salvage value is $100. Straight-line depreciation is to be followed. Interest rates are 12% and the plowing capacity is 1.2 ha/hr [3 a/hr]. What will be the plow's fixed costs plus repair and maintenance per area if it is to be used 100 ha [247] annually?

4.8.   a. The remaining value for the tractor in Fig. 4.8 is estimated to be $30,000 at the end of 1991 with a $1500 R&M cost for an estimated 300 hours of use in 1991. Complete the column in Fig. 4.8 for 1991 assuming a 4% inflation rate.

b. The chisel plow in Fig. 4.7 is expected to have a remaining value of $1500 at the end of 7 years (1991). Use an inflation rate of 4% and complete the column in Fig. 4.7 for a $400 R&M, $826 fuel and oil, $600 labor costs, and 600 acres of use. Assume the 100 hours of chisel plow use is included in the 300 hours of tractor use.

c. What is the $/a cost for chiseling in 1991?

d. Find the average accumulated costs for the plow. In light of this answer, should the farmer repair rather than replace in 1991?

4.9.   A self-propelled combine is used for 200 ha [500 a] annually. Predict the optimum replacement time with 12% interest rate, actual depreciation as in Table 4.2, R&M costs as in Table 4.5, and an obsolescence factor of 0.01.

4.10. Calculate the IRS allowable depreciation for each year of a tractor classed as a 7-year property. Use a 150% depreciation rate with the MACRS method. The tractor is placed in service in August 1998. Use a Section 179 deduction. The equipment dealer was paid $80,000 cash and given a trade-in tractor with an undepreciated value (basis) of $15,000.

OPERATIONS

Operations is the portion of machinery management concerned with optimum adjustment and use of individual machines.

The machinery manager should be concerned with the details of operating equipment. When not the actual operator of the equipment, the machinery manager must supervise its operation. The operator must be instilled with the objective of accomplishing economic return rather than thrilling to the control over power or drifting into a trance with the monotony of it all.

The proper operation of specific machines can contribute to the economy of the farm enterprise just as much as other aspects of good machinery management. The manager may select the correct size of machines, keep costs low, and replace wisely, but all is lost unless the machines are doing their assigned jobs in an efficient manner. The adjustment of harvesting machines to improve their material efficiency is a widely recognized good-management practice. Calibration of distributing equipment and maintenance of tillage elements are also recognized as beneficial practices. But the optimum use of farm machines is not well defined and consequently is not generally thought of as being important.

The philosophy, "If a little is good, more is better," contributes greatly to the overuse and consequent misuse of machines. The lack of understanding of the pertinent crop, soil, and economic facts of production contributes to this attitude. Many times it is an understanding of the system's concept that is needed. The farm enterprises must be defined in terms of a corn production system, a beef production system, etc. When examined from this broad view, it is seen that machines are operated as elements of a system and not as ends in themselves. Machine use is specified as only the amount that contributes to the optimum productivity of the system—no more, no less.

Chapter 5

TILLAGE

Tillage has been defined as those mechanical soil-manipulating actions that nurture crops. The nurturing objective is one of developing a desirable soil structure that promotes seed germination, plant emergence, and root growth. A firm, bare soil surface layer conducts moisture and heat to the seed. Plant emergence is promoted by a crustless surface. Root growth is promoted by an air-porous soil having capillary water and appropriately placed fertility. Nurturing also requires inhibition of competing weeds.

Conventional or clean tillage describes the system where crop residues are disposed of by chopping and incorporating into the tilled soil layer. Such a tillage system has a high-energy, primary tillage operation followed by a later secondary tillage to kill sprouted weeds and prepare a seed bed for the subsequent seeding. Historically, the ideal seed bed was a continuous blanket of fine-crumbled soil free from surface residues. The buried residues promoted trouble-free passage of the planting machines and a reduction in insect pests that overwintered in the residues. Weeds were controlled by deep plowing and mechanical cultivation. With the advent of selective herbicides, insecticides, and residue-cutting furrow openers, some of the objectives of clean tillage are no longer as important.

Reduced tillage refers to any system that requires less soil manipulation than a conventional system used for comparison. The reduced tillage objective saves tractor fuel and time when unnecessary, unproductive tillage operations are eliminated. Combining operations into a once-over trip, strip tilling only the planned rows, and tilling only as deep as is productive are actions to reduce tillage energy and ultimately the cost of production. Chemical weed control may be required.

Conservation tillage is defined as a system that places a premium on reducing soil loss. Since 1985 the U.S. government has provided complying farmers incentives for restricting soil loss to stated amounts. On highly erodible soils, conservation tillage procedures are necessary. The objectives include reduction of both wind and water erosion. Maintaining a soil cover with previous crop residues or producing a soil surface of dense clods meets this objective. Weed control is dependent on chemical herbicides. Conservation tillage objectives are met by a broad range of tillage systems, including the four below:

1. *No-till*, where there are no tillage operations apart from the seeding operation. The seeding machine creates a narrow, tilled strip for receiving the seed. Such seeding may be done in the standing residue from the previous year's crop.
2. *Ridge planting*, which requires the establishment of permanent ridges 15-25 cm [6-10 in.] high. Row crop and even small grain seeding are done on the ridge top, leaving crop residues and wheel track compaction for the furrows. Tillage is needed to reform the ridges seasonally. Seeding is accomplished after scalping the top to remove residues, weed seed, and dry soil. Ridges placed across slopes or on contours are very effective for absorbing heavy rainfall without suffering soil erosion or drowning of young plants.
3. *Mulch tillage*, which includes all conservation tillage systems that leave at least 30% of the soil surface covered with crop residues. The seeding machine must be equipped with tooling that will cut through the residue and deposit the seed into a granular, moist, firm seedbed.
4. *Hard-ground listing*, a system applicable to low rainfall areas. As opposed to ridge planting, the seeding is done in the furrows, where moisture is available and where the young plants are protected from abrasive winds. Listed land is mechanically cultivated to a flat surface and is effective for plant support and weed control without chemicals. If the listing is done in previously worked soil, the tillage is referred to as "loose-ground" listing.

A 1991 survey by J. A. Gliem, T. G. Carpenter, R. G. Holmes, and G. S. Miller of the Ohio State Agricultural Department indicated Ohio farmers were using combinations of tillage systems. The farmers reported using four tillage systems: conventional (75%) conservation (64%), no-till (38%), and ridge till (5%).

Tillage Machines

Tillage absorbs well over half the power expended on the farms in the nation. It is of great economic importance that the machinery manager understand operating characteristics, applicability, and performance of the various tillage machines.

Moldboard plows are primary tillage implements consisting of warped surfaces equipped with cutting edges that crumble and invert the soil. This implement gives best residue coverage and superior pulverization under ideal conditions. It is particularly effective for bringing mature grass and legume fields into cultivation and for control of insects that over winter in crop residues. A *two-way plow* is a moldboard plow having both right- and left-oriented bottoms. This plow maintains a level field and is used where flood irrigation is important. Two-way plows, particularly if tractor-mounted, are very applicable to plowing contours or terraced fields. Their ability to return down the furrow just completed eliminates dead furrows, back furrows, and lost time in finishing irregular lands. A less expensive design uses a specially shaped bottom that can throw either to the right or to the left when rotated in a horizontal plane about the standard. Such a design cuts in half the total number of bottoms needed (Fig. 5.1).

Disk plows use free-turning, concave, disk blades to lift and invert the soil. They do not cover crop residues as completely as do moldboard plows. Standard disk plows have a large, inclined disk mounted on individual standards and are used where the soils are (1) extremely hard or loose, (2) rocky or have many roots, (3) poor scouring, and (4) highly abrasive. Vertical disk plows have all the disks mounted on a common axle. The cutting edge of the disk is in a vertical plane. This implement is a wide-cut, shal-

Fig. 5.1. Five-bottom, two-way moldboard plow.

low-depth, primary tillage machine sometimes used in wheat growing areas (Fig. 5.2).

Listers are primary tillage implements with planting attachments. Only a portion of the soil surface area covered is actually worked. As a result, this type of tillage requires less power and has a greater field capacity than other primary tillage implements (Fig. 5.3).

Middlebreakers (Fig. 5.4) can be described as listers without planting attachments. The bottom appears to be right- and left-hand moldboard plow bottoms joined at their landsides. These implements, also called *middlebusters*, are used in soils with poor internal drainage to make ridges or beds on which to prepare a seedbed.

Subsurface tillage implements rip and pulverize soil without inverting it. These implements are used where residue coverage is not desirable. They are effective in preventing both wind and water erosion.

1. *Subsoilers* (Fig. 5.5) are strongly constructed rippers used for breaking up deep (50 cm [20 in.]), compacted layers of soil to improve their internal drainage.
2. *Chisel plows* (Fig. 5.6) are multistandard rippers designed to operate 25-30 cm [10-12 in.] deep. Sweep tooling may be used but narrow, pointed teeth are more common. Chisel plows are used as primary tillage implements.
3. *Field cultivators* (Fig. 5.7) are constructed more lightly than chisel plows and have closer spaced standards designed to operate from 7-12 cm [3-5 in.] deep. Sweeps of 20 cm [8 in.] and 25 cm [10 in.] are the most common tooling. These implements are used more for weed control and for seedbed preparation than for primary tillage. In dry regions massive single-sweep tooling as wide as

Fig. 5.3. Toolbar-mounted lister with rotating moldboards.

Fig. 5.4. Tractor-mounted middlebreaker.

Fig. 5.5. Subsoiler.

Fig. 5.6. Chisel plow.

Fig. 5.2. Vertical disk plow.

Fig. 5.7. Field cultivator.

Fig. 5.8. Slicing chisel plow with twisted shovels for partial incorporation of crop residues.

2.0 m [7 ft] operating at shallow depths is used to produce the ultimate in weed control with a minimal disturbance of the residue cover on the soil surface. These *blade cultivators* are used for both primary and secondary tillage.

Subsurface tillers leave the soil with surface residues and a rough profile that are very effective in reducing both wind and water erosion. Heavy crop residues can interfere with the operation of subsurface tillers. Most tillers have a staggered standard design to encourage the passage of roots, stalks, leaves, and other surface material through the machine. A combination of disk coulters and chisels with curved shovels is used on the implement in Fig. 5.8 to cut through and partially bury heavy corn crop residues.

Combined tillers are designed to prepare seedbeds in one pass, whether used on no-till or rough-tilled fields. Implements such as those in Fig. 5.9 consist of disk coulters for cutting through crop residues, leveling blades, subsurface tillers, rotating crumblers, and a tine harrow.

Rotary tillers (Fig. 5.10) consist of powered, rotating knives or tines that pulverize the soil. These machines may process the soil over the full machine width or they may be *strip-tillers* that process only a row crop seedbed. The latter can be used for row crop cultivation also. Rotary tillers are especially adept at incorporating chemicals and crop residues uniformly into soil. Fig. 5.11 shows the particle seperation as the rotor turns counter-clockwise while the tiller moves to the right. Light plant material drops immediately to the bottom of the cut. The heavy clods are thrown against the first shield and are directed downward. Fine, light soil particles strike the second shield and land on top of the tilled strip.

Fig. 5.9. Combined tillage implement.

Fig. 5.10. PTO-driven rotary tiller.

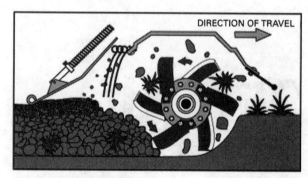

DIRECTION OF TRAVEL

Fig. 5.11. Rotary tiller action.

Moldboard Plow Types

The modern moldboard plow has been developed after centuries of experimentation with plow bottom shapes and materials.

The shape of a moldboard bottom is a compromise among the factors of draft, completeness of inversion, pulverization or breakup of the furrow slice, burying of crop residues, and scouring. Plow bottoms with sharply turned moldboards do a superior job of pulverization but require considerable draft. Long, gently turned moldboards pull easier but only invert the soil slice. Such moldboards can be modified in shape to partially invert the soil and partially bury crop residues for soil conservation reasons. Cohesive or sticky soils require moldboards designed to be uniformly scrubbed by the soil pressure to ensure cleaning or scouring.

Plow bottom parts are subject to rapid wear and are an operating expense that varies with use. Most modern designs include a separate shin (Fig. 5.12) allowing this rapid wearing portion of the moldboard to be replaced economically.

Different materials may be used for each plow bottom part. Cast-iron bottoms have very good wearing characteristics but are too brittle for use in rocky soils. Most plows have steel shares and moldboards. The steel scours easily and the points are not as likely to break upon hitting an obstruction. Steel bottoms may be of two types: solid or soft center. The soft center steel is a metal sandwich having hard, wear-resistant steel in the outer surfaces with a softer, tougher layer inside. The soft center shares and moldboards usually scour more easily than the solid bottoms but are slightly more expensive. Recently plastic coated bottoms have been developed which will provide more thorough scouring in sticky soils.

Implement manufacturers sell more semimounted and mounted plows than trailing plows. Mounted plows are more convenient to maneuver in the field, but a more important advantage is increased traction for the tractor drive wheels. The downward force of the furrow on the plow bottom causes increased tractor tractive effort. Such an effect results when the plow is held *up* to a certain depth by the hydraulic system rather than allowed to "float" in a balanced condition. The downward soil reaction is transmitted through the hydraulic system to the tractor frame with a resulting downward force on the traction wheels.

Even greater traction is gained from using an automatic *traction-assist control system*. In a three-point-

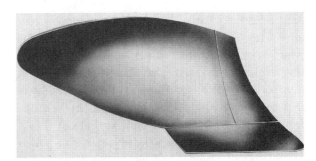

Fig. 5.12. Moldboard plow bottom with expendible shares.

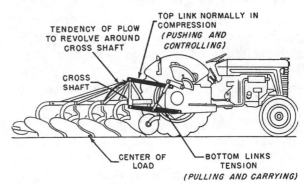

Fig. 5.13. Three-point-hitch link forces.

hitch-mounted plow (Fig. 5.13) the top link works against a spring to actuate the hydraulic lift mechanism. As the pull of the plow increases, the linkage causes the hydraulic cylinder to lift on the lower links resulting in a downward force on the rear wheels beyond that due to weight. It is desirable that the plow is not actually lifted very far, as uneven plowing would result. The sensitivity of the system is adjustable on most tractors. Because individual systems vary widely in design, the operator's manual for the tractor and for the implement should be studied thoroughly.

Hitching Moldboard Plows

The hitching or attachment of moldboard plows to tractors is more complex than for most other tillage implements. The moldboard plow bottom experiences substantial unbalanced side forces and a non-centered resistance to pull. Some of this unbalance is accommodated within the plow itself, but the rest is absorbed by the hitch position. The solution to the hitching problem is to: place the hitch point of the tractor and the plow in the straight line between the *center of resistance* of the plow and the *center of pull* of the tractor. The center of plow resistance is defined as the mean of the individual bottoms' centers of resistance, each located about one-fourth the width of cut to the right of the landside at the junction of the share and the moldboard. The center of pull of the tractor is defined as the point halfway between the driving wheels, and at the height of the front support for the tractor drawbar links.

Ideal conditions occur when the center of pull is placed directly ahead of the center of resistance. Under such conditions no forces tend to pull the front wheels of the tractor to one side. With small plows, even with a drive wheel operating in the furrow, it may be impractical to narrow the tractor drive wheel spacing enough to obtain ideal conditions; then the drawbar hitch points should be placed on a straight line between the center of pull and the center of resistance to minimize the effect of off-center pulls. With large plows the opposite situation may occur– the wheel spacing cannot be adjusted wide enough for ideal alignment of the two centers. To keep the bending forces in the rear axle to a low value, the rear wheel tread should be narrowed and the tractor operated completely on the unplowed land. Further advantages for such operation are the uniform weight distribution on the drive wheels, the uniform tractive effort of each wheel, and the elimination of puddling or packing the soil on the bottom of the furrow. Doubtless it is easier for the tractor operator to steer with the drive wheel in the furrow; but with the dual or large-size tires used on most big tractors and for the previously mentioned reasons, operating out of the furrow and hitching immediately ahead of the center of pull is to be preferred.

Moldboard Plow Adjustments

One of the most important adjustments is the heel and landside clearance effected by the set of the tail wheel. Such a clearance between the landside and the unplowed soil transfers the drag of the heel and landside onto a rolling wheel that reduces the draft of the plow.

Other adjustments are plowing depth and plow levelness. On trailing-type plows the three wheels limit the depth of plowing. Hydraulic control is used to adjust the wheels higher or lower with a consequent change in plowing depth if proper heel clearance is maintained.

The major adjustments on a mounted plow are depth and steering adjustments. To make a deeper furrow, the individual bottoms are tipped forward.

This increased pitch gives greater "suck" and the share penetrates to a new depth. A constant depth results when the plow is leveled. Decreased depth occurs as the plow is tipped backward. Steering to the right or left varies the width of cut of the first bottom. Since a mounted plow hitch must flex to accommodate uneven ground and turns, these steering adjustments must be accomplished without restricting the movement of the hitch links.

Fig. 5.14 shows the average location of forces acting on a mounted plow supported by a traction-assist system. The magnitudes of these forces are approximated by the width of the arrows.

The tractor wheel spacing for large mounted and semimounted plows may be different from similar-size pull-type plows. As shown in Fig. 5.15, it is necessary to attach the tractor to the plow's drawbar (1). If the operator chooses to run the tractor's drive wheel in the plow furrow, the horizontal-pivot pin (3) (the point of pull) may be to the right of a straight-ahead line of pull from the plow's center of resistance. Such a hitch is not undesirable. This hitch has a tendency to rotate the plow counterclockwise (looking down on the plow). As viewed from above, the soil forces acting on the bottom normally produce a clockwise rotation which is counteracted by the tail wheel and the landsides. Therefore, a hitch that produces a counterclockwise tendency may be expected to relieve the cross loads on the tail wheel and landsides. The reduced friction loads on these elements will reduce the draft of the plow. The disadvantages of operating the right drive wheel of the tractor in the furrow are discussed in the section on hitching. Reclamping the pivot pin (3) along the frame bar (2) to be in line with the plow's center of resistance may allow the tractor to be operated out of the furrow.

Semimounted plows have simple hitch adjustments. Fig. 5.15 shows the plow's hitch elements. The plow is free to pivot in a horizontal plane about the pin included in housing (3). Horizontal hitch adjustment is made by shifting housing (3) along bar (2) and/or shifting the plow's drawbar (1). Vertical hitching, suction, and depth control are provided by the tractor's lower hitch link position and by the action of the hydraulic cylinder on the tail wheel. The operator first lowers the front of the plow, producing a pitch extremely favorable for penetration. Then the rear of the plow is dropped and the plow levels out in equilibrium with a desired depth set by limits on the hydraulic control system. The depth of plowing holds rather constant even over undulating soil.

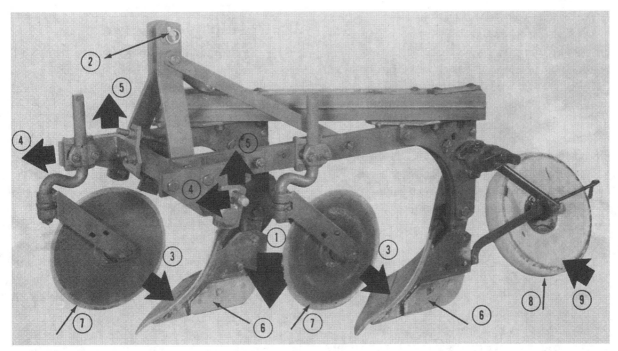

Fig. 5.14. Major and minor forces acting on plow mounted with three-point hitch.

1. Weight of plow
2. Top link force
3. Soil force on bottoms
4. Lower link draft forces
5. Lower link lifting forces
6. Landside horizontal forces
7. Soil forces on coulters
8. Tail wheel vertical force
9. Tail wheel horizontal force

Fig. 5.15. Semimounted plow hitch elements.

A variable width-of-cut plow has several advantages. Fig. 5.16 shows such a plow in its narrowest position for transport. Hydraulics rotate the individual bottoms counter clockwise as viewed from above and the effective width of the plow widens. This adjustment can be made from the tractor cab while operating. Should high-resistance soil be encountered, a reduction in width will allow a momentary match to the power capability of the tractor. Narrowing the width will also produce more soil pulverization. Widening the width of cut reduces crop residue coverage for soil conservation needs. Widening beyond the width of the bottoms leaves a small strip of unplowed ground for erosion control.

Moldboard plows also have trip features which protect the bottoms from damage when striking buried obstructions. Upon impact, the bottoms uncouple and swing back. Shear bolts may need replacing, or the bottom lifted allowing a spring force to reset the bottom, or the reset may be done automatically by springs or hydraulics without stopping forward motion.

Fig. 5.16. Variable width-of-cut moldboard plow.

Draft of Moldboard Plows

The draft or amount of pull required to move a plow is naturally quite dependent on the size of the plow and depth of plowing. A term called *unit draft* is often used as a draft measure considered to be independent of both plow width and furrow depth. Unit draft is defined as the force per cross-sectional area of the plow's action. If a 3-40 cm [3-16 in.] plow has a draft of 10 kN [2248 lb], the unit draft would be 4.17 N/cm² [5.85 lb/in²] if the plow is operating at a depth of 20 cm [8 in.]. (The centimetre is used for length measure as tillage operations lack the precision implied by the millimeter.)

Soil type is a major factor contributing to unit draft; it is well established that the speed of plowing is also very important. Unit drafts can be predicted from:

$$D = C_1 + C_2 S^2 \qquad (5.1)$$

where D = unit draft, N/cm² [lb/in²]

S = speed, km/hr [MPH]

C_1, C_2 = coefficients depending on soil type and unit system

Unit draft equations for specific soils and for plows equipped with high-speed moldboards, coulters, and landsides are

Silty clay (s. Tex.)	7	[10.24]	+ 0.049	[0.185] S²
Decatur clay loam	6	[8.77]	+ 0.053	[0.2] S²
Silty clay (n. Ill.)	4.8	[7]	+ 0.024	[0.09] S²
Davidson loam	3	[4.5]	+ 0.021	[0.08] S²
Sandy silt	3	[4.4]	+ 0.056	[0.21] S²
Sandy loam	2.8	[4]	+ 0.013	[0.05] S²
Sand	2	[3]	+ 0.013	[0.05] S²

These equations are multiplied by 1.07 for an added jointer or coverboard. An increase in soil moisture content of 1% can decrease draft 10%. An increase in the apparent specific gravity of 100 kg/m³ [6.2 lb/ft³] can increase unit draft by 10%.

The amount of wear on shares has some effect on unit draft. Roy Morling (International Harvester) studied the effects of hard surfacing on wear and draft of 35-cm [14-in.] flat expendable shares in dry sandy loam soil. For worn shares he found the draft increased and the suction force decreased as compared to new shares:

Worn Share	Draft % of new	Suction % of new
Regular	103	38
Hard-surfaced on top	102	37
Hard-surfaced on bottom	107	70

These data show that wear has only a minor effect on draft but a great effect on suction and the penetrating ability of a share. Morling traced the decrease of the suction force with use (Fig. 5.17) and found it to be quite rapid in sandy soils. When wear progresses to the point where negative suction occurs, only the plow weight keeps the shares in the soil.

The increase in draft with speed has management implications. The machinery manager is not free to select any combination of plow width and field speed that will utilize the tractor power effectively. A small, half-size plow operating at double speed requires more power than a full-size, low-speed plow.

Fig. 5.18 shows the penalty in theoretical field capacity from high-speed plowing. This diagram was developed by assuming 75 kW [100 HP] was available at the rear axle of a 2-wheel-drive tractor having 8000 kg [17,600 lb] or 78.29 kN [17,600 lb] SRAF on the axle. Various plowing speeds are selected. The silty clay soil (Illinois) data were used in Eq. 5.1 to determine the draft per unit width for each speed. Fig. 2.12 was entered at the actual field speed value and the predicted DRAWBAR PULL/SRAF was converted into maximum plow size for that speed. Theoretical field capacities were then found for that plow width and speed.

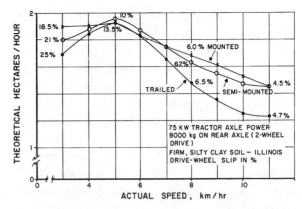

Fig. 5.18. Capacities and tractor drive-wheel slippage for moldboard plows.

The beneficial effects of mounted tillage implements are shown in Fig. 5.18. At all speeds the mounted plows show less slip and the tractor can pull slightly larger mounted plows than trailed plows. A maximum exists (5 km/hr [3.1 MPH]) when the plow is large enough to cover a lot of ground, the wheel slip is low enough to keep tractive efficiency high, and the speed is low enough to avoid high unit draft.

Disk Plow Adjustments

Disk plow adjustments are somewhat different from those for moldboard plows, primarily because disk plows have no landside and thus must absorb more of the side-thrust of the soil on their wheels than moldboard plows. Pull-type plows have provisions for adjusting the tail wheel, furrow wheel, and hitch points to balance the side-thrust between those two inclined wheels (see Fig. 5.2). As a first trial for a standard disk plow, the hitch should be adjusted along a line from the tractor's center of pull to the plow's center of resistance, a point a little below the soil surface and at the average of the centers of the disks. For a vertical disk plow having a wide width of cut, it is quite common to hitch far to the right of the plow's center of resistance so that the tractor pull may counteract the clockwise turning tendency due to the side-thrust of the soil.

The width of cut may be adjusted on standard disk plows by changing the angle of the disks with respect to forward motion. As a class, disk plows naturally swerve into a narrower cut when in hard ground and into a wider cut in soft ground.

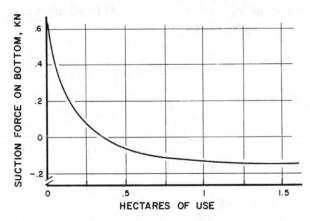

Fig. 5.17. Effect of wear on penetrating ability of expendable shares.

The depth of operation is limited by the wheels or by hydraulically controlled lift arms if the disk plow is mounted. Penetration to this depth depends on the angle from the vertical and weight per disk for the standard plow and on weight alone for the vertical plow. Some soil types require steeper angles for the standard plow to penetrate, others require a shallow angle.

The effect of speed on unit draft for a 66-cm [24-in.] diameter disk with 0.38-rad [22°] tilt and a 0.78-rad [45°] angle from the direction of travel for two soils is:

Decatur clay	$5.2 [7.6] + 0.040 [0.15] S^2$
Davidson loam	$2.4 [3.4] + 0.045 [0.17] S^2$

where S is the forward speed in kilometers per hour [MPH].

Subsurface Tooling Adjustments

Implements designed to till beneath the surface have pointed and balanced tooling. Weight should not be resorted to as a means of securing penetration; the rake angle of the tool (Fig. 5.19) is increased to secure greater penetration. The implement must be adjusted so that when the tooling falls onto the soil

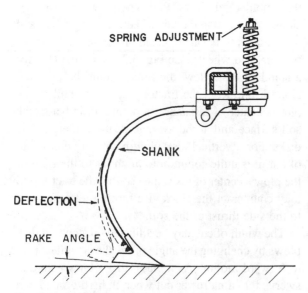

Fig. 5.19. Subsurface tiller construction.

surface an appreciable rake angle exists. As the tool penetrates, the manufacturer's design may cause it to run out of rake angle or suction at the desired depth or else gage wheels are used to prohibit further penetration. Deflection of the shanks due to soil force causes the rake angle to increase. Excess deflection causes increased draft and unnecessarily deep grooves in the seedbed bottom.

Springs absorb shock and permit the tooling to deflect over solid obstructions. The spring force should be adjusted to a level high enough that tool deflection is minimal in normal operations, yet not so high that the tooling fails to trip over obstructions.

Tooling on subsurface tillers can be shovels or sweeps. Deep tillage is accomplished with narrow shovel points. Weed control is best accomplished with sweeps of such a size that adjacent paths overlap; if the overlap is too small, weeds can pass through untouched when the implement slews to one side because of hill slopes or nonuniform soil resistance.

Hitching subsurface tillage implements to tractors, unlike the hitching of moldboard or disk plows, is very simple. The center of resistance is the geometric average resistance points of all the tooling and is below the soil surface. This center should be exactly behind the tractor's center of pull. The advantages of mounted and semi-mounted hitches as described for moldboard plows applies to subsurface tillers also.

Subsurface Tillage Draft

The draft of subsurface tillers depends on both depth of operation and operating speed. The draft of a chisel plow equipped with points or narrow shovels is expected to increase with the square of the depth until the vertical triangular areas of disturbed soil from adjacent tools intersect. PAMI found the draft, D, kN/m [lb/ft] to be

for heavy soils:	$0.61 d + 0.1 S - 1.9$
	$[106 d + 11 S - 130)$
for light soils:	$0.27 d + 0.1 S - 0.5$
	$[47 d + 11 S - 34]$

where d is the depth of penetration in cm [in.] and S is the speed in km/hr [MPH].

Setting a depth of 15 cm [6 in.] as the operational dividing line between chisel plows and field cultivators, the draft D in kN/m [lb/ft] for field cultivators is

for heavy secondary tillage: 0.27 d + 0.1 S − 0.5
[47 d + 11 S − 34]
for light secondary tillage: 0.24 d + 0.06 S − 0.7
[42 d + 2.6 S − 48]

The effect of speed on draft of subsurface tillers was found by J. D. Summers, A. Khalilian, and D. G. Batchelder (Oklahoma State University, 1984). Table 5.1 shows the results for both silt and clay loams. The speeds ranged from 4 to 10 km/hr [2.5 to 6.2 MPH]. Depths ranged from 7 to 20 cm [2.75 to 7.87 in.].

Optimum Widths

The fact that the draft of tillage implements increases with speed means that an optimum width implement exists for any given power level of tractor. The machinery manager who wishes to have high tillage capacity at the *least cost* must balance the cost of large equipment operating at slow speeds against the cost of power for smaller equipment operating at higher speeds. Suppose it is decided that a capacity of C ha/hr is required for a tractor and tiller to do A ha/yr and one wishes to determine the least-cost size of implement and tractor for this area. Since the hours per year are constant, labor costs are not pertinent to this problem.

In Chapter 4 the repair costs are related to hours of use only. In this problem where speeds are allowed to vary, repair is probably not related directly to time. It is quite likely in stony areas that the relationship between tiller speed and the cost of repair is increasing; but to simplify the problem, it will be assumed that repairs for both the tractor and the tiller are proportional to the area tilled. As the area considered, A, is constant, the repair costs are not a factor in the basic question of optimum size-speed relationships. The cost of oil is also assumed to be proportional to area covered and is not considered.

As an example for all tillage, moldboard plowing will be used to develop an equation for optimum-width tillage implements to produce least-cost operations.

$$\$/yr = \frac{(FC\%)}{100} pw + \frac{A}{C}\left(\frac{fuel\ cost}{hr}\right) + (F)\frac{(FC\%)}{100} t\ PTOP$$

where FC% = annual fixed cost percentage

TABLE 5.1. Effect of Speed on Unit Draft of Subsurface Tillers

Soil Type	Tooling	Unit Draft, N/cm² [lb/in.²]
Hollister clay loam	chisel	0.36 [0.52] + 0.023 [0.033] S
	wide blade	3.6 [5.22] + 0.54 [1.26] S
Reinach silt loam	chisel	0.35 [0.51] + 0.019 [0.028] S
	wide blade	4.0 [5.80] + 0.11 [0.26] S
Tabler silt loam	chisel	0.44 [0.64] + 0.033 [0.048] S
	wide blade	6.6 [9.57] + 0.23 [0.54] S

Note: S is km/hr [MPH]; 30-cm [12-in.] chisel spacing, 1.83 m [6 ft] and 3.66 m [12 ft] blade widths.

p = purchase price of plow, $/m [$/ft]

w = width of plow, m [ft]

(F) = fractional portion of tractor's yearly operating time used for plowing, decimal

t = purchase price of tractor, $/kW [$/hp]

PTOP = maximum power take-off power of tractor, kW [hp]

C = effective field capacity, ha/hr [a/hr]

To obtain proper relationships, the fuel cost must be expressed in terms of w and C with units of $/hr.

Fuel cost/hr = f B/(0.96 T H)

where 0.96 = ratio of axle power to PTOP

f = price of fuel, $/L [$/gal]

H = fuel efficiency at the % of max. loading, (Table 2.3), kW·hr/L [HP·hr/gal]

T = tractive efficiency, decimal

B = power at tractor drawbar required to pull plow, kW [HP]

Also,

$$B = wd/C_3 (C_1 + C_2 S^2)S$$

where w = width of plow, m [ft]

d = depth of plowing, cm [in.]

C_1, C_2 = constants from Eq. 5.1

C_3 = 36 [31.25]

S = speed of plowing, km/hr [MPH]

From Eq. 1.1, C = S w e/10. [C = Swe/8.25] Rearranged, the speed of plowing required for a given capacity is

$$S = \frac{10\,C}{w\,e} \quad \left[S = \frac{8.25C}{w\,e}\right]$$

Therefore,

$$B = \frac{Cd}{3.6\,e}\left(C_1 + C_2 \frac{100\,C^2}{w^2 e^2}\right)$$

$$\left[B = \frac{Cd}{3.788e}\left(C_1 + C_2 \frac{68.06\,C^2}{w^2 e^2}\right)\right]$$

$$w = \sqrt[3]{\frac{100 C_2 d}{FC\% \, p\, e^3}\left(\frac{75\,AfC^2}{H} + \frac{0.75(F)(FC\%)\,t\,C^3}{\%L}\right)}$$

$$\left[w = \sqrt[3]{\frac{68.06\,C_2 d}{FC\% \, p\, e^3}\left(\frac{66.7\,AfC^2}{H} + \frac{0.714(F)(FC\%)\,t\,C^3}{\%L}\right)}\right]$$

(5.2)

Substituting this expression for B in the previously derived expression for fuel cost per hour and setting T to be 0.77 (the optimum value for firm soil in Fig. 2.12), the fuel costs per hour can be expressed as

$$\frac{f\,Cd}{2.66\,H\,e}\left(C_1 + C_2 \frac{100\,C^2}{w^2 e^2}\right)\left[\frac{fCd}{3.0\,He}\left(C_1 + C_2 \frac{68.06\,C^2}{w^2 e^2}\right)\right]$$

The fixed costs for the tractor are based on its purchase price per maximum PTOP. The power rating necessary to pull the plow can be estimated from the DBP, the tractive efficiency, and the load factor, %L:

$$PTOP = \frac{Cd}{2.66\,e\,(\%L)}\left(C_1 + C_2 \frac{100\,C^2}{w^2 e^2}\right)$$

$$\left[PTOP = \frac{Cd}{2.8\,e(\%L)}\left(C_1 + C_2 \frac{68.06\,C^2}{w^2 e^2}\right)\right]$$

The cost of plowing per year is the sum of the annual fixed costs for the plow, the annual fuel consumption, and the portion of the tractor fixed costs charged to plowing.

This sum is denoted algebraically by substituting the expressions for PTOP and fuel costs into the original equation. (Repair and oil costs need not be included.) The least-cost plow size is determined by differentiating this cost equation with respect to w and setting it equal to zero (see Appendix B).

Eq. 5.2 gives the expression for the optimum width of plow.

Fig. 5.20 is a plot of Eq. 5.2 for these following constant values for a semimounted moldboard plow in a Davidson loam:

$$C_2 = 0.021 \quad p = 2000 \text{ \$/m} \quad (F) = 0.3$$
$$d = 20\text{cm} \quad e = 0.8\% \quad \%L = 0.8$$
$$FC\% = 16 \quad f = 0.20 \text{ \$/L} \quad t = 400 \text{ \$/kW}$$
$$H = 2.46 \text{ kW·hr/L at 80\% load}$$

The least-cost plow size is obtained by using the following sequence:

1. Select the line representing the annual ha to be plowed (Fig. 5.20).
2. Decide on a dashed line representing the capacity figure.
3. At the intersection of the two lines the plow width, the economical operating speed, and the maximum PTOP of the tractor are determined.

For example, a farm operator has 100 ha of plowing annually and feels that he must have a capacity of 1 ha/hr, C = 1. The least-cost size of the plow is about 1.8 m,

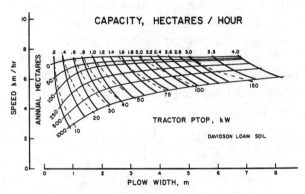

Fig. 5.20. Least-cost plow size, single unit operation.

the best speed is 6.8 km/hr, and the tractor is required to have a maximum PTOP rating of 46.7 kW.

Several general conclusions may be drawn from Fig. 5.20. If the conditions assumed are typical and the plow draft-speed relationship is correct (Eq. 5.1), plow speeds should never be much above 7 1/2 km/hr, and these speeds are profitable only for small acreages. The increase in draft at higher speeds causes the power required and the accompanying fuel costs to rise rapidly. Instead, greater economy is obtained by purchasing larger plows with accompanying larger

tractors and keeping the forward speeds at a lower level.

Also, larger equipment operating at slower speeds is indicated as being more economical as the total area of plowing increases. This conclusion is rationalized by recalling from Chapter 4 that fixed costs per acre decrease as the use increases; thus it becomes more economical to operate the larger equipment. Speeds are slower for the same required capacity, and the economy actually improves even though the larger equipment has a greater purchase price.

Practice Problems

5.1. A 7–40 cm [7-16 in.] plow is pulled by a tractor with an outside wheel clearance dimension of 2.5 m [100 in.]. How far from the furrow edge should the edge of the rear tire be driven if the plow's center of resistance is directly behind the tractor's center of pull?

5.2. A mounted plow attached with three-point hitch linkage swings to the right when in operation, reducing the width of cut of the front bottom. What corrective adjustment is indicated?

5.3. What is the expected draft of a chisel plow 4 m [13.12 ft] in width operating at a depth of 20 cm [7.87 in.] and at 8 km/hr [5 MPH]? Use the data in Table 5.1 for the Tabler silt loam.

5.4. What would be the least-cost size of moldboard plow for 200 ha [494 a] with a capacity of 1 ha/hr [2.5 a/hr] in a sandy loam soil? All other data are the same as in Fig. 5.20.

5.5. The sweep pattern of a subsurface tiller is composed of 3 ranks spaced 91 cm [36 in.] apart. Lateral spacing in each rank is 30 cm [1 ft]. What width of sweep is necessary (wing tip to wing tip) to ensure complete coverage even with 4.5° of implement skewing?

SEEDBED PREPARATION

Seedbed preparation operations are linked to the various tillage systems described in Chapter 5. But regardless of the system, seed must be deposited into a soil condition that promotes germination and growth. This seedbed may be field-wide for broadcast or drilled crops or as narrow as a few centimetres for each crop row.

The primary objective of a machine for seedbed preparation is the reduction of soil particles to a size that establishes, through proper compaction, a close contact around the planted seed to encourage moisture movement into the seed; retains enough soil porosity to drain away free water and permit air movement through the soil; and leaves a rough surface that dries out rapidly and will not germinate weed seeds.

Several different types of tooling can reduce soil particle size, including disks, shovels, tines, and sweeps. To some extent these tools also accomplish soil settling and compaction although rolling is more common to reduce soil clods as well as firm the soil.

Field cultivators (see Fig. 5.7) satisfy most of the objectives for a seedbed in that a fine particle seedbed is established at seeding depth while the surface is left rough and cloddy. Because their draft is somewhat less than comparable width disk harrows, field cultivators are good seedbed preparation machines. They are not as effective in incorporating crop residues into the soil as other implements. But in low rainfall areas a desirable seedbed should have surface residues to reduce wind erosion and to retain snow and rainfall. For areas of higher rainfall, field cultivators can be used for mulch tillage conservation practices. Cultivator sweeps establish a 5-7 cm [2-2.75 in.] bottom to the seedbed. Incorporating wheels behind the sweeps mix any applied chemicals as well as bring some buried residues to the surface.

A combination of subsurface tooling and V-packer wheels is used by the implement in Fig. 6.1. This *roller harrow* is particularly effective in producing a granular, level, firm seedbed. The front rollers crush surface clods. The spring teeth of the harrow portion then bring buried clods to the surface where they can be crushed by the rear roller. This implement is most effective in residue-free seedbeds.

Disk harrows are versatile tillage implements that produce seedbeds intermediate between those of the field cultivator and the roller harrow. Heavy harrows with large diameter disks can do primary tillage operations. Disk harrows are effective in chopping surface residues and incorporating these residues and chemicals into seedbeds.

A typical tandem disk harrow is shown in Fig. 6.2. It has provision for adjusting the angle of the gangs and hydraulically actuated transport wheels to allow road transport and operating depth limits.

Offset disk harrows trail to one side of the tractor hitch point. These disks were originally used for cultivating work under low branches of orchard trees. The sturdy construction of these disks also has made them popular for field seedbeds (Fig. 6.3).

Fig. 6.3. Offset disk harrow.

Adjustments

Depth of operation is the major disk harrow operational adjustment. If limited depth tillage or a shallow incorporation of chemicals is desired, this harrow may be limited in its operating depth by the position of its transport wheels.

Penetration is no problem in plowed ground but is a factor when disk harrows are used for initial tillage in order to cut up and incorporate surface residues into the soil and to level the soil surface. Penetration is increased by increasing the angle of the gang from 0.25 rad [15°] to 0.45 rad [25°]; keeping the hitch points low; increasing the weight per disk blade common weights per blade range from 40 kg [88 lb] to 90 kg [200 lb]; or decreasing the concavity of the blades. Penetration is less with increased forward speed.

Penetration is also affected by size of the disk blades. Large diameter blades will not penetrate as easily as smaller blades with the same weight per disk but they do have the ability to operate deeper. Blade diameters can range from 50cm [20 in.] to 80cm [32 in.].

The spacing of the disk blades on the axle affects penetration. Manufacturers offer spacings from 18 cm [7 in.] to 28 cm [11 in.]. The narrower spacings produce more pulverization, while the wider spacings encourage penetration and trash clearance. Some models of tandem disk harrows are available with wide spacing for the front gangs and narrow spacing for the rear gangs.

Disk blade spacing and the depth of penetration affect the amount of residue remaining on the soil surface. Tests by Deere, as shown in Table 6.1, indicate the relationship.

Uniform penetration of all disk blades is difficult

Fig. 6.1. Roller harrow (4.5-m [15-ft]) produces fine, level, firm seedbed.

Fig. 6.2. Tandem disk harrow.

Table 6.1. Percentage of Original Residue Remaining after Tandem Disking

Disk spacing, cm [in.]	Depth cm [in.]	Small diam. stalks and/or light yields	Large diam. stalks and/or heavy yields
18 [7]	10.2 [4]	25	50
23 [9]	10.2 [4]	30	67
23 [9]	15.2 [6]	17	35

to achieve. Inherently an individual gang tends to ride high at its convex end and low at its concave end because of the soil forces on the blades. Furthermore, as the tractor's center of pull is well above the disk harrow's center of resistance, which is the geometric center of the gangs and below the soil surface, the rear gangs will not penetrate as deeply as the front gangs. The gangs on rigid frame disk harrows are bolted firmly to the harrow frame; thus the gangs are held firmly in place and the only adjustment problem is leveling the frame from front to rear. Rigid harrows excel in leveling the soil surface but do not give a uniform depth of soil pulverization. Flexible disk harrows have the ability to penetrate uniformly over uneven surfaces but need additional adjustments for keeping down-pressure on the convex end of the gangs.

Disk Harrow Draft

The draft of disk harrows increases with speed but decreases as penetration decreases with speed; consequently, a simple rolling resistance coefficient predicts draft at any speed. Robert D. Wismer and Harold J. Luth (Deere and Co.) report coefficients of 1.5, 1.2, and 0.8 for heavy, medium, and light soils, respectively.

Elevated Seedbeds

Seedbeds raised above the field level are used in some areas. Single-row planting is done on a *ridge* and multirow planting is done on a flat-topped *bed* whose surface is 10 cm [4 in.] or more above the adjoining furrow. These practices are used on soils with poor internal drainage to provide a warm seedbed that is drained of free water early in the season. In dry areas, the furrows supply irrigation water to the seedbed. Tractor and implement wheels run in the furrows; thus excessive compaction of the seedbed is avoided. Some automatic steering of machines is possible where a tool carrier guides on the bed or ridge created previously.

Elevated seedbeds are made by middlebreakers (Fig. 5.4), by bedders (Fig. 6.4), and by sled-type tool carriers (Fig. 6.5). The soil usually needs to be preworked by plowing or middlebreaking before being formed into beds. But ridges and beds can be maintained year after year to eliminate the compaction effects of wheels on seedbeds.

Seedbed preparation may also include placement of fertilizers and pesticides. These materials may be broadcast and then incorporated into the soil with tillage.

The advent of preemergence chemical application for weed control has caused the development of implements called *incorporators* that mix herbicides uniformly into the soil where weed seeds will germinate. Incorporation is a once-over operation on previously tilled soil and should leave a level, finely crumbled seedbed. Popular tooling for incorporators are Danish or S tines that vibrate in all directions as they operate at a depth of 7.5-10 cm [3-4 in.] (Fig. 6.6). The S tines are often followed by rotating soil tillers and levelers called crumbler baskets (Fig. 6.7, also shown in Fig. 6.9). Other designs use harrow sections with flexing tines (Fig. 6.8). The herbicide should be applied only a short time before the operation of the incorporator to avoid chemical degradation.

Fig. 6.4. Disk bedder.

Fig. 6.5. Seeding on beds with sled carrier.

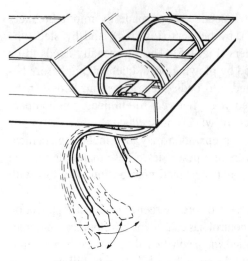

Fig. 6.6. Danish or S tines flex in all directions.

Fig. 6.7. Rotating crumbler baskets follow S tines.

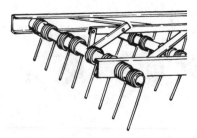

Fig. 6.8. Flex-tine tooling.

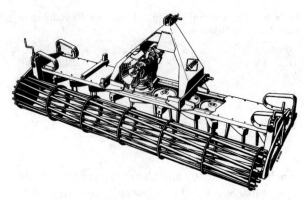

Fig. 6.9. Powered harrow combined with crumbler basket.

The Monsanto Company recommends specific incorporator design and adjustments. To secure uniform incorporation the tines must be spaced to till all the soil. Preferably, S tines should be in three ranks for good residue clearance characteristics. Flexing tines should operate in a nearly vertical position. The flex tines and the crumbler baskets should till 5 cm [2 in.] deep and produce a level seedbed. Forward speeds should be 10 km/hr [6 MPH] or more.

Powered harrows use a tractor's PTO to cause reciprocation, rotation, or oscillation of spike tooling while the harrow is moving forward. Figure 6.9 illustrates one design being manufactured. These implements produce finely broken-up seedbeds for vegetable crops and can also be used for incorporating chemicals. Powered harrows work well in rock- and residue-free fields. They require less draft than comparable fixed spike harrows; but their total overall power requirement is greater, approaching that of rotary tillers. Power and energy requirements increase with depth of operation and degree of soil pulverization. Operating depths can be 18 cm [7 in.] or more.

Practice Problems

6.1. Contrast the different objectives for seedbeds prepared by listing and by bedding.

6.2. Analyze field cultivator performance with respect to:

a. surface conditions

b. seedbed level conditions

c. weed control

6.3. Show with a sketch of forces acting on a pull-type tandem disk harrow why the rear gang tends toward shallower penetration.

6.4. Show by sketch why an offset disk harrow can be pulled off-center for equilibrium operation.

Chapter 7

CULTIVATION

Cultivation refers to tillage operations performed after seeding. The term *cultivator* is usually applied to an implement designed for row crop cultivation, although the field cultivator (Fig. 5.7) is an exception.

While it is possible that chemical means alone will be used for weed control, there is still a present need for cultivators. Although weed elimination is the primary purpose of cultivation, the secondary purposes are

1. Preparing the soil to retain rainfall,
2. Aerating the soil,
3. Incorporating fertilizers and pesticides, and
4. Providing plant support.

Cultivators may be equipped with various types of tools. Shovels are for deep working, soil throwing, and burying and uprooting weeds. Sweeps are shallow depth weed-cutting tools (Fig. 7.1). Other tools are used for special purposes—furrowing shovels, diamond-pointed grass uprooters, and weed knives.

For extremely weedy or grassy fields, *disk hillers* (Fig. 7.2), which throw large amounts of soil though only skimming the surface, may be used next to the row. If the field is infested with viny weeds, it may be profitable to use a *disk cultivator*.

On the *rolling cultivator* pictured above, note the special spoked-wheel tooling that engages the soil surface. These wheels are assembled in gangs that can be angled in the horizontal and vertical planes. The shape of the spoke allows the soil crust to be worked thoroughly as the wheel rotates with the forward motion of the

Fig. 7.1. Cultivator tools. Single-point, double-point, and spear-point shovels; full sweep and right and left half-sweeps. Sweeps may be obtained in sizes running from 120-mm [5-in.] to 450-mm [18-in.] widths.

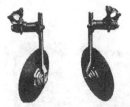

Fig. 7.2. Pair of disk hillers.

cultivator. The gangs can be set to throw considerable amounts of soil into the row for good weed control and plant support. This machine can also be used to reform, and even make, beds for planting.

The conventional row crop cultivator consists of a series of tools mounted on gangs suspended from a toolbar. Some cultivators are front-mounted on tractors to permit the operator to have a clear view of the operation. Large cultivator designs may have quick-hitch rear mounting (Fig. 7.3). They are towed endwise to and from the field on their own transport wheels, which become toolbar gage wheels when cultivating. Typically, these cultivators will have spring-loaded tooling, rolling coulter stabilizers, depth gage wheels, and independent hydraulic lifts for each side to permit improved operation in point rows.

The *rotary hoe* (Fig. 7.4) is a popular and effective cultivation implement. Compared to a row cultivator, it is simple and inexpensive and can have a very high field capacity. It is most effective when the crop is well rooted, the weeds are just emerging, and a light crust has formed on the soil. Speeds may be as high as 16 to 19 km/hr [10 to 12 MPH]. Fig. 7.4 illustrates a 9.5-m [31-ft] rotary hoe with ball bearing hoe wheel mountings, spring pressure on the wheels, and automatic trip action.

The *spike-tooth harrow* can be used both for cultivation and seedbed preparation. It is a simple, inexpensive implement that can have very great effective widths. As a cultivator it is used effectively under the same conditions as a rotary hoe. For seedbed preparation it is used to level and smooth the soil surface.

Other operations are often combined with row crop cultivations. Fertilizer may be added when cultivating. A herbicide is being applied on top of the rows in Fig. 7.5.

Fig. 7.4. Cultivation with rotary hoe.

Fig. 7.3. Row crop cultivator in transport mode.

Fig. 7.5. Herbicide application with cultivation.

Adjustments

The primary problems in adjusting a row crop cultivator are to obtain uniform and adequate penetration of the tools and an optimum throwing of soil into the row for intrarow weed control.

Penetration is obtained in the same manner as for subsurface tillers—increase the forward pitch of the tool for greater operating depth. Some cultivators use spring pressure to aid penetration in hard ground conditions.

Depth of operation may be controlled by using gage wheels to limit depth or by having the tool "run out of suction" as it lowers. The method used depends on whether a parallel lift linkage or a single-point pivot linkage is used in the cultivator gang. In the parallel lift linkages equipped with gage wheels, the tools should be set with a slight suction when in operating position. With the pivot linkages the tools should have no suction when at operating depth.

The amount of soil thrown into the row is a function of speed, depth of operation, and set of the tool. Field adjustments are the most effective.

Special protection for the crop is required for the first cultivation. Metal fenders or shields (Fig. 7.5) are often used and should be adjusted with enough clearance to permit substantial amounts of soil to filter around the crop plant. Rotary hoe wheels have been very effective as shields—small particles are able to filter through the wheel into the row, while larger soil clods are fended away.

Penetration may be increased by adding weight to rotary hoes and to spike-tooth harrows; otherwise, field adjustments to permit trash clearance are the only problems in operating.

Row cultivation damages the crop by pruning the roots if the tooling is operated close to the plant row. Shovel tooling particularly causes damage when set too close to the row and too deep. Serious root pruning is avoided by using shallow sweeps near the row and by widening the spacing. Disk coulters can be used as tooling near the row since they can be set both wide and at a shallow depth and still throw substantial amounts of soil onto the row to smother small weeds. A very direct way to avoid root pruning damage is to use a herbicide spray over the row, as in Fig. 7.5.

SEEDING MACHINES

Seeding machines are a group from a larger class of machines that distribute materials over fields. Management problems of this class include ensuring that the desired application rate is obtained. Rate settings for most current machines are set by hand and left constant for the whole field. But soil conditions may require different rates for different parts of the field. A machinery manager should be sensitive to this requirement and change settings as needed. Rate controllers that can be set from the tractor operator's cab are available. Computer programs based on a satellite navigation system that locates the implement on a stored map of the field can make adjustments automatically and more precisely than the operator. The machinery manager must evaluate the costs and benefits before making an adoption decision.

Whether the materials applied are seed, fertilizer, or chemical sprays, the mechanical operation of the equipment must be understood before adjustments and calibration are attempted. Three general types of rate application machines are

1. Ground wheel driven,
2. Power take-off driven, and
3. Constant delivery.

The ground wheel-driven machines such as corn planters or grain drills are the simplest to operate, since speed variations within practical limits do not affect the application rate. In general, the application rate of ground wheel-driven machinery is varied by changing the gear ratio between the metering mechanism and the ground wheel.

The PTO-driven rate application implements are not affected by changes in speed *within* each forward gear of the tractor, but are definitely affected by changes in the *selection* of forward gears. The application rate of this type is varied by changing either the forward gear selected or the driving gear ratio of the seeder.

The constant delivery applicator (such as a sprayer) is the most sensitive to speed variation. Any slight change in the forward speed of the implement causes a change in the rate of material applied. In this type, the rate is varied by forward speed or by changing the concentration of the material applied.

It is much more important to calibrate the last two types than the first.

Fig. 8.1. A mounted grain drill.

Small-Grain Seeding Machines

Cereal grains, such as wheat, oats, rye, barley, rice, may be seeded by three types of machines.

Drills are seeders that open furrows spaced from 10-30 cm [4-12 in.], meter the seed into the furrow, and firm soil around the deposited seed in one operation (Fig. 8.1). Each seeded row is produced by a separate metering device, furrow opener, and closing mechanism.

Plain drills have only a seeding capability. *Combination drills* (Fig. 8.13) are equipped to distribute both seed and fertilizer. The grass and legume seeding attachment in the figure can be carried by either drill.

Air drills (photo at beginning of chapter) have a large central seed supply and metering device. The seed is delivered to a separate tillage tooling by air flow through individual hoses to each opener.

Field distributors are machines that consist of a seed box with metering devices in the bottom of the hopper. These implements in no way prepare a furrow for the seed. They may be used for fertilizer as well as seed. Compared to drills, these machines are inexpensive to purchase, have higher effective field capacities, and are not affected by field conditions. Unlike drilling, however, a separate seed covering operation is required. A typical field distributor is shown in Fig. 8.2.

Broadcasters are machines that meter materials onto a revolving flanged wheel (Fig. 8.3 and 9.11). This type of seeder is the least expensive to purchase. It has the highest work capacity but is the hardest to calibrate because of uneven distribution from the flanged wheels, unequal seed weight and shape, and difficulties with wind and uneven soil surface. As a seeder this machine must be followed by a seed covering operation.

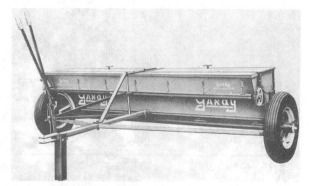

Fig. 8.2. Field distributor.

Fig. 8.3. Tractor-mounted broadcaster applying fertilizer.

Metering Mechanisms

The seed rate controlling mechanisms are volume-metering mechanisms that may be divided into three general types.

Fluted Feed

This universal type of metering device has a delivery system that is varied by changing the drive gear ratio and by exposing different lengths of the rotor to the seed hopper. A positive-type feed, this mechanism is well adapted for use with light and irregular seeds. The lever shown in Fig. 8.4 adjusts the seed gate for different sizes of seed. Seeds as large as corn and beans can be metered by this device.

Internal Run

The smooth, internal-lipped rotor varies its delivery by changing a gear ratio, by using reducers or inserts, or by sliding the rim to greater or less contact with the seed. The internal run feed is noted for its gentle selection of the seed and for even-flow delivery (Fig. 8.5). The seeding rate is changed by sliding the drive shaft through the meter to change the exposure of the wheel to the seed supply. Changing the speed ratio between the drive shaft and the ground dive wheel also changes the seeding rate.

Variable Orifice

This simplest and oldest type of metering mechanism is used primarily on the more inexpensive field distributors. The rate of application is controlled by increasing or decreasing an aperture below the agita-tor in a grain box. While this type of metering device appears to be rather unreliable, it works well *if properly calibrated* (Fig. 8.6).

Air seeding is accomplished by a fluted roller placing seed (and/or fertilizer) into an air stream. Fig. 8.7 indicates the flow of seed around the roller. Air enters from above, picks up the seed, and delivers the seed through tubes to the furrow opener.

All of these metering devices work best when the seed is dry, uniform, and clean and if a mechanical agitator is used in the seed box.

Fig. 8.5. Internal run metering mechanism.

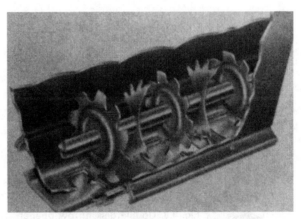

Fig. 8.6. Variable orifice agitator metering device.

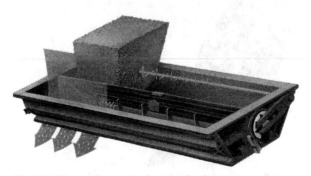

Fig. 8.7. Air seeding metering mechanism.

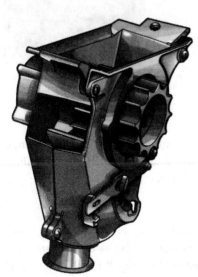

Fig. 8.4. Fluted feed with seed gate.

Furrow Openers-Small Grain

Several different types of furrow openers are available on grain drills to meet the many varying seeding conditions.

Single Disk

This mechanism permits good soil penetration and excellent trash clearance (see Fig. 8.13 and 8.20). A particular type is the *deep-furrow opener* used on drills in the drier areas of the United States. In this case, the seed must be planted deep to find moisture, and the resulting ridges assist both in soil and water conservation and in protection for the young plants from winds.

Double Disk

A double disk gives a uniform depth of placement to the seed in soil that is well worked (Fig. 8.8).

Hoe Opener

Used in minimum tillage operations where good penetration is required, Fig. 8.9 is a hoe opener for both seed and fertilizer. Some special openers have been developed for air seeders. Fig. 8.10 shows a sweep tooling which gives a spread of fertilizer and seed. Fig. 8.11 shows an opener that produces a separation

Fig. 8.10. Sweep opener seeding pattern.

Fig. 8.11. Slant opener to give seed and fertilizer separation.

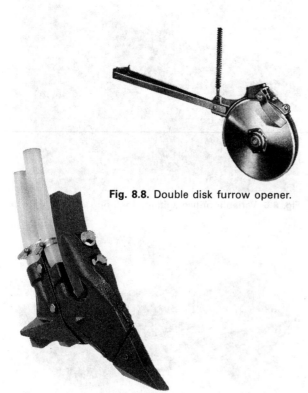

Fig. 8.8. Double disk furrow opener.

Fig. 8.9. Hoe opener seeding pattern.

of seed and fertilizer. Additionally, it can cut through residues and thus avoid an accumulation of residues that may wrap around hoe openers.

Corrugated Roller

Used primarily for legume and grass seeding, seeders using corrugated rollers for furrow openers are commonly called *packer-seeders* (Fig. 8.12). This opener is particularly effective when shallow depth and a firm

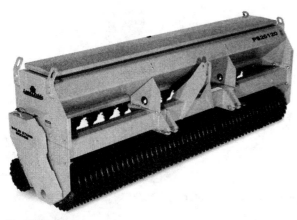

Fig. 8.12. Packer seeder.

seedbed are desired after clean tillage. This opener is very effective for seeding grasses and legumes.

Furrow openers are often arranged in staggered order to allow surface residues to pass through more easily. The need for planting into conservation tillage seedbeds requires additional tooling over that for conventional seedbeds. Additional coulters are most often used to penetrate untilled soil and to cut through crop residues left on the soil surface. *No-till* or *conservation drills* are designed to be sturdy and heavier than conventional drills. The extra weight is used to help penetrate unworked soils and those having a substantial residue cover. Kushwaha, Vaishnav, and Zoerb, Agricultural Engineering Department, University of Saskatchewan, determined the performance of coulters in a straw-covered soil bin (Table 8.1). The 46 cm [18 in.] diameter coulter appears to be an optimum size for a 55 mm [2 in.] penetration. The larger diameter requires greater down force, while the smaller was less effective in cutting the straw. The last column of Table 8.1 indicates the increase in forces when cutting through straw over that for operating in bare soil.

Soil bin work by J. E. Morrison, J. G. Hendrick, and R. L. Shafer, USDA-ARS, reported the draft of a 41-cm [16-in.] diameter coulter in bare soil was

essentially independent of speed in a sandy loam soil but had an increase in draft per depth of 0.22 KN/cm [128 lb/in.].

Covering Devices

The last process of the seeding operation is that of placing and firming soil around the seed. The machinery manager adjusts the covering device to produce a deeper soil cover for large seeds and for dry soil conditions. A firm soil is needed to ensure moisture transfer to the seed but the covering process should not create soil compaction or soil crusting.

Press wheels are used to firm the soil over the planted seed and eliminate air pockets in the seedbed. They may be made entirely of steel but are usually covered with flexible rubber which can flip-off any soil accumulations. Small-diameter, narrow, press wheels can be used to firm the seed in the furrow (Fig. 8.13). Wider, larger-diameter wheels firm a wider band. A popular design uses a press wheel as a gage wheel for the opener to produce an accurate control over the depth of the opener. Such designs are common in row planters.

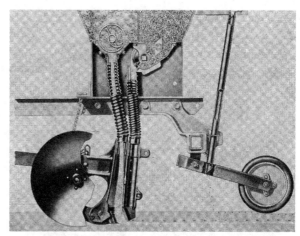

Fig. 8.13. Fertilizer drill with grass and legume seeding attachment having single disk openers and press wheels.

TABLE 8.1. Coulter Performance with Various Straw Residue Densities; Percentage Straw Residue Cut at 55-mm [2-in.] Depth

Coulter Diameter cm [in.]	Residue rates, kg/ha				Range of Increased Force, N [lb]	
	1000	2000	3000	4000	Vertical	Horizontal
36 [14]	86	60	30	17	30-160 [7-36]	45-90 [10-70]
46 [18]	100	100	99	98	35-200 [8-45]	30-100 [7-22]
60 [24]	100	100	40	30	190-300 [43-67]	30-130 [7-30]

Row Planters

Row planters are distinguished from grain drills in that they deposit seed in rows spaced to permit interrow cultivation and they usually have single seed metering.

Like grain drills, row planters may have many different attachments. Fertilizer applicators, either liquid or dry, are often used to provide nutrients for the newly emerged plant. Chemical applicators, liquid or dry, allow herbicide and insecticide treatments at planting time. A granular applicator is shown mounted behind the seed box in Fig. 9.28.

Row planter design has evolved into a toolbar concept in which the individual planting units may be mounted anywhere along the toolbar to give a desired row spacing. Fig. 8.14 is a rear-mounted, toolbar base planter. Note the V-shaped wing preceding the openers. These wings scrape aside residues and dry soil from the top of beds that may be spaced arbitrarily. If the metering mechanism is driven by its associated press wheel, the design is called a *unit planter*. Fig. 8.15 shows such a design. This planting unit is free to move vertically relative to the toolbar through the parallelogram hitch linkage shown to the left in Fig. 8.15.

Fig. 8.15. Unit planter.

Metering Mechanisms

Most row planters may also be distinguished from grain drills because they have single-seed-metering devices. These mechanisms must be able to select as many as 5000 seeds/min that are of variable size and shape; yet these seeds must not be damaged to the extent that germination is destroyed. Tests have shown that 1-2% of the seed may be damaged with mechanical metering.

Both mechanical and air systems are used to meter seed. Fig. 8.16 illustrates a traditional mechanical

Fig. 8.14. Toolbar row planter.

metering mechanism; the seed enters cells in the edge of a flat, circular plate that rotates in the bottom of the seed box. Excess seed is scraped away by the cutoff pawl. The plate then carries the seed to the knockout pawl, which ejects the seed from the cell into the seed tube where it falls down to the furrow opener. The seed must be graded to a uniform size and matched carefully to the planter seed plate for proper operation.

Another mechanical seed selection device is shown in Fig. 8.17. Inclined plates with cup-shaped vanes dip into a pool of seed and carry seed to the discharge point at the top of the plate. A fixed brush ejects the seed into the seed tube. Single seeds are selected if the plate cups can hold only one seed against the pull of gravity as the seed is lifted from the seed pool. Such gravity drop mechanisms can be identified by the housing covering the inclined plates as shown on the planter in Fig. 8.15. The vegetable planter shown in Fig. 6.5 uses the inclined plate mechanism. This mechanism permits some variation in seed size and shape as well as offering a more gentle selection and ejection action.

Air metering use the flow of air through a carefully sized hole in a moving seed plate to trap and hold a single seed until it is released to the seed tube. Fig. 8.18 is a mechanism using air flow to hold a seed in the hole of a vertical, rotating seed plate. As the plate turns counterclockwise, the seed is picked up from the supply pool, is transferred to the 6 o'clock position where the air flow is cut off, and is dropped into the seed tube. The curved seed tube imparts a backward velocity to the seed, which ideally matches the planter's forward velocity to the left. The seed thus has no actual horizontal velocity at impact with the soil surface and is unlikely to bounce and interfere with the desired seed spacing.

The air selection mechanism does not require graded seed although uniform seed will give best results. The absence of cutoff and knockout pawls leads to less seed damage. The vertical plate design permits closer row spacing than the flat, horizontal, circular plate design.

Not all planters use single seed selection mechanisms. In some cases the seed does not exist as a single germ and in others several seeds at a point are required to ensure at least one plant's emergence. *Hill dropping* is a practice where more than one seed is dropped in one place. Multiple seeding is desired to ensure at least one plant at that location or hill, to have multiple plant strength to push through a soil crust, and to provide mutual support for young plants. Either hand or mechanical thinning is usually required

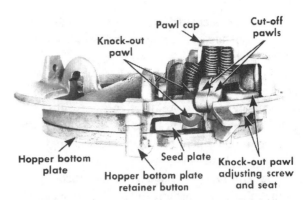

Fig. 8.16. Mechanical seed plate metering.

Fig. 8.17. Inclined plate-metering mechanism with brush ejection.

Fig. 8.18. Air seed-metering design.

after the plants emerge. As a class, vegetable crops are often planted on a volume basis rather than as single seeds.

Several recent developments have permitted some seedings to shift from the volume category to the single seed category. Acid delinting of cotton seed, decorticating of sugar beet seed, and coating of some vegetable seeds permit a more precise single-seed planting of these crops.

The term *precision planting* is sometimes applied to the practice of single seeds planted to stand. One should appreciate that true precision planting also involves placing the individual seed at the correct depth, in the correct environment, and at a proper spacing.

Fig. 8.19 pictures a row planter with a central seed hopper. An air delivery system supplies seed as needed to the individual row metering units. A large central hopper requires less time to fill with high-capacity augers, seed bags, or pneumatic transfer systems.

Furrow Openers

Furrow openers for row planters are similar to those for grain drills. Disk openers with or without adjacent gage wheels (Fig. 8.20) provide effective soil penetration and consistent seeding depth for clean tillage as well as for tillage practices having substantial amounts of surface residue.

Some planter have springs to increase furrow opener penetration by applying a down force. Fig. 8.21 shows a pneumatic bellows used to apply added down force. The air pressure is provided by a hydraulic-powered

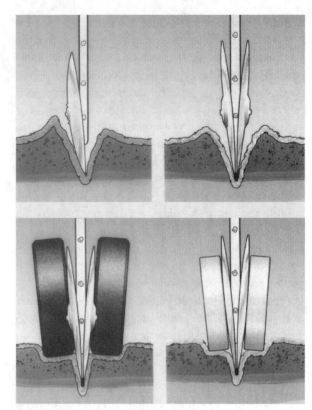

Fig. 8.20. Single disk, double disk openers, rubber and steel gage wheels.

Fig. 8.19. Row planter with a central seed hopper.

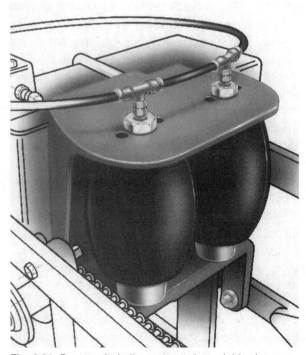

Fig. 8.21. Pneumatic bellows to apply variable down force on furrow openers.

air pump which can be controlled from the tractor cab. This feature allows the operator to secure needed penetration in hard soils or through thick plant residues. The total down force is limited to the total weight of the seeder.

Covering Devices and Press Wheels

Mechanisms to cover a seed and firm the soil around it after it has been deposited in the furrow are of many types. Press wheels of many designs are used for row planters (Fig. 8.22). The indented center press wheels firm the soil on the side of the seed row but leaves the portion over the seed loose to not impede plant emergence. See also Fig. 9.13.

Seeder Adjustments

Seeder adjustments consist of

1. Row spacing
2. Rate setting
3. Depth of seed placement
4. Amount of soil coverage
5. Amount of soil compaction

The operator's manual describes how the adjustments are made but it is up to the machinery manager to determine when and how much. To do this, soil conditions, moisture, temperature and fertility, anticipated weather, and growth requirements of the seed must be known. These data are then transformed into definite adjustment settings.

The forward speed of the seeder is an adjustment that is completely at the discretion of the operator. To obtain high field capacities, fast field speeds are

desirable; yet the penetration of furrow openers and the rate of delivery from metering mechanisms usually decrease with an increase in ground speed. High ground speeds are not objectionable if allowance is made for the negative effects and a constant speed is maintained. Speeds up to 12 km/hr [7.5 MPH] are possible without significant reduction in seeding rates.

One of the most important factors affecting the performance of ground-driven seeders is the effective rolling radius of their drive wheels. The inflation pressure in pneumatic tires is generally recognized as influencing the rolling radius. The rolling radius is measured as the vertical distance from the center of the axle of the drive wheel to a level, firm ground surface. This radius passes through the deflected portion of the loaded tire and is sometimes referred to as the squash-radius. The weight on the tire affects the amount of tire squash or deflection. Since seeders are inherently distributing machines, their weight is constantly changing. It is most desirable to have only constant weight on the drive wheels; but if such is not the case, the rolling radius should be determined under average weight conditions.

Seeder Performance Tests

The performance of a planter may be checked in the laboratory by powering the drive wheels of the positioned planter. The seed is then caught and examined.

Calibration of Small-Grain Seeders

A small-grain seeding machine is a *volume* seeder. As a result of this particular characteristic the rate of application is affected by height of grain in seed box,

Fig. 8.22. Six designs for row planter press wheels.

the agitation of the seed, and seed characteristics such as size, shape, density, and moisture content. The factors are usually minimized by the design of the seeder, but for best results the seeder should be calibrated with the seed to be planted.

Ground wheel-driven drills and distributors are so easy to calibrate that no farmer should begin to seed without first checking the seeder's performance. Simply jack up the seeder, turn the drive wheel enough turns to represent a known area, collect and weigh the seed, and finally convert the data into mass per area and compare with the recommended rates.

For PTO-driven machines the same general procedure is followed. The primary difficulty is in establishing a relationship between PTO speed and ground travel. The ground speed of the tractor must be measured with the PTO running at the correct field working speed. The relationship between PTO speed and forward travel is often indicated on the tractor's tachometer.

Broadcasters are most difficult to calibrate in that not only is the amount of grain seeded important but also the distribution of the seed on the ground. Operating the broadcaster while stationary on a level, wide concrete floor will indicate the distribution pattern and the quantity delivered—*if* the broadcaster is operated at the same PTO speed to be used in the field.

An example solution to a small-grain seeding problem is presented below. The unit-factor system is most helpful in calculating all calibration problems.

A drill with an effective width of 3 m [10 ft] is to seed 75 kg/ha [66 lb/a]. Each wheel drives one-half of the drill. The tires have a squash-radius of 350 mm [14 in.] when the seed box is half full. A 1% wheel slip is assumed. The rate lever is set at 75 kg/ha. One side of the drill is blocked up and that drive wheel is turned 50 revolutions at field speed. What is the expected field seeding rate if 1 kg [2.2 lb] is collected in a pan beneath the drill?

$$60\frac{kg}{ha}=\frac{1\,kg}{50\,rev}\times\frac{1\,rev}{2.2\,m}\times0.99\times\frac{2}{3\,m}\times\frac{10,000\,m^2}{ha}$$

$$\left[52\frac{lb}{a}=\frac{2.2\,lb}{50\,rev}\times\frac{1\,rev}{7.3\,ft}\times0.99\times\frac{2}{10\,ft}\times\frac{43,560\,ft^2}{a}\right]$$

The correct rate lever setting would appear to be 75/60 [66/52], about 1.25 times the test setting, or 94 kg/ha [82.5 lb/a]. But this setting *should be tested*

since the performance of the drill may not be linear with the lever setting.

Row Planter Calibration

The calibration of a row planter involves selecting the correct seed plate and checking its performance. Seed producers and planter manufacturers work together to recommend the correct plate for a particular seed. In general, the machinery manager's responsibility is merely to follow these recommendations.

The characteristics of most seeds are such that one must overseed to get an expected emergence. In a series of tests in Illinois, H. P. Bateman found that 32 varieties of hybrid corn had an 88.8% emergence and 81.8% ears per population of seed planted.

The following problem serves as an example of the type of calculations required in determining the rate settings for a planter.

Corn is to be planted in 1-m [39-in.] rows and spaced 18 cm [7.09 in.] in the row. The planter wheels have a rolling circumference of 2.2 m [86.6 in.]/rev, and they drive a cross-shaft through overrunning clutches at double the wheel rpm. The 6- and 10-tooth sprockets on the cross-shaft may drive any of the 18-, 14-, 12-, 10-, 8-, and 7-tooth sprockets on the plate drive shaft. The 38-cell seed plate turns at a ratio of 12/40 of the plate drive shaft speed, (Fig. 8.23). (The wheels drive through overrunning clutches to permit the planter to make turns when operating.) Which sprocket combination is most appropriate for the desired seeding rate, assuming a 100% cell fill and a 2.6% drive-wheel slip?

The number of seeds to be planted in 2.2 m [86.6 in.] = 12.22. Required ratio of seed plate revolutions to 1 planter wheel revolution = 12.22/(38 × 0.974) = 0.33. Let X = necessary speed ratio of the plate drive shaft to the cross-shaft.

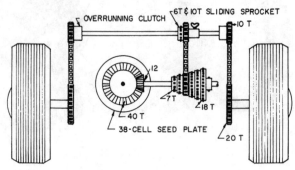

Fig. 8.23. Ground-driven seed plate drive.

$$\frac{0.33 \text{ plate rev}}{1 \text{ wheel rev}} = \frac{2 \text{ cross} - \text{shaft rev}}{1 \text{ wheel rev}}$$

$$\times \frac{X \text{ plate drive rev}}{1 \text{ cross} - \text{shaft rev}} \times \frac{12 \text{ seed plate rev}}{40 \text{ plate drive rev}}$$

$$X = \frac{0.33 \times 40}{2 \times 12} = 0.555 \text{ or approx. 5 to 9 ratio}$$

The sprocket speed ratios are in reverse ratio to their numbers of teeth; therefore, the 9 to 5 ratio is matched exactly with the combination of the 18-tooth and 10-tooth sprockets.

If only 90% of the cells fill because of a fast plate speed, what sprocket ratio is required for the same plant population?

To produce the same pant population per area, the seed cell travel velocity has to be increased as if to provide a plant spacing that is 90% of the previous 18 cm [7.09 in.]. The planting pattern will not be uniform as 10% of the spacings between the planted seeds in the row are expected to be 34.4 cm [12.76 in.]. The cells exposed in 2.2 m [86.6 in.] of planter travel will be 12.22/0.9 or 13.58. Required ratio of seed plate revolutions to drive wheel revolutions = 13.58/(38 × 0.974) = 0.367. Thus

$$X = 0.367 \times \frac{40}{12} \times \frac{1}{2} = 0.612$$

The 6- and 10-tooth sprockets with a ratio of 0.6 most nearly match the X value of 0.612.

Using the 6- and 10-tooth sprockets, the theoretical seed spacing would be

$$\frac{16 \text{ cm}}{\text{cell}} \left[\frac{6.33 \text{ in.}}{\text{cell}} \right] = \frac{220 \text{ cm} [86.6 \text{in}]}{0.974 \times 1 \text{ wheel rev}}$$

$$\times \frac{1 \text{ wheel rev}}{2 \text{ cross} - \text{shaft rev}} \times \frac{1 \text{ cross} - \text{shaft rev}}{0.6 \text{ plate drive rev}}$$

$$\times \frac{40 \text{ plate drive rev}}{12 \text{ plate rev}} \times \frac{1 \text{ plate rev}}{38 \text{ cells}}$$

The actual seeded population would be

$$\frac{1 \text{ cell}}{0.16 \text{ m}} \times \frac{1}{1 \text{ m}} \times \frac{10,000 \text{ m}^2}{\text{ha}} \times \frac{0.9 \text{ seeds}}{\text{cell}}$$

$$= 56,250 \text{ seeds} / \text{ha}$$

$$\left[\frac{1 \text{ cell}}{6.3 \text{ in.}} \times \frac{1}{39 \text{ in.}} \times \frac{144 \text{ in.}^2}{\text{ft}^2} \times \frac{43,560 \text{ ft}^2}{\text{a}} \times \frac{0.9 \text{ seeds}}{\text{cell}} \right.$$

$$\left. = 22,976 \text{ seeds} / \text{a} \right]$$

Field tests of planters show substantial variability of seed spacing in a row. Even if the metering is accurate, variation in seed fall through the seed tubes and variable seed bounce in the furrow cause error. In a South African test by B. D. Boshoff (Department of Agricultural Technical Services) a finger-type metering planter was set to space seeds at 16.9 cm. It actually averaged 17.1 cm with a standard deviation of 7.3 cm. The complete range was 3.3–52.4 cm.

Monitoring equipment has decreased, but not eliminated, the need for calibrating row planters. Seed counters in the seed tubes and distance-traveled sensors enable microprocessors to present information while the seeding operation is under way. A failure or reduction in seed count for any row will trigger both a buzzer and a light indicator. Plant populations and seed spacing in the row can be presented on the readout panel. Variations in seed size, drop trajectory, dust and other particles, and high seed flow rates all lead to miscounting. High seeding rates in which two closely spaced kernels are sensed by the counter as one seed leads to a common undercounting. The machinery manager may wish to calibrate the monitor for each seed type and for high seeding rates. Operating the row planter at shallow depths on a bare, hard soil produces a display of seeds that are easy to count over a measured distance.

Practice Problems

8.1. In a full-width calibration test of an 18×200 mm [18×8 in.] grain drill, 30 turns of the drive wheel yielded 3 kg [6.6 lb] of oats. If the rolling circumference of the wheel is 3.7 m [12.14 ft], what is the seeding rate? Assume 1% drive-wheel slip in the field.

8.2. A planter plate turns one-half revolution for each revolution of the ground wheel. The ground wheel has an effective circumference of 2 m [78.74 in.] and the plate has 16 cells. What is the spacing in the row?

8.3. The fertility of a particular field is such that the corn yield is at a maximum at 37,500 plants/ha [15,000 plants/a]. The average survival is 85%. The row spacing is 760 mm [30 in.]. What would seed spacing be?

8.4. Seeders sometimes must operate around curves. If the seeder's metering mechanisms are driven by press wheels in a unit-planter configuration, the seeding rate remains constant across the seeded width. Describe the seeding rate pattern across the width for a PTO-driven, central metering configuration.

Chapter **9**

CHEMICAL APPLICATION

Chemicals are applied to fields and plants as fertilizers; as killers of weeds, brush, insects, nematodes, fungi, and rodents; and as defoliants and growth regulators. The use of herbicides has replaced much of the mechanical cultivation done formerly. Chemicals may be applied with attachments to tillage machines and seeders rather than with single-purpose chemical applicators.

Fertilizer Application

Fertilizers may be applied at various times, in various ways, and with various machine methods. Bulk fertilizer and lime applications are usually made before primary tillage. The applicators may be owned by individual farmers but often are owned by custom operators or fertilizer sales organizations. In the latter case, the applicator is loaned or rented to the farmer.

Controlled amounts of fertilizer may be added at seeding time with mechanisms attached to seeders. This fertilizer is placed near the seed and is called a *starter fertilizer*. The practice of adding fertilizer to fields with growing plants is called *side dressing* and is done with attachments to cultivators and with anhydrous ammonia applicators.

Whatever the practice, the farm manager is concerned that a uniform application be made at a proper rate and at a proper position with respect to the seed or plant. Fertilizers may be applied in gaseous, liquid, or solid forms. The applicator's construction depends on the form in which the fertilizer is applied.

Dry Fertilizer

The special characteristics of dry commercial fertilizer lead to the following general requirements for fertilizer-applying machinery:

1. Easy to clean
2. Provide a wide range in the rates of application
3. Mechanical agitation desirable
4. Corrosion-resistant parts

Various mechanisms for metering dry fertilizer have been developed over the years. Aggressive mechanisms are needed to crumble the occasional caked materials and to deliver rates unaffected by depth or weight of the material in the supply bin. Plastic bins and metering mechanisms avoid the chemical corrosion common to steel parts. There are several common designs.

Augers are screw conveyors or portions of a helix over the outlet tube. These mechanisms (Fig. 9.1) agitate the material and produce a strong flow. Application rates are changed by varying the rotational speed of the auger with respect to ground speed.

Belt conveyors are used to apply large quantities of fertilizer. The belt carries the fertilizer from the bottom of the hopper past a variable height scrape-off plate to the delivery tube. (Fig. 9.2 and 9.11).

A *variable orifice with rotating agitator* is typical in full-width field distributor machines. The horizontal rotor acts as an agitator and a feeder for the orifice. As a class, these machines may also be used as small-grain seeders as shown in Fig. 8.6.

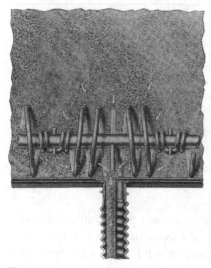

Fig. 9.1. Auger type of feed.

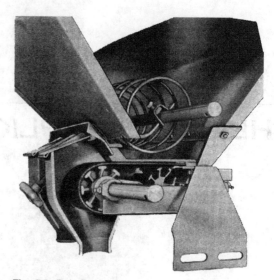

Fig. 9.2. Belt feed.

Liquid Fertilizer

The main advantage of liquid fertilizer is the possibility of reduced labor in handling since either pumps or gravity-flow systems may be used. These fertilizers may be applied with dedicated machines, but attachments to row planters are also popular (Fig. 9.5).

One popular metering mechanism for liquid fertilizer is a gravity-flow, constant-head metering device (Fig. 9.3). The tank should be filled to the top so that a minimum of air space remains. An airtight cap is installed and a vacuum forms above the liquid almost immediately after the off-on valve is opened. As soon as the vacuum becomes great enough for atmospheric pressure to support the mass of all the liquid above the bottom of the vent tube, a constant head occurs across the metering orifice and a constant flow results regardless of the fluid level in the tank. The vacuum gradually decreases as the tank empties and as air bleeds in through the vent tube. The clear plastic liquid level tube indicates to the operator the need for refilling. Refilling should occur before the level drops to the bottom of the vent tube as the delivery rate is no longer constant at lower levels. This mechanism meters at a constant rate with respect to time alone. The forward speed of a machine equipped with such metering must be constant for a constant application rate over a field.

The *squeeze pump* (Fig. 9.4) overcomes the disadvantage of having to operate the gravity-flow applicators at constant forward speed. Small volumes of liquid are trapped as the rollers, on the rotor, press against the tightly held plastic tubes. As the rotor turns, the trapped volumes move along the tube and

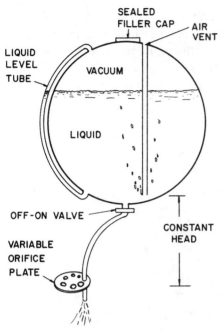

Fig. 9.3. Constant-head metering.

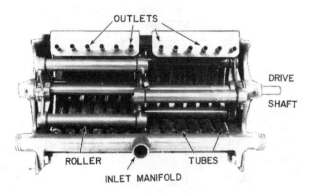

Fig. 9.4. Squeeze pump.

Fig. 9.5. Liquid fertilizer attachment to a row planter.

are released to the furrow opener. The amount delivered per time depends on the speed of the rotor and the size of the tube. The amount delivered per distance depends on the gear ratio between the ground wheel and the rotor. Fig. 9.5 shows a squeeze pump installation driven by the ground wheels of the planter. The plastic delivery tubes lead to separate double disk openers that place the fertilizer in a specific relationship to the seed row.

A variable-stroke, positive-displacement group is also used for applying liquid fertilizers. Fig. 13.13 shows such a pump operated by a chain from a small drive wheel that bears against a transport wheel. This pump has a provision for varying its stroke and thus its output. Its delivery rate is directly proportional to the forward advance of the

applicator and to the stroke setting. A constant rate per area is expected at any field speed. Such pumps are more expensive than squeeze pumps and much more expensive than the gravity-flow metering method.

Gaseous Fertilizer

Anhydrous ammonia is the important gaseous fertilizer used in agriculture. It must be handled in tanks and by transfer equipment capable of withstanding pressures as high as 1.7 MPa [250 psi]. Although equipment for handling ammonia is very similar to that for LP gas, brass fittings must not be used for anhydrous ammonia equipment.

The transfer of liquified anhydrous ammonia from supply to implement tanks must be done with care. If a two-hose system is used, the liquid-carrying hose is connected to the lowest outlets of the two tanks. The second hose is called the vapor return line and conducts the vapors from the empty tank to the supply tank. A vapor-operated transfer pump is mounted on top of the tank in Fig. 13.12. This device uses some of the return vapor to power a pump, which transfers the rest of the vapor more rapidly than natural flow. Filling time is thus much shorter than that when using just the pressure differential between the two tanks. Usually less than 1% of the vapors are exhausted to the atmosphere.

A machine to apply anhydrous ammonia is shown in Fig. 9.6. This applicator really meters a liquid since ammonia under pressure remains in the liquid form until the pressure is released. The gas must be released at least 10 cm [4 in.] below the soil surface where the gas can be trapped by soil moisture. A narrow opener, commonly called a knife opener, is required to produce a slit in the soil to minimize the escape of gas. Two covering blades are used (Fig. 9.6) to seal the slit. Provisions are made to isolate the gas tube from the opener to reduce freezing of soil on the opener. This refrigeration effect occurs because of the ease of vaporization of ammonia. Fig. 9.7 shows a common type of knife opener.

In the simplest metering mechanism, the pressure in the tank forces the ammonia out through a restricting orifice. But a pressure regulator is required to maintain a constant pressure or head drop across the orifice, as tank pressures will vary with temperature and with the amount of liquid in the tank. Such a pressure-regulated device is diagrammed in Fig. 9.8 and is mounted on top of the tank in Fig. 9.6. The desired metering rate is set by adjusting the position of the variable orifice-metering valve. When the off-on valve is opened, liquid flows from the tank through the metering orifice and past the balance valve. As the pressure drops in the tank, the pressure on the right side of the diaphragm decreases and the spring force opens the balance valve until the pressure drop across the metering valve is as before. If the pressure in the tank should increase because of a rise in temperature, the diaphragm moves to the left, partially closes the balance valve, and restores the pressure drop across the metering valve. Constant flow is maintained. As with gravity-flow liquid metering, a constant forward speed is required for the applicator to achieve a constant application rate per area.

A positive displacement pump for metering anhydrous ammonia is shown in Fig. 9.9. The long rope-controlled lever operates an off/on valve. The stroke of the pump is set by adjusting the marked dial. The heat exchanger on the left keeps the fluid in a cool and liquid state for accurate metering. As with other ground-driven applicator pumps, the amount applied per area is independent of machine travel speed. However, the pump method of metering is more expensive and requires more maintenance than the pressure-regulated orifice method.

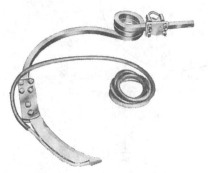

Fig. 9.7. Knife opener.

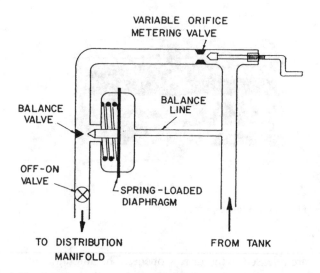

Fig. 9.8. Pressure-regulated orifice.

Fig. 9.6. Anhydrous ammonia applicator.

Fig. 9.9. Variable-stroke anhydrous ammonia pump.

Operations

American farmers apply large quantities of fertilizer each year. Several hundred kilograms per hectare [pounds per acre] are quite common. Such quantities are no longer efficiently handled in sacks. Bulk handling methods are used for filling the fertilizer boxes on seeding machines (Fig. 9.10) as well as for filling a fertilizer broadcaster (Fig. 9.11). While broadcasting fertilizers may not be as efficient in fertilizing row crops as precision placement methods, the time saved during the critical seeding operation may more than make up for any economic losses due to poor placement. The machinery manager must be sure that bulk handling equipment purchased to fill the boxes on seeders is capable of being operated by one person. No economy is experienced if an extra person is required to operate an unloading auger from the truck cab.

Adjustments

The rate of application and the placement position in the soil are the major adjustments for fertilizer applicators. Disk and knife openers are usually used as furrow openers for placing both dry and liquid fertilizers. A range in adjustment permits several alternative depths and spacings relative to a seed row.

The rate adjustment is most often accomplished by varying an orifice opening. Sometimes a combination of orifice opening and a speed change for the impeller-agitator is used to produce a change in rate.

The weight in a fully loaded fertilizer hopper can cause a significant increase in both the rate and depth of placement. The machinery manager seeking precision farming should require the operator to make appropriate adjustments. Modern second and third level instrumented fertilizer distributors make such adjustments routine.

Fig. 9.11. A dry fertilizer broadcaster and a liquid sprayer.

Fig. 9.10. Attached fertilizer-distributing auger used to fill planter's fertilizer boxes.

Calibration of Fertilizer Distributors

The calibration of fertilizer applicators is difficult because of the variability in the action of metering mechanisms and the variability in fertilizer particle size. P. H. Southwell and J. Samuel (University of Guelph, Ontario) tested three types of metering mechanisms with a fertilizer described to be in the middle range in regard to both particle size uniformity and moisture absorption and found the variations shown in Table 9.1. They noted that the actual delivery was 128–195% of the manufacturer's calibration. They concluded that variations in manufacturing account for most variations in mechanisms, that variations in fertilizer particle size account for most of the down-row variations, and that hopper size has some influence on metering variation. They propose that manufacturers of volume-metering mechanisms report the bulk density, kg/m³ [lb/ft³], of the material used in their calibration.

The techniques used to calibrate small-grain seeders apply equally well for dry fertilizer. Weights of fertilizers are caught while operating the distributor for a significant time or over a simulated distance.

Broadcast fertilizer distributors are especially variable in performance. Fig. 9.12 shows a plot of percentage of application rate variation across the swath of a single flinger wheel broadcaster as reported by Swedish tests. Urea was applied at an average rate of 270 kg/ha [240 lb/a]. The vertical dashed lines indicate the effective width of the pattern and define the amount of overlap required. The upper solid line shows the effective rate when overlap is included.

Anhydrous ammonia applicators are very difficult to calibrate. One must rely on keeping an account of the amount used while the machine is in operation and must make adjustments accordingly. Distribution from the openers can be checked for uniformity by using water rather than ammonia in a performance test.

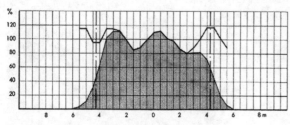

Fig. 9.12. Application pattern for fertilizer broadcaster.

TABLE 9.1. Variations in Fertilizer Application as a Percentage of the Mean

Factor	Gravity-Flow, Variable Orifice with Agitator	Star Wheel Metering	Wire Auger Metering
Variation between mechanisms			
average	25.5	17.1	11.0
range	17.7-31.6	8.1-22.5	5.7-16.7
Variation down rows			
average	13.1	8.0	8.4
range	2.2-20.2	5.4-14.0	2.3-13.9
Overall performance variation	24	18	12

Liquid fertilizer applicators operating on a gravity principle require a special calibration technique that is also representative of sprayer calibration; that is, the delivery is not proportional to distance traveled but to time elapsed. Forward speed of the machine is a fundamental measurement that affects the application rate. Speed must be held constant if the application rate is to be uniform. An example calculation of this time-rate type of calibration is considered here.

A liquid fertilizer distributor is being checked for application rate. A container catches 1.5 kg [3.3 lb] of 30% N solution in 36 s from one outlet tube. What is the rate of N applied if the forward speed is 7 km/hr [4.375 MPH] and the machine is 6 row with outlet tubes spaced 1 m [39 in.] apart?

$$64.3\,\frac{\text{kg}}{\text{ha}} = \frac{1.5\,\text{kg sln}}{36\,\text{s}} \times \frac{0.3\,\text{kg N}}{1\,\text{kg sln}} \times \frac{3600\,\text{s}}{1\,\text{hr}}$$

$$\times \frac{1\,\text{hr}}{7\,\text{km}} \times \frac{10{,}000\,\text{m}^2}{1\,\text{ha}} \times \frac{1}{1\,\text{m}}$$

$$\left[57.4\,\frac{\text{lb}}{\text{a}} = \frac{3.3\,\text{lb sln}}{36\,\text{s}} \times \frac{0.3\,\text{lb N}}{1\,\text{lb sln}} \times \frac{3600\,\text{s}}{1\,\text{hr}} \right.$$

$$\times \frac{1\,\text{hr}}{4.375\,\text{m}} \times \frac{1\,\text{mi}}{5280\,\text{ft}} \times \frac{43{,}560\,\text{ft}^2}{\text{a}}$$

$$\left. \times \frac{1}{39\,\text{in.}} \times \frac{12\,\text{in}}{\text{ft}} \right]$$

Field Sprayers

Chemical application by sprayers is a common field operation in crop production. In the Midwest 80% of the annual row crop acreage is sprayed with herbicides and nearly half is sprayed with insecticides.

Sprayers may be designed as attachments to tractors and planters (Fig. 9.13); or they may be designed as individual implements, either as self-propelled or as PTO-driven trailed machines (Fig. 9.14). The necessary components of any sprayer are a tank with agitator and strainer, a pump, a filter, a pressure regulator, valves, piping, and nozzles with dirt screens (Prob. 9.1). The farm field sprayer differs from others in that it uses nozzle pressures that are just high enough to give adequate coverage without causing excessive spray drift. Orchard and grove sprayers use high-velocity fans to blow the spray instead of using booms as do field sprayers (Fig. 9.15). These sprayers use high nozzle pressures, have large supply tanks, and are equipped with high flotation tires to permit wet weather operation.

Specifications for an all-purpose field sprayer are

1. High clearance for row crops
2. Long, light, flexible booms, adjustable in height
3. Hand boom for stationary applications
4. Noncorrosive construction to enable the sprayer to be used for liquid fertilizers as well as for sprays
5. Boom selector control valve
6. Accurate ground speed indicator
7. Flexible connections on the nozzle drops from the boom

Fig. 9.13. Row planter with a spray attachment.

Fig. 9.15. Orchard sprayers use air blast to deliver mist of spray materials.

Fig. 9.14. Effective, high-capacity field spraying requires good management.

Farm sprayer *pumps* may be either positive or nonpositive displacement types (Fig. 9.16, 9.17).

A. *Positive*: Only a small amount of liquid leaks past actuator—very high pressures are possible.
1. *Plunger*: Common piston-cylinder arrangement with one-way inlet and outlet valves.
2. *Gear*: Pair of gears operating in a close-fitting housing. *External* uses two (usually bronze) spur-tooth gears; *internal* uses one spur gear and one internal gear.

3. *Vane*: Spring-loaded vanes slide radially in the rotor. The vane tips follow the eccentric housing and trap fluid between the rotor and the housing.

B. *Nonpositive*: Produce maximum pressures of from 500 to 700 kPa [70 to 100 psi].
1. *Diaphragm*: Small stroke plunger pump in which the liquid is sealed from the rest of the pump by a rubber disk extending over the plunger to the cylinder walls. These pumps are particularly suitable for handling highly corrosive fluids.

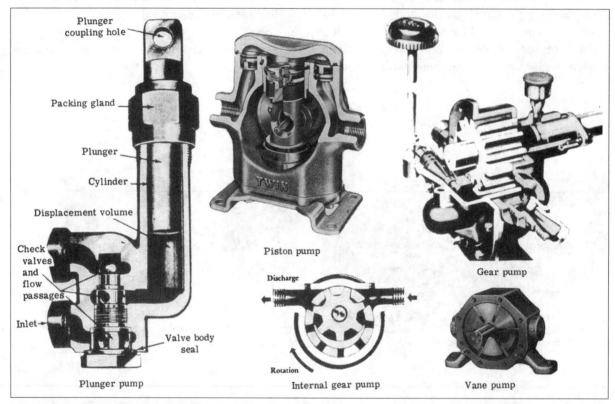

Fig. 9.16. Positive displacement sprayer pumps.

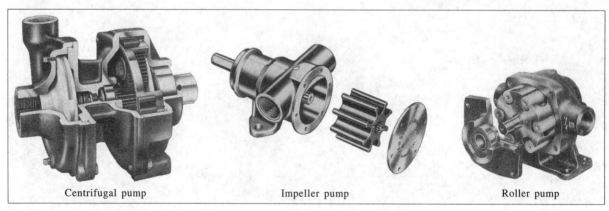

Fig. 9.17. Nonpositive displacement sprayer.

2. *Impeller*: Flexible neoprene rubber rotor bears against a cam surface.

3. *Roller*: Slotted rotor having loose-fitting rollers in the slots that bear against the cam surface of the outer housing.

4. *Centrifugal*: High-speed rotor flings the fluid to the periphery by centrifugal force. These pumps have become increasingly popular in recent years because of their ability to handle wettable powders and abrasive materials. Since high speed is essential, these pumps may have step-up gear ratios or be driven by high-speed hydraulic motors.

The *pressure regulator* is a mechanism to maintain any preset pressure by bypassing some of the liquid to the tank (Fig. 9.18, 9.19). Field sprayers usually depend entirely on this bypass flow for agitation of tank liquids. Wettable powder materials require as much as 50 L/min [13 gal/min] bypass flow at 200 kPa [29 psi] to keep the particles adequately suspended in a 750-L [200-gal] tank.

1. *Manual*: Adjustable spring-loaded ball rises off its seat when the pressure increases. With the relief valve type, the pump works against pressure when the boom outlet valve is closed; not so with the hand-unloaded type.

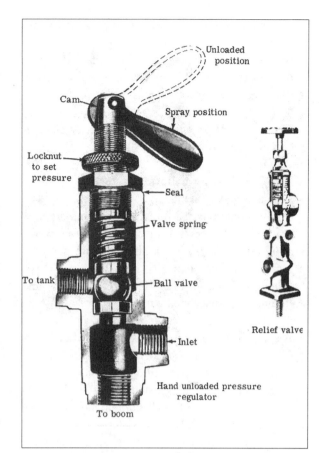

Fig. 9.18. Manual pressure regulators.

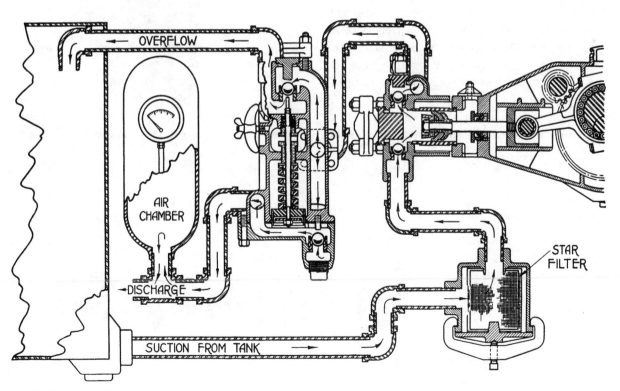

Fig. 9.19. Automatic unloading pressure regulator operation.

2. *Automatic unloading*: Plunger-operated release valve maintains constant pressure until all boom outlets are closed, whereupon a bypass circuit is opened to the tank. The pump, when unloaded, does not have to work against pressure.

Nozzles (Fig. 9.20) should be protected from clogging with a removable fine mesh screen.

1. *Flat fan*: Used for uniform coverage applications such as weed spraying.
2. *Hollow cone*: Gives a fine mist for complete coverage of plants being sprayed for insect control.
3. *Solid cone*: Used when a high-pressure penetrating spray is needed.
4. *Flooding*: Used for liquid fertilizers, slurries, and liquids with suspended particles.

A spraying system is easily assembled after securing the component parts. Solving Problem 9.1 will demonstrate how the parts fit together.

Check valves can be used at the nozzle (Fig. 9.21) to give drip-free shutoff. A diaphragm closes a valve as soon as the line pressure drops.

Nozzle performance changes as spray materials erode the nozzle tip. Brass tips show wear about one-third as fast as aluminum; stainless steel and some of the new plastic tips show wear only one-quarter as fast as brass. Nozzle wear can be significant in only 50 hr of use, depending on the abrasiveness of the spray material. The machinery manager should test nozzle performance periodically for changes in flow rate at the spraying pressures used and for changes in spray pattern owing to nozzle tip wear.

A *rotary spray applicator* uses centrifugal force rather than hydraulic pressure to break liquids into small droplets. These rotary applicators, also called controlled droplet applicators (CDA), produce more uniform droplet sizes; and spray drift is reduced because of the absence of very fine droplets. According to T. N. Jordan (Purdue University), the greatest advantage of CDA is that the spray volumes applied can be eight to ten times less than those for hydraulic nozzles. Fig. 9.22 presents the construction of a CDA

Fig. 9.21. Flooding nozzle equipped with check valve.

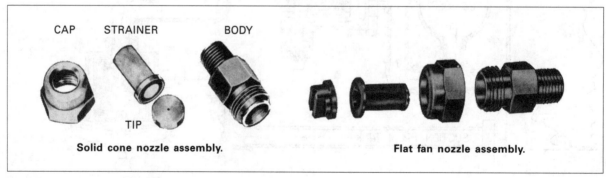

Fig. 9.20. Nozzle assemblies.

and shows the flow of liquid with small arrows. Smaller droplets are produced with an increase in rotary velocity.

Solenoid-operated boom valves (Fig. 9.23) use 12-V electric current from the tractor battery to start and stop flow to the desired booms. Such an arrangement permits control of spraying from remote locations such as the inside of a tractor cab.

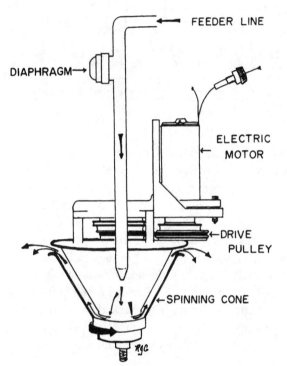

Fig. 9.22. Controlled droplet applicator.

Fig. 9.23. Solenoid-operated boom valves.

Calibration of Sprayers

Societal environmental concerns as well as economics require precision placement of only enough chemical to accomplish pests control. Yet obtaining complete control may involve over-application. S. E. Law and R. D. Oetting (University of Georgia) and S. C. Cooper (Electrostatic Spraying Systems) state that conventional spraying technology depending on gravity and spray droplet inertial forces often achieves less than 50% deposit of the total spray on the plant targets. They estimate the actual quantity reaching the insect or disease pest can be as low as 0.01% of the total spray volume. These and other investigators are developing turbulent air pattern and electrostatic charge technology to achieve penetration of plant canopies and more uniform deposit on the plant parts, particularly on the undersides of leaves. Determining the effectiveness of spraying, eliminating over-spraying, and calibrating the sprayer are important duties of the farm machinery manager.

Perhaps no other farm implement requires as much care in calibration as the farm sprayer. For complete comprehension of the problem the action of all the variables must be understood. The factors affecting the eventual spray concentration per acre are

1. Pressure and delivery of pump
2. Speed of forward travel
3. Water-spray ratio in tank
4. Height of boom
5. Nozzle spacing
6. Concentration of spray materials

Each of these factors has to be considered when solving this example problem.

1. The pump delivers 20 L [5.3 gal]/min of spray at 300 kPa [43.5 psi] to ten nozzles.
2. The water-spray concentrate ratio in the tank is 10:1. (In this presentation the term *spray* refers to the mixture of water and *spray concentrate*. The spray concentrate is composed of *active ingredient* [A.I.] and *carrier*.)
3. The sprayer travels at a constant forward speed of 8 km/hr [5 MPH].
4. When the sprayer nozzles are spaced 75 cm [30 in.] apart on a boom and carried 50 cm [20 in.] off the ground, the application is uniform.
5. The liquid purchased consists of 0.5 kg [1.1 lb] of A.I. in each 15 L [4 gal] of spray concentrate.

If the recommended rate for killing a certain species of weed is 1.2 kg/ha [1.05 lb/a] of A.I., will the weeds be killed?

The unit factor system is very helpful in making this calculation. Only one nozzle and its associated swath need be considered in spraying rate problems. Use both of these techniques to find the amount of A.I. actually applied by the sprayer.

$$0.606 \ \frac{\text{kg A.I.}}{\text{ha}} = \frac{2 \text{ L spray}}{\text{min}} \times \frac{1 \text{ L conc}}{11 \text{ L spray}}$$

$$\times \frac{0.5 \text{ kg A.I.}}{15 \text{ L conc}} \times \frac{60 \text{ min}}{1 \text{ hr}} \times \frac{1 \text{ hr}}{8 \text{ km}}$$

$$\times \frac{1}{0.75 \text{m}} \times \frac{10,000 \text{ m}^2}{\text{ha}} \times \frac{1 \text{ km}}{1000 \text{ m}}$$

$$\left[0.525 \ \frac{\text{lb A.I.}}{\text{a}} = \frac{0.53 \text{ gal spray}}{\text{min}} \times \frac{1 \text{ gal conc}}{11 \text{ gal spray}} \right.$$

$$\times \frac{1.1 \text{ lb A.I}}{4 \text{ gal conc}} \times \frac{60 \text{ min}}{\text{hr}} \times \frac{1 \text{ hr}}{5 \text{ mi}}$$

$$\left. \times \frac{1}{2.5 \text{ ft}} \times \frac{\text{mi}}{5280 \text{ ft}} \times \frac{43,560 \text{ ft}^2}{\text{a}} \right]$$

This answer is about half the recommended rate. Any of the following adjustments could be made:

1. Travel at half speed while keeping the pump speed constant.
2. Reduce the water-spray concentrate mixture from 10:1 to 4.5:1.
3. Purchase a double-strength spray concentrate.

Overlap of the spray patterns from individual nozzles is required to obtain the uniform coverage of a field needed when applying preemergence herbicides. No general recommendation as to the amount of overlap can be made because of the distinctive patterns from different types of nozzles. Values can range from 30% to 100% when the percentage overlap is defined as the ratio of the width of the area double-covered by adjacent nozzles to the spacing between the nozzles on the boom.

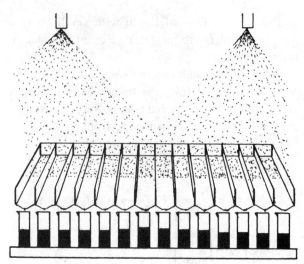

Fig. 9.24. Checking pattern overlap with calibration stand.

The machinery manager must check the total spray pattern for uniformity when total area coverage is desired. Uniformity can be checked by spraying into a nozzle calibration stand (Fig. 9.24). The overlap of adjacent nozzles should produce the same application rate (indicated by the height of spray in the collection tubes) as in the center of the nozzle pattern. If the overlap rate is low, the boom must be raised relative to the ground surface. Lacking a calibration stand, the pattern can be roughly checked by operating the sprayer over a dry clean surface. Excess application shows up as wet streaks in the pattern.

Sanitation and Maintenance

The machinery manager should take special care to avoid residues left in sprayers. In addition to causing deterioration of the sprayer, these residues may contaminate the sprayer for future use.

Pesticide wastes can present environmental problems. They are potentially harmful to humans, livestock, and wildlife in addition to being detrimental to noncrops because of drift and runoff. Various Environmental Protection Agency (EPA) divisions have issued guidelines and regulations important to machinery managers of chemical application equipment, who should refer to these publications for details of approved practices and procedures.

Since complete detoxification of waste pesticide materials is seldom possible, good managers reduce the volume of waste generated and dispose of it in a safe manner. Determine areas to be treated, volume to be mixed, and applicator's performance to reduce amount of leftover solution. Mix spray materials and clean equipment in the field when possible. Avoid producing concentrations of equipment rinse water that may run off into streams or seep into groundwater.

Containers with residual pesticide are classified by EPA as special wastes and must be taken to special or hazardous waste landfills. Only containers that have been triple-rinsed can be recycled or taken to a general refuse landfill. Container rinse water should be poured into the applicator tank and not on the ground.

In addition to cleanup requirements, sprayers must be thoroughly rinsed before using a different spray. A mixture of 0.6 kg [1.3 lb] of detergent/100 L [26 gal] water should be circulated through the bypass for 30 min and then sprayed through the nozzles. For 2,4-D and similar compounds, a 2% solution of household ammonia in water is needed. Rinse with clean water and then spray out a tankful of water with 3-4 L [1 gal] of light oil added to the top of the water in the tank.

A new design of sprayer mixes the spray concentrate with the water while spraying instead of requiring a whole tank of pre-mixed spray. This approach avoids having unused spray for disposal at the end of the field operation. A variable delivery pump, usually ground driven, injects the spray concentrate into the flow of water from the main pump to the booms. Calibration of this design requires the additional testing of the injection pump delivery at various speeds. Since this pump operates against the spraying system pressure, a restricting valve maintaining that pressure is required for the outlet of the injector test apparatus.

Many tall weeds in an established crop can be controlled if translocatable herbicides can be applied selectively to the weeds (Jordan, extension weed specialist, Purdue University). They must be applied during vigorous growth periods of the plant for maximum effect. The following designs isolate the herbicide from the crop: The *recirculating sprayer* directs spray patterns horizontally into a catch basin that collects the spray not left on the weeds (Fig. 9.25). The *rope-wick applicator* brushes the foliage of tall weeds. Rope wicks are attached to the lower side of a capped pipe and extended into the pipe to contact the stored herbicide (Fig. 9.26); capillary action wets the exterior portion of the wick. Forward speed is limited by the speed of capillary flow. Similar applicators may use sponge bars or carpet rollers instead of wicks.

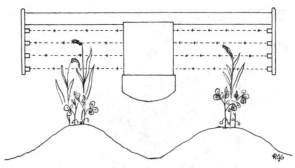

Fig. 9.25. Recirculating sprayer design.

Fig. 9.26. Rope-wick applicator.

Dry Chemical Application Equipment

The farm machinery manager can elect to apply chemicals in a dust or granule form instead of in a spray form. These dry chemical compounds weigh less, require no mixing, can be used in simple machinery, and in some cases may be more effective chemically than the water-diluted sprays. Sprays, however, are easier to apply uniformly.

Dusters compete with sprayers as insect and plant disease control machines. The use of high-capacity power dusters is quite common for many horticultural crops. Fig. 9.27 illustrates the components of a power duster. Constant fan speed and constant forward speed are required for uniform distribution. Notice that effective agitation is provided in the dust hopper.

Granular applicators can substitute for sprayers as control machinery for weed, insect, and soil pests. These applicators have small ground-wheel-driven metering mechanisms such as the variable orifice (Fig. 8.6) or fluted feed (Fig. 8.4). As typical application rates are 10 kg/ha [9 lb/a] or less, this applicator's metering mechanism is much smaller and requires more careful adjustment than other typical farm distributing equipment. Accurate calibration is very important, as slight changes in application rate may cause chemical damage to the crop, be wasteful of materials, or be completely ineffective.

Fig. 9.28 shows a safe handling procedure for granules. The granules are purchased in a container having a special valve closure. This valve fits over a similar

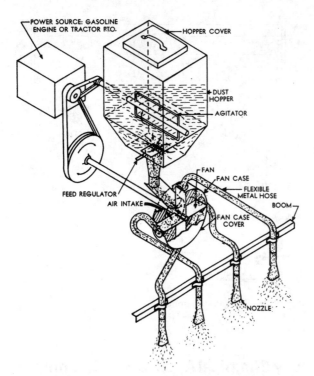

Fig. 9.27. Schematic drawing of typical power duster.

Fig. 9.28. Closed handling system for granules.

valve on the implement's granule hopper and filling proceeds without waste or exposure to the air or to the operator.

Band application of granules, where only a narrow strip over the row is covered, is practiced to save on herbicide costs. Some confusion can occur in reporting application rates for this practice. In this presentation, all application rates will be assumed to be on a band basis; that is, the rate is that existing within the area covered by the band. The band application rate and the field application rate are the same; the total amount of chemical purchased will be different.

Adjustments

The metering adjustments for granular applicators and power dusters are similar in nature to all other variable orifice devices. In granular applicators the levelness of the hopper, the size range of the granules, the depth and compaction of material in the hopper, and perhaps the agitator speed all contribute to variations in the output from the metering mechanism. Furthermore, for herbicide application, even distribution across the width of the outlet as well as in the direction of travel is needed. Only the best calibration techniques will give predictable results.

Calibration

As the amount of granules applied is very small, either actual or simulated test areas should be large enough to obtain a significant sample. If inexpensive weighing equipment is used, the sample collected should not weigh less than 0.5 kg [1 lb]. Example calibration computations follow.

It is known that 1 kg/ha [0.9 lb/a] of A.I. will give preemergence weed control. Granules containing 20% A.I. are to be applied behind a planter having 1-m [39-in.] row spacing.

1. How many kg [lb] of granules should be purchased to treat a 100-ha [247-a] field if
   a. the granules are applied solidly over the field?

$$500\frac{\text{kg}}{\text{field}} = \frac{100\,\text{ha}}{\text{field}} \times \frac{1\,\text{kg A.I.}}{\text{ha}} \times \frac{1\,\text{kg granules}}{0.2\,\text{kg A.I.}}$$

$$\left[1112\frac{\text{lb}}{\text{field}} = \frac{247\,\text{a}}{\text{field}} \times \frac{0.9\,\text{lb A.I.}}{\text{a}} \times \frac{1\,\text{lb granules}}{0.2\,\text{lb A.I.}}\right]$$

   b. the granules are applied in 25-cm [10-in.] bands centered over the rows?

$$125\frac{\text{kg}}{\text{field}} = \frac{500\,\text{kg}}{\text{field}} \times \frac{0.25\,\text{m}}{1\,\text{m}}$$

$$\left[285\frac{\text{lb}}{\text{field}} = \frac{1112\,\text{lb}}{\text{field}} \times \frac{10\,\text{in.}}{39\,\text{in.}}\right]$$

2. To check the application rate when distributing in bands, what quantity of granules should one applicator meter out in 100 m [328 ft] of forward travel?

$$12.5 \, gm = \frac{1 \, kg \, A.I.}{ha} \times \frac{1 \, kg \, granules}{0.2 \, kg \, A.I.}$$

$$\times \frac{100 \, m}{1} \times \frac{0.25 \, m}{1} \times \frac{1 \, ha}{10,000 \, m^2}$$

$$\times \frac{1000 \, gm}{1 \, kg}$$

$$\left[0.45 oz = \frac{0.9 \, lb \, A.I.}{a} \times \frac{1 \, lb \, granules}{0.2 \, lb \, A.I.} \right.$$

$$\times \frac{328 \, ft}{1} \times \frac{10 \, in.}{1} \times \frac{ft}{12 \, in.}$$

$$\left. \times \frac{a}{43,560 \, ft^2} \times \frac{16 \, oz}{lb} \right]$$

Pesticide Regulations

The U.S. Congress passed the Federal Environmental Pesticide Control Act, which regulates the use of pesticides, in 1972. In the language of the act, the term pesticide includes defoliants, desiccants, and growth regulators. All pesticides must be classified as either for *general* or *restricted* use. Restricted-use pesticides are those that can harm the environment or the person making the application even when used as directed.

A person must be certified to use a pesticide classified for restricted use. Farmers applying pesticides to their own land become certified by attending a training session. Commercial or custom operators must pass a written examination that checks their competency in a number of areas.

1. Understanding labels and labeling information and the classification of pesticides (general or restricted use)
2. Knowing the causes of pesticide accidents and how to guard against injury
3. Realizing the need for protective clothing and equipment
4. Recognizing the symptoms of pesticide poisoning and being able to administer appropriate first aid treatment
5. Knowing how to handle and store pesticides properly
6. Being aware of the influence of pesticides on the environment
7. Being able to identify common pests to be controlled
8. Being familiar with pesticide formulations
9. Knowing the common techniques of application
10. Knowing how to calibrate application equipment

The states are to administer examinations, training sessions, and the regulatory program. Individual states may have more stringent regulations than does the federal program.

Practice Problems

9.1. Complete the schematic sprayer diagram on page 148 by sketching hose connections and filter positions.

9.2. A row planter with openers spaced 760 mm [30 in.] apart is equipped with a dry fertilizer applicator and operates at 8 km/hr [5 MPH]. A bag is placed over one outlet and the planter is operated. In 100 m [328 ft], 4 kg [8.8 lb] of material was caught. What is the application rate?

9.3. Sketch the shape of the wetted areas underneath flat fan, hollow cone, and solid cone nozzles.

9.4. At 300 kPa [43.5 psi] a flat fan nozzle has a 1.5-rad [86°] included angle as the spray pattern emerges from the tip of the nozzle. At such an angle, adjoining spray patterns should overlap about 30% of their effective width to give uniform coverage for weed spraying. If the nozzles are spaced 760 mm [30 in.] apart on the boom, what should be the boom height above the ground surface?

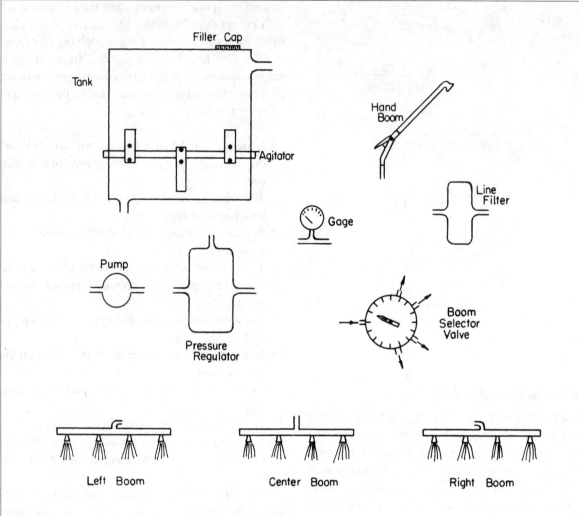

Filler Cap

Tank

Agitator

Hand Boom

Gage

Line Filter

Pump

Pressure Regulator

Boom Selector Valve

Left Boom

Center Boom

Right Boom

9.5. A sprayer travels at 7 km/hr [4.375 MPH]. The solution in the tank is prepared by putting 20 L [5.3 gal] of spray concentrate containing 0.1 kg/L [0.83 lb/gal] of A.I. into a 200-L [53-gal] tank and filling it with water. The nozzles of the sprayer are spaced 760 mm [2.5 ft] apart and each delivers 2 L [0.53 gal] spray/min at 200 kPa [29 psi] pressure. What is the application rate of A.I. at that pressure?

9.6. An airplane flying at 3-m [10-ft] elevation has an effective spray pattern width of 10 m [32.8 ft]. Its travel speed is 200 km/hr [124 MPH]. The A.I. application rate should be 0.8 kg/ha [0.7 lb/a]. The spray concentrate contains 0.2 kg/L [1.66 lb/gal] of active ingredient and is

mixed evenly with water. What must be the sprayer pump output rate?

9.7. A sprayer delivers 2L [0.529 gal]/min from each nozzle. The nozzles are spaced 1 m [39.37 in.] apart. The spray concentrate contains 50 g/L [0.42 lb/gal] of A.I. The required application rate is 1 kg/ha [0.89 lb/a]. If the forward speed is 6 km/hr [3.75 MPH], what should be the spray concentrate-water mixing ratio?

9.8. An applicator places granules in a 25-cm [10-in.] band over rows spaced 760 mm [30 in.] apart. The necessary application rate for weed control is 1.5 kg/ha [1.34 lb/a]. What should be a metering mechanism's delivery in 500 m [1640 ft]?

Chapter 10

GRAIN HARVESTING

The mechanization of grain harvesting has been a longtime objective of farmers. The cutting and threshing of small seeds by hand methods were extremely tedious. Before recorded history, people learned to use reaping tools and animal treading to increase the productivity of small-grain harvest. Evidence of the use of a wheeled reaping cart in ancient Gaul has been found. Horse-drawn reaping machines and stationary threshers date from the early 1800s in the United States. An important early application of steam power was for the stationary threshing, separating, and cleaning of food grains. Attempts to develop a field machine to accomplish all the small-grain harvest in one operation dates from the 1830s; but not until after the advent of lightweight, powerful, high-speed internal combustion engines could the reaper, thresher, separator, and cleaner be combined effectively into a single field machine referred to simply as a *combine*. The modern combine is one of the most important machines in mechanized agriculture. It has been adapted to the harvest of more than 100 food, feed, and processing grain crops.

Small-Grain Harvesting

As discussed in Chapter 1, the farm machinery manager is most interested in an economic material efficiency for machines; that is, a material efficiency that may not be 100% in any aspect but that produces the greatest return above the costs of operating the machine.

Small-grain harvesting is a good example of compromise machinery management. If a combine is adjusted to thresh out every grain, the severity of threshing will likely damage the grain and overload the cleaning section with ground-up straw. The material efficiency of the combine is actually reduced. One must also expect reduced income because of the lower value of damaged grain, increased grain loss from the cleaning section, and reduced forward speed made necessary because of the overloaded cleaning section. Clearly, a compromise adjustment that allows some grain to go through the combine unthreshed is an optimum adjustment, since the overall return is increased.

The ability to recognize and evaluate compromise solutions is a valuable trait of the harvesting machinery manager who must understand the detailed operation of the machines, be able to check their performance, and then arrive at adjustments or operating procedures that produce the greatest economic return.

Combines are classified according to use:

1. *Hillside*: Having provision to level the body of the combine on sidehill slopes (Fig. 10.1)
2. *Prairie*: Not having such a provision

And they are classified according to power:

1. *PTO*: Power taken from a pulling tractor (Fig. 10.2)
2. *Self-propelled*: Power provided by single engine that operates both the machine and its driving wheels (Fig. 10.1 and 10.3)

Mechanisms

The mechanisms included in a combine accomplish five general functions:

1. Cutting
2. Feeding
3. Threshing
4. Separating
5. Cleaning

Cutting

The *cutting* operation is accomplished with a cutter bar and a reel (*Cb* and *Rl*, Fig. 10.3). The reel may be a plain bat reel as in Fig. 10.1 or a pickup reel with cam controlled teeth as shown in Fig. 10.3. A pickup reel can be effective in lifting lodged crops above the cutter bar.

A *floating* cutter bar is illustrated in Fig. 10.4. This cutting device is loosely suspended beneath the combine's header and flexes to follow the contour of the ground. The low cutting height, 2.5 cm [1 in.], is

Fig. 10.1. Self-propelled hillside combine that automatically levels front to rear as well as side to side.

Fig. 10.2. Large PTO combine with windrow pickup.

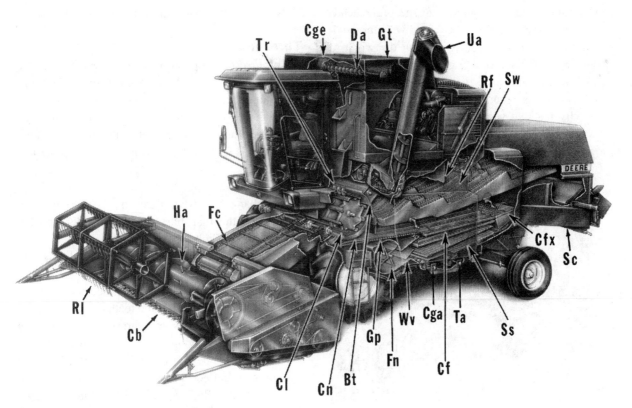

Fig. 10.3. Cutaway view of self-propelled combine.

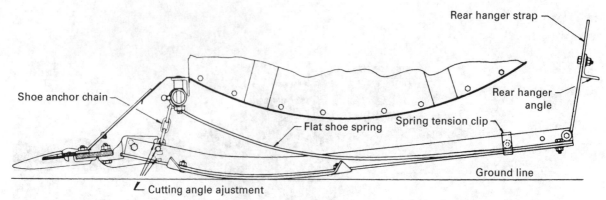

Fig. 10.4. Floating cutter bar construction.

especially advantageous in harvesting soybeans, which often pod close to the ground. A floating cutter bar is especially valuable for wide-cut combines that are seldom able to maintain a uniform low cut in even the most level of soybean fields.

Windrowing the crop before combining is practiced in many areas. Windrowing is expected to increase the ripening and drying rates of the crop to permit earlier harvest. A self-propelled windrower is shown in Fig. 10.5. A windrower has the same elements as a combine's small-grain head—a reel, a cutter bar, and a conveyor. A grain windrow should be laid on stubble at least 20 cm [8 in.] high so that air can circulate under the windrow. The grain heads should be exposed on top of the windrow. After windrowing, the crop is immune to all but the worst of storms. Fig. 10.2 shows a combine equipped with a special pickup head used for windrowed grain.

In many areas windrowing is used only if the crop is heavily infested with green weeds. S. G. Huber

Fig. 10.5. Self-propelled windrower.

(Ohio State University) and R. G. White (Michigan State University) report that total losses when combining weedy crops directly may be as much as 13% of the gross yield compared to 4% for clean fields. Windrowing can dry out the green weeds and lessen their bad effects during combining.

Feeding

The *feeding* operation consists of mechanisms that distribute and deliver the crop material to the thresher in a steady, uniform flow. A large diameter auger (Ha, Fig. 10.3) conveys the cut material away from the cutterbar and to a central conveyor (Fc, Fig. 10.3) that feeds the material to the threshing mechanism.

Threshing

The *threshing* operation is accomplished with a cylinder (Cl, Fig. 10.3), or a rotor (Fig. 10.7) working against curved, grated sections called *concaves* (Cn, Fig. 10.3). Material flow is perpendicular to the axis for cylinders and parallel to the axis for rotors. Two designs for threshing cylinders are rasp bar and spike tooth.

1. Rasp bar (Fig. 10.3). Threshing is accomplished between the rasp bars on the cylinder and the concaves. Much of the threshed grain escapes through the openings in the concaves and falls onto the grain pan (Gp, Fig. 10.3).
2. Spike tooth. Threshing takes place between the teeth of the cylinder and the teeth of the concaves. This design provides an aggressive threshing for rice and crops having tough straw and weeds.

The beater behind the cylinder (Bt, Fig. 10.3) is used to deflect the material from the cylinder. The retarder flap (Rf, Fig. 10.3) knocks down thrown grain from the beater.

Separation

The *separation* operation removes the threshed grain from the mixture of plant parts that comes from the threshing section. The straw walkers (*Sw*, Fig. 10.3) are long units mounted on two crankshafts. Rotation of the crankshafts agitates the material and the heavier grain fails through a grate onto the bottom pan of the straw walker. From there the grain slides down to the grain pan, *Gp*. Most of the threshed grain is expected to escape from material-other-than-grain (MOG) at the concave grate. The separating mechanisms are used to take out the last 20% or so of the threshed grain entrapped in straw, pods, husks, or stalks.

Fig. 10.6 illustrates a design which has a pre-separating cylinder, a main threshing cylinder, a beater, and twin rotary separators. The pre-separator cylinder removes the easily threshed grain which reduces the load on the downstream mechanisms and increases overall capacity.

Rotary combines (Fig. 10.7) use large single or double rotors to replace both the threshing cylinder and the separator. The rotor is enclosed by a cylindrical grate that has vanes to direct flow in an axial

Fig. 10.6. Two cylinder threshing mechanisms.

spiraling motion out the rear of the machine. The centrifugal force imparted by the rotation causes a force on the kernels many times greater than the force of gravity and produces rapid separation of grain from the MOG. The physical volume required for the separating operation of a rotary combine can be substantially less than that for the traditional straw walker design of similar capacity.

Fig. 10.7. Combine with dual threshing and separating rotors.

Also, the rotary design may damage grain less. The threshing function occurs over a large arc of contact with the cylinder and along its full axial length. Larger cylinder-concave areas permit less aggressive machine adjustments for the same degree of threshing.

Cleaning

The grain pan (*Gp*, Fig. 10.3) is a common flow point for threshed grain, unthreshed grain, unhulled kernels, unthreshed heads, chaff, bits of straw, and dirt that enter the *cleaning* section of the combine. Cleaning is accomplished by:

1. Screening out the larger particles with the chaffer and sieves,
2. Blowing out the lighter particles with a fan blast, or
3. Screening out the smaller particles in a recleaner.

Fig. 10.3 shows the cleaning mechanisms. The uncleaned grain is collected from the grain pan, *Gp*, and from the grain return chutes of the straw walkers and dumped on the upper sieve or chaffer, *Cf*. Light materials and chaff are blown off the chaffer and shoe sieves, *Ss*, and out the back of the combine by the fan, *Fn*. A wind vane, *Wv*, can be adjusted to direct the air flow to keep the sieves clean. Clean threshed grain falls through the sieves into the clean grain auger, *Cga*, and is conveyed by the clean grain elevator, *Cge*, to the grain holding tank, *Gt*. A distributing auger, *Da*, levels the grain when filled. Unthreshed grain moves over the sieves and drops through the chaffer extension, *Cfx*, into the tailings auger, *Ta*. These *tailings* are returned by the tailings return elevator to the cylinder for rethreshing. The distributing auger, *Tr*, spreads the returned material evenly over the cylinder.

A *recleaner* is expected to remove all material smaller than the grain harvested. It may be a device as simple as a fine screen around the clean grain auger or a screen along the bottom of the clean grain elevator, or as sophisticated as a powered auger device located above the grain tank. The auger conveys the grain through a cylindrical screen. Fine material is separated from the grain as it passes over the screen. If only dirt is to be removed, the machinery manager may allow the screenings to fall to the ground. But if weed seeds are the major component of the screenings, they should not be allowed to scatter back onto the land.

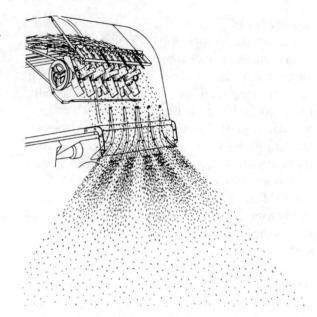

Fig. 10.8. Crop residue chopper attachment for combine.

Some farmers wish to chop the residue coming out the back of a combine (*Sc*, Fig. 10.3) before dropping this material back onto the land. Figure 10.8 shows the action of a straw chopper mounted in the back of a combine. The chopped material is scattered by deflector vanes as it leaves the combine.

Losses

There are five losses relating to the combining of grains.

1. *Shatter loss*: Grain lying on the ground or out of reach of the cutter bar.
2. *Cutter bar loss*: Grain lost due to rough handling by the cutter bar. If windrowing is practiced, the cutter bar loss includes the loss for the windrower's cutter bar and the combine's pickup attachment. The loss is about the same for both methods.
3. *Threshing loss*: Grain lost out the rear of the combine in the form of unthreshed heads.
4. *Separating loss*: Grain lost out the rear of the combine in the form of threshed grain.
5. *Cleaning loss*: Loss in value of the crop due to the presence of foreign matter in the grain tank.

Results in Table 10.1 indicate that harvesting losses from combines can be substantial. Weed-free fields

TABLE 10.1. Typical Combine Losses in Midwestern U.S. (percentage of gross yield)

Crop	Shatter and Cutter Bar Loss	Cylinder Loss	Separating and Cleaning Loss	Total Loss
Oats	1.5	0.8	2.0	4.3
Wheat	3.4	0.1	0.6	4.1
Soybeans	5.8	0.6	0.7	7.1

of dry, near-mature, straight-standing grain should allow combine losses to be less than 3% total loss. A good machinery manager should understand the nature of the various losses and the capability of the modern combine for adjustments to minimize loss.

Adjustments to Reduce Loss

Shatter loss is not the responsibility of the machine other than that the capacity of a particular machine may not be high enough to harvest all the grain before a significant amount of grain is lost due to wind action; to bird, rodent, and insect damage; and to the natural tendency of the crop to shatter as it dries.

According to S. G. Huber and R. M. White, cutter bar losses usually are greater than other tosses for the correctly adjusted, direct-cut combine. These losses are due to low-lying material passed over by the cutter bar, to grain shattered onto the ground by reel and guard action, and to material carried over or fumbled by the reel. The least loss is expected for crops that stand upright.

Crops such as soybeans that head close to the ground are especially subject to loss because the cutter bar cannot be operated low enough. Losses of 1% per additional 2.5 cm [1 in.] of cutter bar height are typical for soybeans. Automatic cutter bar height controllers are available to maintain a constant height of cut. These controllers respond to signals from fingers (Fig. 10.9) or radar-like sensors that sense the soil surface and use hydraulic power to adjust the cutter bar height and tilt for cross slopes. Cultural practices that leave the soybean field as level as possible will significantly decrease cutter bar losses.

A special row crop head has been developed for harvesting soybeans. Similar in appearance to the row crop head for corn, it uses intermeshing rubber loops on the gathering belts to clamp the crop stalks rigidly while they are being cut with a rotary cutter and being transported to the feeder auger. Shock and vibration of the stalks is reduced and less shattering of the

beans from the pods is expected. Forward speeds can be as high as 8 km/hr [5 MPH] without experiencing great loss if the belt loops are synchronized with the forward travel of the combine.

Improper reel adjustment is one of the greatest sources of loss when harvesting cereal grains. J. R. Goss, R. A. Kepner, and L. G. Jones (University of California) compiled the data presented in Fig. 10.10 over a three-year period of barley harvesting. The top two solid curves show the effect of having a fixed-bat reel too low and perhaps too far forward for upright grain. The pickup reel was also misplaced for standing grain. The heights indicated are for the lowest position of the bat or tooth tips of the pickup reel with reference to the level of the cutter bar knife. The distance forward is the horizontal distance between the reel axis and the tips of the cutter bar knife sections. Goss et al. recommended that a fixed-bat reel should be 15–25 cm [6–10 in.] forward and at a height such that the lowest position of the bats is a little below the lowest heads. A pickup reel, when used in lodged crops, should be lower and a little farther forward. With either type of reel, a *real index* (ratio of tip speed at the reel bat to forward speed of the combine) of 1.25 to 1.5 gave consistently satisfactory performance. Some machines have a variable-speed drive to the reel giving the operator maximum control over the reel index.

Cylinder losses should be less than 1% and inversely related to the severity of threshing. However, damage to the grain limits the extent to which severe threshing may be employed. A compromise must be reached. The operator's manual of each combine indicates a range of cylinder speeds for specific crops.

Fig. 10.9. Finger sensors for cutter bar height control.

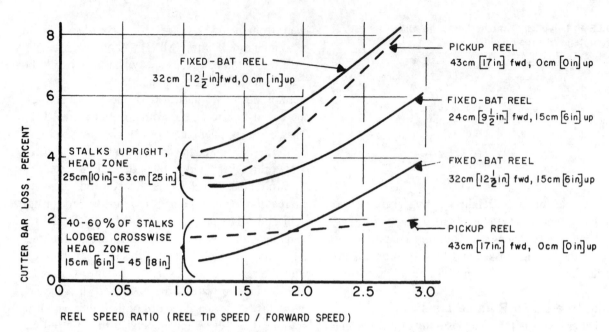

Fig. 10.10. Losses associated with reel adjustment.

Typical tip speeds are

Wheat and small grains	30–36 m/s [100–120 ft/s]
Sorghums and soybeans	18–23 m/s [60–75 ft/s]
Corn	12–15 m/s [40–50 ft/s]
Peas, edible beans	8–15 m/s [26–50 ft/s]

The concave clearance depends on the amount of throughput as much as on the physical size of the seed. In some instances good threshing is obtained by widening the concave clearance and then increasing the throughput. The increased volume of material cushions the seed from damage without reducing threshing effectiveness. However, such a practice may overload the separating section.

The severity of threshing is increased with increased cylinder speed and decreased concave clearance. Such adjustments are indicated if the crop is tough because of high moisture content.

Separating losses can be divided into two categories reflecting their point of origin: shoe losses and separating mechanism losses (straw rack, straw walkers, rotor). Separating mechanism losses, still called *rack losses,* should be less than 1% of yield or the combine may be overloaded. Overloading reduces the ability of the separator section to work grain through the large mass of MOG. While most combines are designed with a concave grate under the cylinder or rotor to encourage separation, at least 10% of the separation is expected to take place in the separator section. Fig. 10.11 relates loss to throughput from data compiled

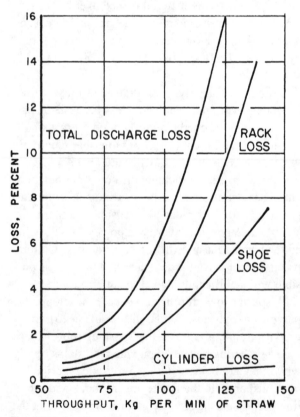

Fig. 10.11. Losses with heavy throughputs of barley.

by Goss et al. To reduce load on the separator, reduce throughput by decreasing forward speed or by raising the cutter bar to lessen the amount of straw in MOG. PAMI experience suggests that raising the cutter bar from 20 cm [8 in.] to 36 cm [14 in.] in small-grain harvest will increase the throughput capacity by 50%. If the crop can be cut as high as 56 cm [22 in.], the throughput capacity can be increased 90% above that for the low cut.

Shoe losses should also be less than 1%. Losses are due to overloading and sieve plugging. Proper airflow should keep the sieves clean by floating the trash off the sieves and out the back of the combine. With clean sieves and no overload, the only loss would be those few light grains blown over the tailgate. Fig. 10.12, prepared from data by Goss and others, indicates a definite optimum setting for fan inlet openings when working in barley. While the absolute settings may be different for other crops, the general shape of the curve is thought to be representative. Note the increased volume of tailings with increased airflow.

Impact boards placed behind the tailgate have been used to indicate instantaneous loss. The frequency of grain hitting the board is conveyed electronically to the operator.

Cleaning losses are associated with the shoe of the combine, but unfortunately they are not really lost since they appear as straw, weeds, etc., in the grain tank. In a correctly adjusted combine, cleaning is improved by closing down the shoe sieve and/or the chaffer sieve; but the problem may be one of overthreshing. The separation of grain from the rest of the throughput is easier if the straw, weeds, etc., are not broken up to work their way down onto the shoe sieve.

Operations

Forward speed is probably the most important factor in optimizing the performance of a combine harvester. Several investigators have determined that total losses increase rapidly as forward speed increases. Because of overloading, rack losses, particularly, rise with an increase in speed. The increase in rack loss appears to be directly proportional to speed and can amount to 4% of the total yield as speed is increased from 3.2 km/hr [2 MPH] to 5.6 km/hr [3.5 MPH] in heavy-yielding grain. Hydrostatic drives are used on SP combines to permit precise control of forward speed.

Both speed and weed infestation increase harvest losses for soybeans. W. R. Nave and L. M. Wax (USDA, Urbana, IL) found that slower speeds reduced losses in fields with heavy, green weed infestation (1 weed/ 30 cm [12 in.]) (Fig. 10.13). While cutter bar losses were somewhat less in weed-free fields, the cylinder and separating losses were reduced to 1/10 of those in the weedy plots. The cylinder and separating losses for the weedy plots were cut in half by delaying harvest until after a weed-killing frost.

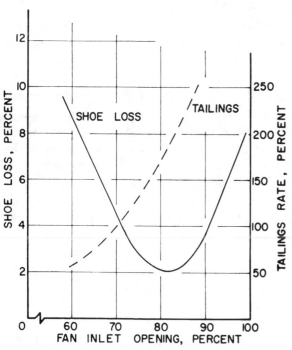

Fig. 10.12. Fan inlet opening and shoe losses.

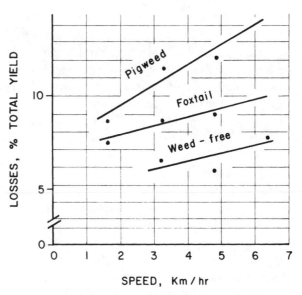

Fig. 10.13. Combine harvest losses in weedy and weed-free soybeans.

Moisture content of the crop is a very important item in deciding when to combine. High-moisture crops can be harvested with combines, allowing the farm manager some options in operations timing since he can artificially dry the harvest. However, field and combine losses are minimized in a rather narrow moisture range for most crops. Additionally, combining at a high moisture may damage grain physically. Optimum ranges for some small grains are wheat, 13–15% wet basis (W.B.) moisture; soybeans, 12–14% W.B.; and grain sorghum, 25–30% W.B. (See the section Drying Machinery, Operations, in Chapter 12 for a discussion of wet basis moisture.)

The time of harvesting with combines is a major decision for the machinery manager. In an extensive study of soft red winter wheat combining, W. H. Johnson (Agricultural Engineering Department, Ohio State University) found that the losses due to delayed harvesting are as shown in Fig. 10.14. Such losses amount to 13.5 kg/ha [12 lb/a] each day of delay. On the other hand, if the grain is to be stored it must be 14% or less W.B. moisture or be dried artificially. Johnson found that the well-adjusted combine's material efficiency was at an optimum at about 20% grain moisture. For an average year, harvesting at 20% will produce about 70 kg/ha [62 lb/a] more, which can help pay for the drying costs.

Soybean harvesting is especially sensitive to the time of harvest. Some varieties are prone to self-shatter as time passes. The combine operation should be planned with this loss pattern in mind.

R. W. Harper (Agricultural Engineering Department, University of Illinois) investigated the economic loss associated with untimely harvesting of Harosoy and Shelby soybeans (Fig. 10.15). Marketing penalties for high moisture, grain damage, and foreign matter were included in the study. The least costly harvest occurred during a 10-day period, after which the increased loss per day per area averaged 0.4% of total yield for Shelby and 1.1% for Harosoy varieties.

Harper found that losses were not associated directly with moisture content. Harvesting should begin after the grain first drops to 13% W.B. moisture, as losses increase afterward even though rain or dew may cause wide fluctuations in grain moisture.

The efficient operator will keep a constant watch over both the grain tank and the tailings. The cleaning performance of the shoe is indicated by the appearance of the grain in the tank, while the general "health" of the combine is indicated by the tailings. The quantity of tailings should be small but never

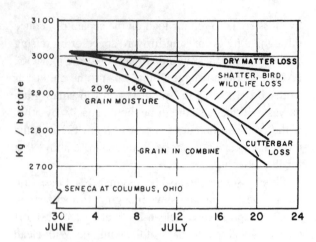

Fig. 10.14. Losses in wheat yields with delayed harvest.

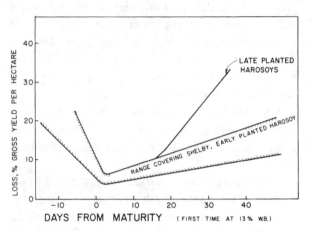

Fig. 10.15. Timeliness losses in soybean harvesting.

zero. Ideally over half the tailings should be threshed grain, as this circumstance will indicate that the shoe sieve is not open too far. Perhaps one-fourth of the tailings should be unthreshed grain, which would indicate that the cylinder and concaves are not operating too severely. The remaining quarter should be chaff and other small bits and pieces to indicate that the airflow through the sieves is not excessive. Instrumentation is available to give the operator a sense of the volume of the tailings.

Some combines deliver tailings for recleaning only, others only for rethreshing, and still others permit the operator to select where the tailings are deposited. A particularly desirable option is a screen in the tailings elevator that will pass the threshed grain into the cleaning area while retaining the unthreshed grain for the cylinder.

Finally, the efficient operator must check the ground behind the combine to get an estimate of losses and material efficiency.

Sample Calculation

An example of the calculations required for determining a combine's material efficiency should be helpful.

A 5 m [16.4 ft] self-propelled combine is harvesting seven 760-mm [30-in.] rows of soybeans. Before combining, a strip 1 m [39.37 in.] long is carefully gleaned across the full 7-row width. All the remaining residue is cleared away. The combine is operated normally until the cutter bar passes over the cleared area. The combine is stopped, allowed to clean out, and then reversed without any of the discharge falling on the cleared area. The cutter bar losses show up as newly dropped beans on the cleared area and as unthreshed pods still clinging to the stubble. A short distance down the field a 10-m [32.8-ft] net yield test distance is marked off parallel to the rows. A container is placed under the grain outlet in the tank as the combine moves past the start of the test distance and is removed as it passes the mark at the end of the distance. The combine discharge is observed to fall over a 2-m [78.74-in.] width. All beans except those pods clinging to the stubble are gleaned from the 2-m [78.74-in.] discharge test length. The following data are obtained.

1. Initial gleanings from the cleared strip, 20 g [0.044 lb]

2. Loose beans gleaned from the cleared strip after the cutter bar has passed, 25 g [0.055 lb]

3. Beans shelled from pods picked from stubble left in the cleared area after the cutter bar has passed, 50 g [0.11 lb]

4. Grain caught in the container, 12 kg [26.4 lb]

5. Loose beans gleaned from the discharge test area, 2 m × 2 m [6.56 ft × 6.56 ft], 40 g [0.088 lb]

6. Beans shelled from pods picked out of the discharge (does not include those on the stubble), 10 g [0.022 lb]

Shatter loss

$$37.6\frac{kg}{ha} = \frac{0.020\,kg}{1\,m \times 5.32\,m} \times \frac{10,000\,m^2}{1\,ha}$$

$$\left[33.4\frac{lb}{a} = \frac{0.044\,lb}{3.28\,ft \times 17.5\,ft} \times \frac{43,560\,ft^2}{1\,a}\right]$$

Cutter bar loss

$$141\frac{kg}{ha} = \frac{(0.025 + 0.05)\,kg}{1\,m \times 5.32\,m} \times \frac{10,000\,m^2}{1\,ha}$$

$$\left[125\frac{lb}{a} = \frac{(0.055 + 0.11)\,lb}{3.28\,ft \times 17.5\,ft} \times \frac{43,560\,ft^2}{1\,a}\right]$$

Threshing loss

$$9.40\frac{kg}{ha} = \frac{0.01\,kg}{2\,m \times 5.32\,m} \times \frac{10,000\,m^2}{1\,ha}$$

$$\left[8.35\frac{lb}{a} = \frac{0.022\,lb}{6.56\,ft \times 17.5\,ft} \times \frac{43,560\,ft^2}{1\,a}\right]$$

Total loose beans from discharge gleanings
= 40 g [0.088 lb]

Subtract shatter loss beans,

$$\frac{20\,g}{1\,m \times 5.32\,m} \times (2\,m)^2 = -15\,g$$

$$\left[\frac{0.044\,lb}{3.28\,ft \times 17.5\,ft} \times (6.56\,ft)^2 = -0.0330\,lb\right]$$

Subtract cutter bar loose beans,

$$\frac{0.025\,kg}{1\,m \times 5.32\,m} \times = (2\,m)^2 = -18.8\,g$$

$$\left[\frac{0.055\,lb}{3.28\,ft \times 17.5\,ft} \times (6.56\,ft)^2 = -0.0412\,lb\right]$$

Net beans for separating loss
= 6.2 g [0.0138 lb]

Separating loss

$$15.5\frac{kg}{ha} = \frac{0.0062\,kg}{(2\,m)^2} \times \frac{10,000\,m^2}{1\,ha}$$

$$\left[13.97\frac{lb}{a} = \frac{0.0138\,lb}{(6.56\,ft)^2} \times \frac{43,560\,ft^2}{1\,a}\right]$$

Net or tank yield

$$2256\frac{kg}{ha} = \frac{12\,kg}{10\,m \times 5.32\,m} \times \frac{10,000\,m^2}{1\,ha}$$

$$\left[2003\frac{lb}{a} = \frac{26.4\,lb}{32.8\,ft \times 17.5\,ft} \times \frac{43,560\,ft^2}{1\,a}\right]$$

Gross yield of the field

$$2460\frac{kg}{ha} = (37.6 + 141 + 9.4 + 15.5 + 2256)$$

$$\left[2184\frac{lb}{a} = (33.4 + 125 + 8.35 + 14 + 2003)\right]$$

Material efficiency

$$\frac{2256}{2460} = 0.917 \qquad \left[\frac{2003}{2184} = 0.917\right]$$

The respective losses are calculated as percentages of gross yield.

$$\text{Shatter}: \quad \frac{37.6}{2460} \qquad \left[\frac{33.4}{2184}\right] = 1.53\%$$

$$\text{Cutter bar}: \quad \frac{141}{2460} \qquad \left[\frac{125}{2184}\right] = 5.73\%$$

$$\text{Threshing}: \quad \frac{9.4}{2460} \qquad \left[\frac{8.35}{2184}\right] = 0.38\%$$

$$\text{Separating}: \quad \frac{15.5}{2460} \qquad \left[\frac{14}{2184}\right] = 0.63\%$$

Obviously the cutter bar loss is excessive. The cutter bar should be lowered further even if it might occasionally run in the dirt. Saving even half the cutter bar loss might pay for increased cutter bar repair.

Considerable time and effort is involved in making a material efficiency test; furthermore, several tests should be made in representative parts of the field. Smaller test areas would speed up the test but at a sacrifice in accuracy. The use of laydown frames and specially built discharge catching devices are most efficient. There are no shortcuts, however. A material efficiency test is a bother, but the time is well spent if the combine is used over very many acres of a valuable crop.

Corn Harvesting

Corn is unique among cereals in that its grain may be stored and used either as kernels or as ears. A different machine is required for each type of harvest. Combines have been used since the mid-1950s for harvesting corn in the shelled form. In areas where corn is grown as a cash-grain crop, almost all the corn is harvested with a combine.

Corn Pickers

Corn pickers have dropped dramatically in popularity as combines were adapted to corn harvest. Seed corn growers and some livestock farmers harvest ear corn to meet their special needs.

1. *Self-propelled*: Picker having its own engine, transmission, and drive wheels.
2. *Mounted*: Component parts are mounted on the tractor framework. The picker is operated by the tractor's PTO shaft.
3. *Trailed*: PTO-driven picker with its weight carried primarily on its own carriage wheels.

Mechanisms

The initial function of the corn harvester is to pick the ear of corn without shelling the kernels or gathering stalk material. The *snapping rolls* (Fig. 10.16) are designed to straddle the standing row of corn and remove the ears with a pair of contra-rotating rolls that pull the stalk downward while pinching off the ear. The ear is then delivered onto associated *husking rolls* (Fig. 10.16) or onto a separate husking bed. The husks of the ears are removed and the ears delivered to a trailing wagon.

Under dry conditions, kernel shelling occurs at both roll mechanisms. Kernels shelled at the snapping rolls are lost to the ground. Kernels shelled at the husking bed are collected, cleaned with an air blast, screened for dirt particles, and delivered to the ear corn elevator.

Corn Combines

Corn combines (Fig. 10.17) differ little from small-grain combines. The essential difference is the con-

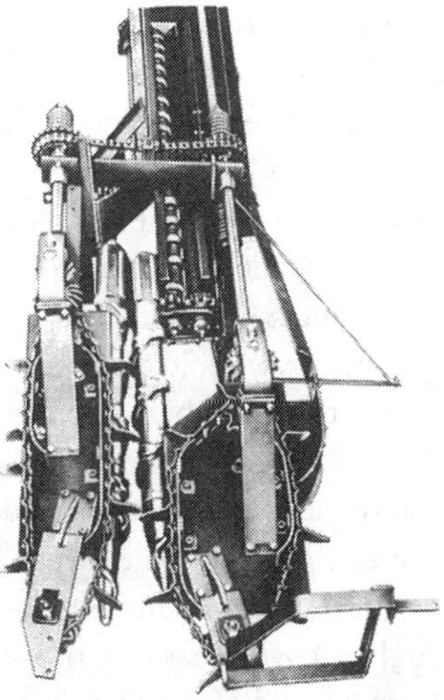

Fig. 10.16. Separate snapping rolls and husking rolls.

Fig. 10.17. Self-propelled combine equipped with corn head.

struction of the row crop head that replaces the cutter bar head. The ear-snapping mechanism for a combine corn head (Fig. 10.18) differs from that for conventional corn pickers. Stripper plates keep the ear from being pulled down into the rolls and causing kernel shelling and loss. The snapping rolls can be very aggressive and will work well in the higher stalk moistures encountered by corn combines.

Losses

The losses relating to corn harvesting are

1. *Loose ear loss*: Ears separated from the stalk and lying on the ground out of reach of the gathering mechanism.

2. *Picker ear loss*: Ears left on the ground that were attached to the stalks but escaped the snapping rolls while being picked up by the gathering points.
3. *Shelled corn loss*: Shelled corn lying on the ground as a result of shelling occurring at the snapping rolls.

For combines, two additional losses occur.

1. *Threshing loss*: Kernels lost because they were not shelled from the cob.
2. *Separating loss*: Kernels lost that were not separated from the husks, silks, etc.

The evaluation of the material efficiency of a picker should include:

1. Amount of foreign matter in the wagon
2. Amount of shelled corn in the wagon

In the case of a combine it should include

1. Amount of foreign matter (cobs, trash, etc.) in the tank
2. Amount of damaged kernels in the tank

G. E. Ayres, C. E. Babcock, and D. O. Hull (Iowa State University) checked the field losses of 84 Iowa combines (Table 10.2). They found that most custom operators were more skillful than owner-operators in operating combines. They observed, without finding any explanation, that afternoon losses were half the morning losses and operating speeds used had no effect on losses. They found that a 50-mm [2-in.] error

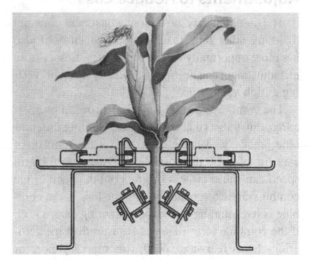

Fig. 10.18. Corn head snapping rolls and stripper plates.

TABLE 10.2. Corn Combine Field Losses (kg/ha [lb/a] shelled corn)

Loss	Mean	Range
Missed ears	94 [84]	0-598 [0-532]
Snapping roll shelling	56 [50]	7-308 [6-274]
Separating	44 [39]	0-384 [0-342]
Cylinder	38 [34]	0-453 [0-403]
Total machine	232 [207]	35-1742 [31-1551]
Preharvest dropped ears	131 [117]	0-629 [0-560]
Total field	363 [324]	70-1742 [62-1551]

in matching the header row spacing to the planted row spacing caused over 450-kg/ha [400-lb/a] increased loss. The extent of the ranges above the mean losses shows that only a few operators experienced serious losses.

A cause-and-effect relation between total losses and the amount of lodging of the stalks is shown below. Lodging was not a serious complaint as less than half the combines were operating in fields that had as much as 6% lodged stalks.

% lodged stalks	Loss kg/ha	[lb/a]
0	245	[218]
1–5	339	[302]
6–10	321	[286]
11–15	440	[392]
16–25	478	[426]
36–40	1157	[1030]

Adjustments to Reduce Loss

The combine may have more adjustment potential than any other farm machine; consequently, it also has more opportunity for misadjustment. Losses from misadjustment can be especially costly when harvesting a high value crop.

The combine operator is in direct control of operating adjustments that affect losses. Guiding the combine precisely over the row reduces the number of ears fumbled by the corn head. Excessive forward speed can cause (a) loss of grain out the back of the combine and (b) greater picker ear loss if the combine is overrunning the header gathering chains (i.e., if the combine forward speed is faster than the gathering-chains rearward speed). Improper cylinder or rotor speed will cause loss of kernels attached to the

cobs if too slow and cracked corn damage if too fast. Slow engine speed will cause inadequate separation and cleaning. These adjustments are operating parameters that are recommended by the combine manufacturer and modified by the machinery manager.

Other adjustments are made as a result of monitoring combine performance. The adequacy of most adjustments is determined from field observation and measurements taken after the combine has passed. A concentration of shelled corn on the ground in the row may mean the stripper plates in the header are spaced too widely for ear size and the ears are pulled down into contact with the rolls. The plates should be spaced only enough to permit passage of stalks. A wider than normal spacing will improve throughput for a dense crop with thick stalks and large ears.

Cracked kernels and foreign matter in the bin indicate overthreshing. Rasp bar speeds greater than 15 m/s [50 ft/s] and concave clearances less than 1.6 cm [5/8 in.] will damage kernels and break cobs. However, greater clearances and slower speeds may not completely remove high-moisture kernels from the cobs.

Single kernels scattered randomly out the back of the combine indicate separating and cleaning losses. Adjustments in cleaning corn are different from those for other grains because the corn kernel is larger and heavier. Larger screen and sieve openings and greater quantities of fan air flow are needed. The tailings return system, so important for small grains, has little utility for corn. A kernel left attached to a large cob piece will be discharged. A more severe threshing setting that grinds the cob into small pieces will tend to damage grain, put much of the cob in the grain tank, and overload the combine with recycled material. Increased fan output may be able to blow out small cob particles while retaining the corn kernels.

Power Requirements

Large combines need to be equipped with powerful engines. Yet typical loading on combine engines may be 50% or less. Maintaining constant threshing and separating speeds is so important that large engines are needed to obtain quick recovery from a drop in speed due to an occasional overload. Fig. 10.19 shows measured power requirements for a corn combine equipped with a 3-row head, 760-mm [30-in.] spacing, and a hydrostatic propulsion drive for infinite forward speed control. Field conditions were dry and firm soil surface, 23% W.B. kernel moisture,

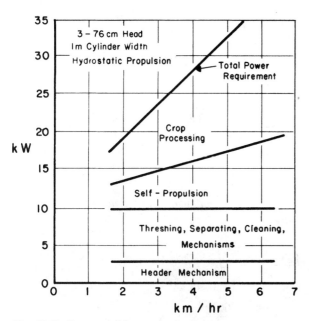

Fig. 10.19. Corn combine power requirements.

the effect of moisture content on the efficiency of conventional snapping rolls. A minimum loss occurs near 25% kernel moisture. At high moisture contents the loss is characterized by crushed ears and at lower moisture contents by shelling loss at the snapping rolls.

Kernel moisture, % W.B.	Loss, % of total yield
30	6
25	3
20	4
15	7
10	10

Proper matching of snapping roll spacing to row spacing becomes increasingly important with the multirow corn heads used on fields with narrow-spaced rows. For example, assume that an 8–75 cm [8–30 in.] combine corn head is to be used on 70-cm [27.6-in.] rows. If the operator guides on one of the inner rows, the opposite outside snapping rolls will be 20 cm [7.9 in.] off the row center.

For small grains, kernel moisture is the prime indicator for timing the harvest; in corn harvesting, the stalk and cob moisture contents must also be considered. Damp and tough stalks withstand the action of snapping rolls much better than dry and brittle stalks. Harvesting is most timely if the kernel moisture is low but the stalks still retain considerable moisture.

The shelling of the corn kernels from the cobs is significantly affected by cob moisture. D. E. Burrough and R. P. Harbage (Agricultural Engineering Department, Purdue University) found a linear relationship between the amount of kernels left on the cob and the cob moisture content. Losses became insignificant if shelling was delayed until the cob dried down to 30% W.B. C. S. Morrison (Deere and Company) reports that shelling loss for combine cylinders is also a direct function of cob moisture and levels below 47% W.B. are required for low cylinder loss.

W. H. Johnson, B. J. Lamp, J. E. Henry, and G. E. Hall (Agricultural Engineering Department, Ohio State University) found that after the kernels dry to 35% W.B. there is no further accumulation of dry matter and ear corn harvesting can proceed. In fact, losses from a corn picker are minimum at this moisture level. Such a moisture level is too high, however, for a shelling type of corn harvest. "Invisible losses" occur, partly due to broken kernel tips that remain in the cob and partly due to unexplained dry

and 8.8 t/ha [140 bu/a] crop yield. The plotted lines and the labels show accumulated power requirements. Intervals shown for the header and the threshing, separating, and cleaning mechanisms are essentially their no-load power requirements, which do not vary with forward speed. Both the propulsion and crop processing power requirements increase with forward speed and the resulting increased throughput.

Operations

Good combine operators can keep losses down to 3% by being sensitive to crop conditions and to individual combine component performances.

Slow forward speed can reduce snapping roll losses and losses related to the overloading of the rest of the harvester. Yet the desire for high field capacity urges the operator to go faster.

The moisture condition of the crop is an important factor in optimizing operations. Ear corn may be safely stored in slatted cribs at 18–20% W.B. moisture but shelled corn, as with other grains, cannot be stored for very long above 13% moisture. The material efficiency of combines is best at about 25% kernel moisture; consequently, combined corn must either be dried for storage or sealed in airtight storage if the combine is to operate at its optimum performance point.

W. H. Johnson and B. J. Lamp (Agricultural Engineering Department, Ohio State University) report

matter losses. The proper operating range for combining appears to be from 20% to 30% with 25% an optimum value.

Damage to the shelled corn kernel should be considered in optimum harvesting decisions. Kernels having a cracked seed coat are vulnerable to mold attack. At Ames, Iowa, USDA research has shown that field-shelled corn can often be as much as 30% damaged and these damaged kernels will deteriorate three to four times as fast as undamaged kernels. In addition, the small particles from damaged kernels (called fines) interfere with airflow through stored grain. Damaged grain should be discounted for farm storage as it is for commercial trade. Lamp and Johnson found that cage shellers caused minimum kernel cracking at 12% moisture. The typical combine cylinder had minimum damage at 16% moisture and a significant increase in kernel cracking at both higher and lower moisture levels. The machinery manager can afford to spend some time in adjusting a shelling mechanism to give a damage rate as low as possible for the circumstances.

Determining the material efficiency of corn harvesters is somewhat easier than for small-grain harvesters, as the kernels are larger and more easily seen and picked up. Areas of at least 0.005 ha [0.012 a] should be used for checking ear losses, 0.0005 ha [0.0012 a] for shelled losses.

The calculations are similar to those for small-grain harvesters. One difference is the need to express ear corn and shelled corn on the same basis. For estimation purposes, ear corn will yield 78–80% of its mass as shelled corn.

Example

A 6-row combine is tested for material efficiency in corn with rows spaced at 1 m [3.28 ft]. The loose ear loss, the picker ear loss, and the net yield are measured over a 10-m [32.8-ft] length. The rack and sieves discharge onto a canvas strip measuring 3 m [9.84 ft] long and 2 m [6.56 ft] wide. The shelled loss occurring at the snapping rolls is found beneath the canvas. The following amounts were collected.

Loose ears	0.5 kg	[1.1 lb]
Picker ears	1 kg	[2.2 lb]
Shelled loss	90 g	[0.2 lb]
Cylinder loss	46 g	[0.1 lb]
Separating loss	64 g	[0.14 lb]
Net yield	45 kg	[99 lb]

Loose ear loss

$$83.3\frac{\text{kg}}{\text{ha}} = \frac{0.5\,\text{kg}}{6\,\text{m}\times10\,\text{m}} \times \frac{10{,}000\,\text{m}^2}{\text{ha}}$$

$$\left[1.06\frac{\text{bu}}{\text{a}} = \frac{1.1\,\text{lb}}{32.8\,\text{ft}\times19.68\,\text{ft}} \times \frac{43{,}560\,\text{ft}^2}{1\,\text{a}} \times \frac{1\,\text{bu}}{70\,\text{lb}}\right]$$

Picker ear loss (mathematically double the loose ear loss)

$$166.7\frac{\text{kg}}{\text{ha}} = \frac{83.3\,\text{kg}}{\text{ha}} \times 2$$

$$\left[2.12\frac{\text{bu}}{\text{a}} = \frac{1.06\,\text{bu}}{\text{a}} \times 2\right]$$

Shelled loss

$$150\frac{\text{kg}}{\text{ha}} = \frac{90\,\text{g}}{2\,\text{m}\times3\,\text{m}} \times \frac{10{,}000\,\text{m}^2}{1\,\text{ha}} \times \frac{1\,\text{kg}}{1000\,\text{g}}$$

$$\left[2.4\frac{\text{bu}}{\text{a}} = \frac{0.2\,\text{lb}}{6.56\,\text{ft}\times9.84\,\text{ft}} \times \frac{43{,}560\,\text{ft}^2}{1\,\text{a}} \times \frac{1\,\text{bu}}{56\,\text{lb}}\right]$$

Threshing (or Cylinder) loss

$$26.6\frac{\text{kg}}{\text{ha}} = \frac{46\,\text{kg}}{3\,\text{m}\times6\,\text{m}} \times \frac{10{,}000\,\text{m}^2}{1\,\text{ha}} \times \frac{1\,\text{kg}}{1000\,\text{g}}$$

$$\left[0.4\frac{\text{bu}}{\text{a}} = \frac{0.1\,\text{lb}}{9.84\,\text{ft}\times19.68\,\text{ft}} \times \frac{43{,}560\,\text{ft}^2}{1\,\text{a}} \times \frac{1\,\text{bu}}{56\,\text{lb}}\right]$$

Separating loss

$$35.6\frac{\text{kg}}{\text{ha}} = \frac{64\,\text{g}}{3\,\text{m}\times6\,\text{m}} \times \frac{10{,}000\,\text{m}^2}{1\,\text{ha}} \times \frac{1\,\text{kg}}{1000\,\text{g}}$$

$$\left[0.56\frac{\text{bu}}{\text{a}} = \frac{0.14\,\text{lb}}{9.84\times19.68} \times \frac{43{,}560\,\text{ft}^2}{\text{a}} \times \frac{1\,\text{bu}}{56\,\text{lb}}\right]$$

$$7.5 \frac{t}{ha} = \frac{45\,kg}{6\,m \times 10\,m} \times \frac{10{,}000\,m^2}{1\,ha} \times \frac{1\,t}{1000\,kg} \left[119.3 \frac{bu}{a} = \frac{99\,lb}{19.68\,ft \times 32.8\,ft} \times \frac{43{,}560\,ft^2}{1\,a} \times \frac{1\,bu}{56\,lb} \right]$$

Summary (shelled corn base for kg/ha)

Item	kg/ha [bu/a]	%
Loose ear loss (× 0.78)	64.9 [1.06]	0.83
Picker ear loss (× 0.78)	130.0 [2.12]	1.66
Shelled loss	150.0 [2.40]	190
Threshing loss	25.6 [0.40]	0.32
Separating loss	35.6 [0.56]	0.45
Net yield and material efficiency	7500.0 [119.30]	94.84
Gross yield	7906 [125.84]	100

Tire loading for combines is described as cyclic in that grain tank weight varies from zero to 8200 kN [18,000 lb] for the largest grain tanks. The Tire and Rim Association has developed a special load rating for the tires used in cyclic service (Table 2.2).

Practice Problems

10.1. Sketch the path of: (1) cut material to the cylinder ===, (2) straw - - - - - - - - - - - - , (3) clean grain - x - x - x - x - x, (4) tailings - o - o - o - o - o, and (5) weed seed ————————for a combine in wheat.

Recleaner

Cutter bar

Elevator

Feeder   Grain tank

Weed
seed
sack

Cylinder   Concaves   Grain pan   Elevator

Beater

Clean
grain
auger

Chaffer   Shoe sieve

Grain
return

Straw rack   Chaffer   Tailings
                extension   auger

Straw spreader

Ground Surface

10.2. What combine adjustment is indicated by each of the following conditions?

a. The tailings elevator is carrying a large quantity of material consisting of 80% clean wheat, 15% unthreshed heads, and 5% chaff and small bits of weed stem and straw.

b. The soybeans in the grain tank are clean but about 50% are split.

c. The small amount of material going through the tailings elevator consists of large pieces of unthreshed heads and straw (no threshed grain). What condition might be found in the grain tank?

d. A combine working in oats has a rather small quantity of tailings—about 10–20% clean grain, a small amount of chaff, and the residue in unthreshed heads.

e. A combine working in oats shows a great deal of finely chopped straw both in the clean-grain tank and in the tailings.

10.3. How much loss in kilograms per hectare [bushels per acre] must a $20,000 SP windrower save to break even with its annual costs? The value of the crop is $0.11/kg [$3/bu]. Assume the windrower operates at 7 km/hr [4.375 MPH] and has a 5-m [16.4-ft] width of cut, an 85% field efficiency, a $7/hr operator labor charge, a 10-year life and a $3.50/hr fuel and oil cost. Use an interest rate of 8%. Express the answer in terms of the amount of area covered per year.

10.4. Early work in Ohio indicated that the additional shoe loss for combines on slopes is $0.001, 23 s^4$, where s is the percentage of slope. If a hillside combine is priced at $30,000 more than a prairie or flatland model, above what average slope would it be more economical to own the hillside combine? Assume the gross yield to be 1 t/ha [29.7 bu/a] of $0.11/kg [$3/bu] grain and the annual use each year to be 200 ha [494 a] at a field capacity of 1.5 ha/hr [3.7 a/hr]. The repair costs for the leveling mechanism alone require 0.5% of the purchase price annually for each 100 hr of use. Assume a 10yr. life and a 12% interest rate.

10.5. One combine header design for soybean harvest uses a row crop harvesting head instead of a cutter bar. A comparison field test indicated 80 beans lost per m^2 [8 per ft^2] for the cutter bar and 26 beans per linear m of row [8 beans per ft] for the row harvest head. The row spacing is 60 cm [23.6 in.]. The soybeans have the average mass per Appendix I, and the value of the crop is $0.22/kg[$6/bu]. What dollar savings would the row harvester generate for 200 ha [474 a] annual use?

10.6. Make a table of the mean losses in Table 10.2 expressed as a percentage of total field loss.

10.7. Compare the load rating of an 18.4–38 bias ply tire for a tractor drive wheel with that for a self-propelled combine at a common inflation pressure of 180 kPa [26 psi].

Chapter 11

FORAGE HARVESTING

Forage harvesting problems are different from other harvesting operations in several ways:

1. All forages are intended for animal consumption, usually after preservation.
2. Forage harvesting involves handling a voluminous, low-value material; and
3. Unlike grains forages have special storage problems.

The first point implies that the mechanization of forages should not be limited to field operations but logically extends from the field, through storage, and to the animal. Do not select forage harvesting equipment until you consider the complete mechanization of forage handling.

The second statement indicates that the worker's 100-W body is not economically efficient in handling forages by hand. However, before purchasing forage handling machinery, the individual operation must be evaluated economically. Complete mechanization may be too expensive for small operations.

Finally, it is suggested that both haymaking and ensiling may be thought of as an art rather than a science because of the many uncertainties in each process. The proper selection and operation of forage machines will depend greatly on the eventual answers to such problems as maintaining hay and ensilage quality, or reduction of spoilage and undesirable bacteria activity.

Forage preservation varies widely in different sections of the nation—from open-air haystacks to elaborate air-tight self-feeding silos. The machinery requirements are as varied as the storage methods.

One practice that is gaining prominence is the green feeding of forages. The animals are not allowed to graze pastures, but instead the pasture is mechanically brought to the animal. The most efficient utilization of green forages is realized with this practice.

As world food needs increase, forage harvesting may need to change in character. Instead of planting grass and legume crops specifically for animal feed, agriculture of the future may have to use grain and foodstuff by-products (straw, leaves, stalks, cobs, tops, husks, pods, etc.) as animal feed. If this occurs, the present forage harvesting machinery will have to be adapted to these new forages. Harvesting operations would evolve into a complex system maximizing the profitability of both the food and feed from the crop.

Forage preservation is accomplished by many methods. Classification of these methods is by moisture content, as in Table 11.1. The column labeled "From storage" includes storage losses *and* field losses. Note that minimum dry matter losses occur when using the low-moisture silage method of preservation.

Losses in forage harvesting are substantial but often unappreciated because they are difficult to observe. The machinery manager should be aware of five types of losses.

1. *Cutting losses* are those where whole plants are missed or a high stubble results, usually because of a lodged crop.
2. *Clipping losses* are particles such as leaves and short stems that filter out of machines because of their small size.
3. *Shatter losses* are those leaves lost because of rough handling of the dry plant.
4. *Fermentation losses* occur even during the field operation if the moisture and temperature are right for microbiological growth.
5. *Leaching and bleaching losses* are losses of nutrients due to rainfall and sunshine acting on the cut material while it is still in the field.

These losses may be reduced with skillful operation, adjustment, and scheduling of forage harvesting machinery. The next several sections discuss the details for specific machines.

Mowers

The benefits gained from study of the mowing machine are not limited to that implement alone, but may be applied to all farm machines using a cutter bar: combines, windrowers, and field harvesters.

TABLE 11.1. Forage Classification

Moisture Content % W.B.	Name	Dry Matter, kg/t [lb/T]		% Loss	Equipment Needs
		In field	From storage		
80-70	silage, green chop	200-300 [400-600]	150-240 [300-480]	22	forage harvester, flail harvester, sealed storage with drainage
70-60	wilted silage	300-400 [600-800]	240-340 [480-680]	17	mower, windrower, forage harvester, sealed storage with drainage
60-40	low-moisture silage, haylage	400-600 [800-1200]	350-500 [700-1000]	15	mower, conditioner, windrower, forage harvester, sealed storage
40-20	high-moisture hay	600-800 [1200-1600]	500-650 [1000-1300]	18	mower, conditioner, windrower, forage harvester, baler, dryer, dry storage
20 and less	hay, stover, straw	800+ [1600+]	650+ [1300+]	20+	mower, conditioner, rake, baler, forage harvester, sweep rakes, stackers, dry storage

Mowers are classified by construction as (1) mounted—rear, mid, or forward; (2) semimounted; and (3) trailed.

A mower is essentially a cutter bar with the necessary drive train from the tractor's PTO to the knife drive. Fig. 11.1 shows a rear-mounted mower connected to the tractor through the 3-point hitch.

Mechanisms

The *guard* is the fundamental unit of a cutter bar (Fig. 11.2). Common spacing for guards on a cutter bar is 7.6 cm [3 in.].

A cutter bar may be likened to a series of scissors in which the serrated sections of the knife and the ledger plate sections of the guards act as the two halves of a pair of scissors. The effectiveness of cutting with a pair of scissors or a cutter bar depends on the sharpness of the parts and the clearance between the parts. In the cutter bar this clearance is held to a minimum by the knife clip. The knife, composed of steel triangular-shaped knife sections riveted to a steel strap, terminates in a rounded head at the inner end that is given a reciprocating motion by a crank rod. The knife strap is held forward against the crop by replaceable and adjustable wear plates. These wear plates absorb the thrust of the crop on the knife (Fig. 11.2).

In addition to picking up and guiding the material into the knife, the guards also provide a shield for the knife sections as they come to a complete stop when they reverse direction.

A divider and swathboard are used on the outer end of the cutter bar to clear a path next to the standing crop, allowing the inner shoe a free path on the next swath.

Fig. 11.3 shows a double guard made of steel which has cutting edges on the bottom of the top lip and on the top of the bottom lip. The knife sections are bolted alternately upside down to match the cutting edges of the guard (Fig. 11.4). This design eliminates knife clips and ledger plates. The cut-out portion of the knife sections reduces weight (and consequently power requirements) and helps to clean out crop residues.

Tractor-powered mowers are equipped with slip clutches and hitch release devices. Both of these features are designed to protect the mower parts from breakage if some obstacle enters the guards or catches the cutter bar. The cutter bar is raised and lowered by hydraulic power.

Fig. 11.1. Three-point hitch, rear-mounted cutter bar mower.

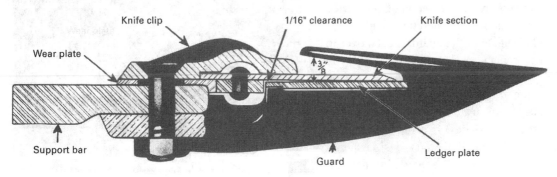

Fig. 11.2. Cross section of cutter bar.

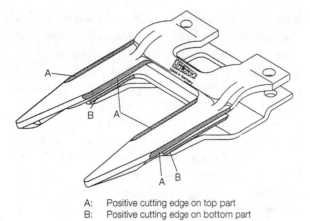

A: Positive cutting edge on top part
B: Positive cutting edge on bottom part

Fig. 11.3. Steel double guard with four cutting surfaces.

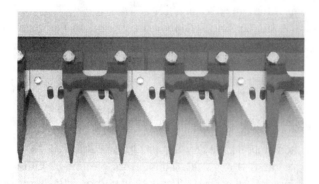

Fig. 11.4. Cutterbar composed of the guards in Fig. 11.3.

Adjustments

One basic adjustment is to vary the cutter bar and knife relationship until the knife sections are exactly centered behind the guards at the end of the stroke. When this condition exists, the mower is said to have correct *register*. Some recent models of mowers do not have centered register. Instead, the stroke is stopped with the tip of the section barely covered by the edge of the guard. On such mowers the stroke does not equal the guard spacing.

Other adjustments are (1) the height of cut, varied by adjustable shoes at the inner and outer ends of the cutter bar; and (2) angle of the guards, nosed over to pick up a down crop and backward to ride over crop residues and rough ground.

The slip clutch, if adjustable, should be loosened to the point that it will just operate without slipping in the heaviest crop growths. If a rubber V-belt is used instead of a slip clutch, the belt tension should not be so tight that it will be unable to slip in emergencies.

The adjustment that the operator has the most control over is forward speed. Cutter bar knife speed is related to PTO speed, but the driver is free to choose a tractor forward speed that can overload the cutting mechanism. All the crop area between two adjacent guards traversed between consecutive strokes is bent forward in a clump before being cut between the knife edge and the ledger plate. Excessive forward speed increases the size of the clump and places stress on the knife, requires more power, and leaves a long stubble for portions of the crop. PTO speed must be kept at rated while forward speed must be limited to the ability of the cutter bar to cut cleanly.

Operations

Cutter bar mowers have not kept pace with other farm machines with regard to increased field capacity. In general, capacities are increased by using wider cuts or faster speeds but in both cases this mower is very definitely limited. Cutter bar lengths have not grown much beyond 3 m [10 ft] with 2-m [7-ft] lengths being the most common. A size barrier seems to have been reached as to the amount of unsupported cutter bar that can be economically used.

Increased knife speeds would increase the field capacity of mowers, but a speed of about 800 cycles/min seems to be the present practical limit. Only balanced knife drives permit even these speeds. Fig. 11.5 pictures a balanced knife drive with linkages that insure straight-line motion.

Maintaining the cutting edges of the knife sections contributes greatly to clean cutting and low

Fig. 11.5. Knife drive linkage to give balanced, straight-line motion from the pitman rod drive.

TABLE 11.2. Effect of Dullness on Cutting Energy

Edge Thickness of Blade, mm	[in.]	Cutting Energy kW·hr/t	[HP·hr/T]
0.05	[0.002]	0.74	[0.9]
0.10	[0.004]	0.94	[1.15]
0.15	[0.006]	1.15	[1.40]
0.20	[0.008]	1.33	[1.62]
0.25	[0.010]	1.44	[1.76]
0.30	[0.012]	1.50	[1.80]

power requirement. W. J. Chancellor (Agricultural Engineering Department, Cornell University) found in laboratory tests that energy requirements varied with the dullness of the section's edge as in Table 11.2. Chancellor also reported that the smooth-edged blades required less energy for cutting than the underserrated blades.

Similar work by R. P. Prince, W. C. Wheeler, and D. A. Fisher (University of Connecticut) indicated that the cutting energy required for alfalfa, relative to a sharp blade, was 2 1/2 times for a medium blade (0.13-mm [0.005-in.] edge) and 3 1/3 times for a dull blade (0.25-mm [0.01-in.] edge).

Rotary Mowers

Rotary mowers, also called rotary cutters and choppers, have replaced cutter bar mowers in many instances. These mowers shear plant materials as a result of high-speed impact of a knife or hammer without need for a shear bar. The rotor may be a single blade rotating in a horizontal plane (Fig. 11.6) or a series of knife-tipped arms rotating in vertical planes (Fig. 11.7). These rotary machines have almost completely replaced cutter bar mowers for weed mowing, brush cutting, and plant residue chopping.

Rotary mowers having multiple rotors (Fig. 11.8) can replace the cutter bar mower in hay fields. They are not as limited in forward speed as the cutter bar mower and are superior in cutting through lodged and tangled crops. A hood gives the operator protection from flying debris. Replaceable swinging knife sections are attached to gear-driven disks. These machines are often called *disc mowers.*

Cutter bar mowers are inherently more efficient than flail mowers (Fig. 11.31) or rotary mowers (Fig. 11.6) for forage harvesting. H. D. Bruhn (Agricultural Engineering Department, University of Wisconsin) measured the amount of clippings (material passing through a 5-cm [2-in.] screen) after various hay cutting operations (Table 11.3).

Fig. 11.6. Horizontally rotating, single-blade mower cutting brush.

Fig. 11.7. Chopping residues with vertical-plane rotary cutter.

Fig. 11.8. Forage mower with multiple rotors.

TABLE 11.3. Clippings from Mowers

Treatment	% of Total Weight
Cutter bar mower	4
Mower followed by:	
crusher at 6.0 km/hr [3.7 MPH]	8
crusher at 4.3 km/hr [2.7 MPH]	12
Flail mower	
at 10-11 km/hr [6-7 MPH]	14
at 6.4-8 km/hr [4-5 MPH]	20
at 4.0-4.8 km/hr [2.5-3 MPH]	30

Although the rotary machines absorb considerably more power than do the cutter bar mowers (Table 2.4), the advantages of simple construction, reduced and easy maintenance, and higher field capacities make the rotary machines attractive. One disadvantage of rotary machines is the ever present chance that solid objects may be picked up and thrown a long distance. Such events damage the machine and present a hazard to bystanders.

Adjustments

In most operations, horizontal rotary mowers should operate in a tipped-forward position. Such reduces the amount of re-cutting of the material and is a power conserving measure. If a mulched material is desired, operating the knife plane horizontally or even tipped back causes the desired re-cutting. The rotary cutter needs to be balanced for smooth operation. During re-sharpening of the blades, attention must be given to preserving equal knife weights.

Rakes

In haymaking, a rake is used to form special windrows that are an important part of field curing. The properly made windrow has the small-stemmed, quick-drying, leafy portions of the plant surrounded by the coarse-stemmed, slow-drying part of the plant. This arrangement encourages uniform drying of the hay crop.

Rakes are also used in gathering materials together for later operations. The form of the windrow is not important for such an operation as raking straw for baling.

Rakes are made in two general forms.

1. The conventional *side-delivery rake* uses an angled, spring-toothed beater to sweep the hay into a windrow. This type is normally four barred and ground wheel driven. A special bar positioning mechanism is needed to keep the bar teeth always pointed down.

Some studies indicate that the distance traveled by the hay to a windrow is directly related to the amount of leaf loss. Therefore, a more sharply angled rake has been developed, the *oblique reel head rake*. The distance traveled by the hay to get to the windrow is less than that for the previous design. Figure 11.9 shows two oblique reel head rakes operating in parallel to produce a large windrow for large balers.

2. A second type of rake is the *finger wheel rake*, which consists of several large, spring-toothed wheels in contact with the ground surface and arranged in echelon. As the rake is moved forward the wheels turn, dragging the hay to one side and forming a windrow. This rake works very well on rough ground, as each wheel is independently mounted and free to follow the soil contour (Fig. 11.10).

The wheel rake is sometimes front mounted. Other types of rakes may be trailed, semimounted, or completely mounted.

Rakes are driven by ground wheels, PTO, or by hydraulic motors, as in Fig. 11.9.

Fig. 11.9. Left- and right-hand hydraulic-powered rakes producing a windrow for large balers.

Fig. 11.10. Finger wheel rake.

Adjustments

Rakes have only limited provisions for adjustment. In the side-delivery rake the reel height and the angle of the teeth are important. The reel should be set to just clear the ground surface. The angle of the teeth determines the form of the windrow—tightly roped or loose and fluffy. A compromise must be made in the windrow form: for quick drying the windrow should be loose; for even feeding into balers or other harvesters, the windrow should have continuity and be tightly roped.

Operations

The direction of travel of the side-delivery rake should be the same as the preceding mowing operation. The top parts of the plant ideally fall backward over the mower's cutter bar and lie pointing in a direction opposite to mower travel. The rake bars then reach the top parts of the plant first and roll them into the center portion of the windrow. The coarser, slower drying parts of the plant are placed in the outer layer of the windrow where they are subject to the direct rays of the sun and to the most airflow. The optimum windrow rests on raked ground, not on unraked hay.

For leafy plants such as alfalfa and clovers, raking should be completed before the moisture content of the whole plant drops below 40%, the approximate point where leaf shatter occurs.

The wheel rake should be used only in clean fields, as any previous crop residues will be placed in the windrow along with the hay.

The timeliness aspects of raking are most important as revealed in research work by J. B. Dobie, J. R. Goss, R. A. Kepner, J. H. Meyer, and L. G. Jones (Agricultural Engineering Department, University of California). Hay was deliberately raked at a moisture level between 10% and 15% and compared with hay raked between 40% and 50%. Hay left in the swath dried at a faster initial rate than that in the windrow, but by baling time there was no significant difference in moisture content. Baling at two moisture levels was also included in the study. Results are given in Table 11.4. The losses were even greater when measured on a crude protein basis.

Rain is a risk to field drying hay. Even with careful planning, cut forage may get wet before it can be made into hay. *Tedding* is the operation that fluffs damp hay to encourage faster air drying. Some side-delivery rakes can be operated with reverse rotation to provide a tedding action. Other machines are designed

TABLE 11.4. Hay Yields with Different Practices (moistures given as % W.B.)

Raked at	Baled at	Yield kg [lb]	%
40%	15%	1330 [2920]	100
40%	6%	1270 [2800]	96
12%	15%	990 [2170]	74
12%	6%	865 [1900]	65

to ted only. Fig. 11.11 shows a tedder operating on windrows and on hay left as a mower swath. A windrow inverter (Fig. 11.12) is effective for gently overturning windrowed hay to dry out the damp bottom portion.

Hay Conditioners

If forage is to be field cured for hay, several days may be required before the sun and wind action reduces the moisture content to a safe 20%. This time may be accelerated by the use of a *hay conditioner* as shown operating in the photo at the beginning of the chapter. Conditioning refers to the mechanical rupturing of the plant stems to accelerate air drying. The

Fig. 11.11. Hay tedder.

Fig. 11.12. Windrow inverter.

hay conditioner machine is in effect a combination mower and rake which prepares a rapid-drying windrow. PAMI tests show that in 18 hours an alfalfa-brome mixture was reduced 15 moisture points beyond that for mowing alone.

Hay conditioning mechanisms differ in detail but all produce a crushing, abrading effect on the plant material. Fig. 11.13 illustrates the roller-crusher mechanism used behind a cutter bar. A similar arrangement uses a pair of engaging toothed rolls to break, or crimp, the stems rather than crush them. Figure 11.14 shows a flail mechanism behind a disc mower.

These machines may be PTO-operated, pull-type or self-propelled. The self-propelled hay machine is a windrower with crushing attachment (Fig. 11.15).

Different mechanisms produce varying drying rates and legume leaf loss. Work done in Michigan by C. A. Rotz and D. J. Sprott (U.S. Department of Agriculture) found the fastest drying rate from flail machines operating with a thin material flow. But flail machines had a 6.2% loss of leaves, which was about double that for other designs. Table 11.5 reports leaf loss for the several conditioner designs.

Adjustments

Work done on alfalfa by G. E. Fairbanks and G. E. Thierstein (Agricultural Engineering Department, Kansas State University) shows that the fastest drying rate is with the roll-type conditioner following a conventional mower (Fig. 11.16). Any machine that

TABLE 11.5. Haymaking Losses, Dry Matter

Equipment	Loss, % of Total
Conditioner and raking loss (by design)	
Disc mower and rolls	4.70
Disc mower and flail	6.43
Cutterbar and rolls	3.95
Baler Losses	
Pickup	2.17
Bale chamber	
Rectangular	2.79
Round, variable chamber	3.83
Round, fixed chamber	10.89

Fig. 11.15. Self-propelled windrowers.

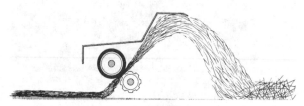

Fig. 11.13. Hay crushing principle.

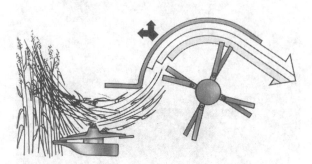

Fig. 11.14. Hay conditioning using a disc mower and flails.

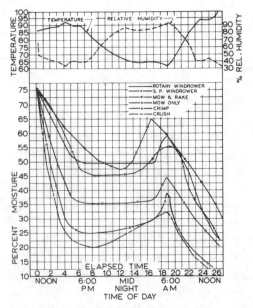

Fig. 11.16. Drying rates for various hay conditioning treatments.

windrows the material while conditioning it seems to lose any drying time advantage over unconditioned hay lying in a mower swath.

Roll-type conditioners have adjustment provisions for varying the pressure between the rolls, height of operation, and sometimes the spacing between the rolls. In general, the greater the roll pressure the faster the drying rate. The height of operation should be as high as necessary to avoid stones and other debris, yet low enough to pick up all the cut material.

Forward speed can be as fast as the mowing mechanism will allow.

Balers

Balers are the most popular of today's haymaking machinery. It is estimated that about 90% of the nation's hay is baled. The packaged hay lends itself to many varied types of hay storage. Baled hay is the major type of hay considered on the commercial market. In addition to haymaking, balers are often used for packaging crop residues, especially small-grain straw. Also, balers can process green forage as silage bales.

Rectangular Balers

Balers producing rectangular bales are designed as self-propelled machines (Fig. 11.17) and as PTO-operated pull machines (Fig. 11.18, 11.19). These machines produce bales weighing up to 36 kg [80 lb] that can be handled by man. Larger bales weighing up to 600 kg [1320 lb] are produced by the baler Fig. 11.20 and are intended for off-farm market. Mechanical handling is required for these bales. Balers may be designed to pick up windrows to one side of the tractor (Fig. 11.18) or to pick up the windrow straddled by the tractor (Fig. 11.19). The latter design provides a straight-through material flow.

Mechanisms

The baler includes a pickup device to take windrowed hay to the baling chamber. This pickup may be either PTO or ground driven. In some instances a precompression chamber is used as a preliminary

Fig. 11.17. Self-propelled baler.

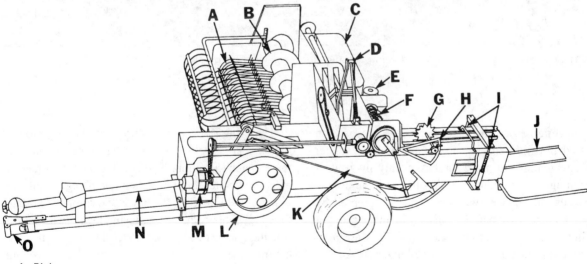

A. Pickup	F. Knotter	K. Compression chamber
B. Feeding auger	G. Star wheel	L. Flywheel
C. Twine box	H. Metering arm	M. Slip clutch
D. Wad board	I. Bale density cylinder	N. PTO shaft
E. Hydraulic density control	J. Bale chute	O. Hitch

Fig. 11.18. Components of a PTO-driven, side-feed, rectangular baler.

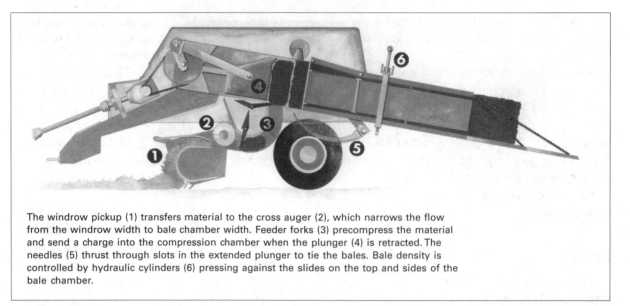

The windrow pickup (1) transfers material to the cross auger (2), which narrows the flow from the windrow width to bale chamber width. Feeder forks (3) precompress the material and send a charge into the compression chamber when the plunger (4) is retracted. The needles (5) thrust through slots in the extended plunger to tie the bales. Bale density is controlled by hydraulic cylinders (6) pressing against the slides on the top and sides of the bale chamber.

Fig. 11.19. Flow through a bottom-feed rectangular baler.

Fig. 11.20. Large rectangular baler and trailing accumulator for bales as large as 600 kg [1320 lb].

compacting device. In many balers the hay is then stuffed into the compression chamber by a feeder fork, which should place the hay evenly in front of the plunger. The plunger presses the hay against the previous slug of hay, which is held compressed by hay retainers. Most balers have a knife on the plunger acting against a stationary shear bar. This arrangement gives a clean slice for each plunger stroke.

One of the chief functions of a baler is the tying of the bales with either twine or wire. A star wheel, whose teeth penetrate the bales, releases the tying mechanism clutch after rotating a predetermined amount to give a definite bale length.

The tying cycle occurs during the short period of time the hay is held compressed when the plunger is in the extended position. The needles must be timed to enter the knotter through slots in the plunger while the plunger is still holding the hay in the compressed state. Fig. 11.21 shows the needle laying lower twine alongside the upper twine in the knotter. The knot is tied and cut from the strands retained in the twine holder. As the needles retract a second knot is tied to give two knots in each upper side strand of twine around the bale.

A wire baler has a tying device that twists the wires together instead of forming a knot. A twister mechanism design is shown in Fig. 11.22.

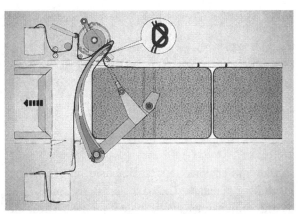

Fig. 11.21. Twine tying action for a baler.

Fig. 11.22. Loop-type wire twister. The wire ends are looped over a rotating finger (1) which makes the twist. This looped twist (2) is pulled off the finger by the advance of the bale.

Adjustments

The forward speed of the baler is perhaps the greatest responsibility of the operator. A baler is more sensitive to the rate of feeding than any other implement. Smooth, continuous, even feeding of hay into the bale chamber will result in firm rectangular bales. The form of the bale depends on the width and thickness of the windrow.

Bale density is controlled by compressing the movable sides of the bale chamber. Manually adjusted spring force or hydraulic force is used. The hydraulic control can be designed to give automatic control over bale density. The star wheel (G, Fig. 11.18) senses bale density and operates a valve to either relax or increase the pressure on the cylinders (I, Fig. 11.18) that constrict bale passage.

As with all cutting mechanisms, the knife and shear bar must be kept sharp and have limited clearance for neat, low-power cutting. Most shear bars and knives are replaceable. The clearance may be adjusted by the adjustable shear bar or by adjustable plunger guides.

Constant checks should be made on the needle action. Broken needles are sure to result if either their alignment or their timing is faulty.

Standards for baler twine have been established by the ASAE to promote uniform and reliable performance. The machinery manager should check the specifications of any purchased twine. Table 11.6 shows the required minimum strengths for the four classifications of twine or any substitute material that may be used.

Tension is equally important for the wire twisting mechanism. Additionally, the hardness of the wire must be annealed to acquire a pliability and toughness suitable for bale tying. The ASAE standard calls for tensile strengths of 345 to 483 MPa [50,000 to

TABLE 11.6. Twine for Automatic Balers

Classification	Min. Knot Strength		Min. Tensile Strength	
	N	[lb]	N	[lb]
Light	222	[50]	445	[100]
Medium	334	[75]	667	[150]
Heavy	445	[100]	890	[200]
Extra heavy	556	[125]	1117	[250]
Ball Dimension Categories	Height, mm	[in.]	Diameter, mm	[in.]
I	254 ± 13	[10.0 ± 0.5]	254 + 0 -25	[10+0-1]
II	305 ± 13	[12.0 ± 0.5]	286 + 0 - 25	[11.25+0-1]

Source: ASAE Standard S315.2.

70,000 psi] with a 12% permanent elongation in 254 mm [10 in.] length. Table 11.7 lists the dimension standards.

As with all technical specifications, the machine operator's manual should be consulted.

Operations

Handling hay in bales still involves a great deal of hand labor but progress is being made in reducing this labor. Fig. 11.23 illustrates a baler equipped with a *bale thrower*, which randomly fills a wagon and eliminates the need for hand-stacking the bales. Fig. 11.20 shows the use of a bale accumulator. Instead of randomly dropping the bales all over the field, as many as four can be carried and dropped together at a convenient site.

The importance of baling legumes at the right time is reported in Table 11.4. Leaf loss is minimized by baling at the highest moisture content consistent with safe storage. In dry areas, nighttime baling is practiced to take advantage of the more humid conditions.

The forward speed of the baler should be inversely proportional to the windrow size to keep the feed rate at a constant level. The proper feed rate is the amount that permits the feeding mechanism to present a uniform quantity of hay in front of the plunger before each stroke, resulting in compressed slugs of uniform thickness. If the feed rate is not related to the feeder mechanism's adjustment, the individual slugs will be thicker on one side. The bale will bow as it leaves the compression chamber and, if extreme, will slip its ties and have to be rebaled.

Fig. 11.23. Bale thrower saves labor of loading wagon.

TABLE 11.7. Wire for Automatic Balers (14¹/₂ gage, annealed)

Length per Coil,		Coil Dimensions, mm [in.]		
m [ft]		Outside	Inside	Width
960 [3150]		251 [9.875]	76.2 [3]	95 [3.75]
1981 [6500]		336 [13.25]	206 [8.125]	152 [6]

Source: ASAE Standard S229.6.
Note: Dimension tolerances are ± 12.5 mm [± 0.5 in.]

Bale dimensions and densities are important to a machinery manager. The bale's cross section is fixed by the dimensions of the bale chamber. These dimensions are quoted to indicate the size of the baler. The most popular size is 0.356 m × 0.457 m [14 × 18 in.] but other sizes are available. The theoretical length of a bale can be adjusted by the operator, but actual length can be from 92.5 to 108% of the theoretical length because of variation in feed rate from light to heavy. Bale densities depend on material baled and adjustment of density control. In PAMI tests, alfalfa hay densities averaged 207 kg/m³ [12.9 lb/ft³] while wheat straw averaged 114 kg/m³ [7.1 lb/ft³].

Losses occur at pickup and in the baling chamber. In good conditions (heavy windrows, high moisture content), total alfalfa leaf loss should be less than 4%, but it can be as high as 5% at the bale chamber with overdry hay.

Typical field capacity is 170 bales/hr. Throughputs expected are 8–10 t/hr [8.8–11 T/hr] for alfalfa, 2–4 t/hr [2.2–4.4 T/hr] for the lighter wheat straw.

See Fig. 1.22 for typical PTO requirements for different feed rates. PAMI recommends a tractor size with double the average PTO requirement to operate balers on soft and hilly fields.

Round Balers

Round balers roll a windrow into a large cylinder (Fig. 11.24). These machines are the core of a one-man, low-cost, hay harvest system.

The bales may be up to 1.7 m [5.5 ft] in length and have variable diameters up to 2.0 m [6.5 ft]. They may weigh over 900 kg [2000 lb] and can be wrapped for several revolutions by twine or for fewer revolutions with a full-width plastic net (Fig. 11.25). These bales can shed moderate amounts of rainfall and may be left in the field until more urgent operations have been completed. The bales can be fed directly to livestock in the field, transported and stored with tractor power (Fig. 11.26), placed in pipe racks in feed lots, or processed into chopped hay with tub grinders (Fig. 11.27).

Fig. 11.24. Round baler ready to expel bale.

Fig. 11.26. Large round bales are moved with a tractor loader.

Fig. 11.25. A plastic net may be used instead of twine for wrapping round bales.

Fig. 11.27. Round bale processor shreds hay into feed bunks.

Mechanisms

There have been two designs used for round balers. Fig. 11.28A shows the formation of the bale for the pre-compression, variable-volume design. The hay is lifted, passed through compression rolls, and fed onto the rolling bale held between a series of rotating flat belts. Spring-loaded idlers resist the increase in bale diameter to produce a dense bale. Upon accumulating enough material, the baler's forward motion is stopped and the bale is expelled to the rear as hydraulic cylinders raise the rear gate.

Fig. 11.28B shows the formation of the bale in the open-throat, constant-volume design. As opposed to the variable bale chamber design, the series of flat belts, chains, or roller sets define a fixed volume.

The bale starts with the hay tumbling loosely in the chamber. When the chamber fills, the compression occurs from the outside to the inside. The bale has a characteristic star-shaped pattern in the less dense center and a high density at the circumference.

Measurements at PAMI show a range in throughput from 5 to 13 t/hr [6 to 14 T/hr]. Densities range from 15 to 22 kg/m^3 [9.4 to 14.3 lb/ft^3]. Alfalfa leaf loss can be as high as 7% at pickup and 25% in the baling chamber in overdry hay. Under optimum conditions, total losses can be as low as 5% when working with heavy windrows of conditioned hay at the maximum moisture content that allows safe storage. Bale chamber loss can be minimized with high feed rates that reduce the time a bale is in the bale chamber.

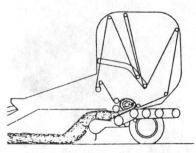

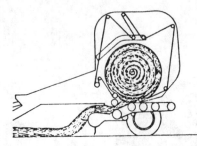

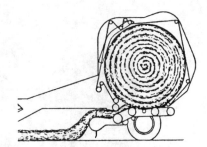

A. Variable-volume compression with pre-compression rolls.

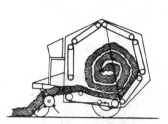

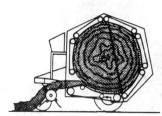

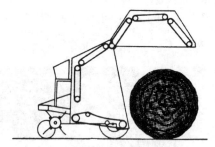

B. Open-throat, constant-volume design.

Fig. 11.28. Designs for round balers.

Even-feeding of the windrow across the complete width of the bale is important to produce a cylindrical bale. The ideal windrow would be of uniform depth and exactly as wide as the bale chamber. Lacking such, the operator must weave back and forth over the narrow windrow to build the bale evenly. The bale monitor in Fig. 3.8 senses the bale formation and advises the tractor driver to steer right or left to form the more uniform bale required for structural integrity.

The power recommended to operate round balers is quite variable. PAMI recommends a maximum operating load factor of only 1/3 because of wide fluctuations in the total power required. The PTO power requirements at initiation of the bale can be as low as 5 kW, but as the bale mass increases the power needed rises to as much as 20 kW for a 900-kg bale. Of course, the rolling resistance of the baler also rises as the bale accumulates mass.

In tests by R. G. Koegel, R. J. Straub, and R. P. Walgenbach of the University of Wisconsin (Table 11.5), significant differences were found in dry matter loss from the various baling systems. Total system losses ranged from 6.1% to 27.1% of total dry matter. An average of 15% dry matter loss translates to about 30% of the nutrient loss since most of the loss for alfalfa is leaves. Cutter bar and roll conditioners and rectangular balers had the least losses.

Forage Harvesters

The field forage harvester permits bulk handling of forage crops, thus replacing the more laborious unit methods.

Forages were originally chopped for making ensilage because in a chopped form the air could easily be excluded by packing. This reason for chopping is still valid but another reason has been added. Chopping forages transforms them into a homogeneous material that can be handled mechanically, even to the point where they may be "untouched by human hands."

Forage harvesters may be trailed, semimounted or self-propelled machines. Figure 3.7 shows the harvest of windrowed hay by a self-propelled forage harvester. Fig. 11.29 shows a PTO-driven, pull-type machine equipped with each of the three interchangeable heads used in forage harvesting.

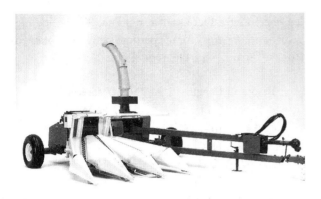

A. Row crop head.

B. Cutter bar head.

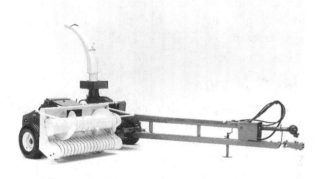

C. Windrow pickup head.

Fig. 11.29. Forage harvester with three interchangeable heads.

Mechanisms

Fig. 11.30 shows the chopping cylinder that rotates against a close-clearance shear bar. The cylinder both cuts and throws the chopped material into a trailing wagon. Any of the three heads shown in Fig. 11.29 are used to feed this cylinder.

Row crop attachments for field harvesters use gathering chains to feed the stalks into a cutter bar and then into the feed rolls. The tops of the stalks are restrained, thus row crops are fed into the chopping mechanism butt end first.

Many harvesters have a clutch and reversing mechanism in the feed roll and conveyor power drive that is especially useful in preventing the chopper from becoming choked. Modern instrumentation can detect tramp metal and automatically disengage the clutch.

Adjustments

As in all cutting mechanisms, two very important maintenance measures are to keep the units sharp and the clearance between the two parts at a minimum. The harvesters have a replaceable shear bar; in some cases different sides of the bar may be presented as a cutting edge. A sharpening device mounted on the harvester can be applied to sharpen the cylinder in place while using tractor power for rotation. Modern design and instrumentation allow for sharpening the cutter and adjusting the clearance from the operator's cab. Sharpening after only a few hours of use can save significant amounts of energy. New Holland reports a blunt edge on the knife doubles the power requirement at the same feed rate, a clearance as much as 0.6 mm [0.02 in.] will double the power requirement even with a sharp knife, and the combination of that clearance and a blunt knife will triple the power required.

Length of cut is adjusted by changing the feed rate per knife cut. A variable-speed gearbox in the drive from the cylinder to the feed rolls provides changes in the gear ratio to produce changes in length of cut. Dynamic balance of the cutter must be maintained.

Two very important functions for the operator are to adjust the forward speed and watch for rocks and

Fig. 11.30. Cylinder or helical knife cutter.

tramp metal. Forward speed determines the through-put. High performance and fuel efficiency are obtained by keeping forward travel as fast as power availability will allow. A variable speed, hydrostatic ground wheel drive is particularly useful for adjusting forward speed to its optimum. Rocks damage cylinder knives and can be stopped at the feeder mechanism by a quick reversing device controlled by the operator. When metal detectors are used, the feed rolls are automatically stopped before the metal damages the cutter. The operator is alerted, reverses the feed mechanism, and clears the metal from the feeder.

Operations

Chopping forages is one of the most power-consuming operations on a farm. The machinery manager must be assured that such power expenditure is economical. The greatest contributors to the high-power requirement are the speed of the cutter mechanism and the shortness of cut. Manufacturers have slowed cutter speeds and increased the number of knives on the cutters to save power yet retain capacity. If the choice arises, the machinery manager should change the feed ratio to get a specific length of cut at a slower speed rather than remove knives to get the same cut at a higher speed.

A good management rule for forage harvesters is to keep the length of cut as long as possible. Unfortunately, finely chopped material is superior to coarse because it packs more tightly for good silage making, is easier for most types of conveyors to handle, and is not sorted out when fed to livestock. Chopped hay need not be cut finely for packing, but it must be fine enough to pass the other two requirements. PAMI tests of several forage harvesters show that doubling the length of cut increased capacity per energy input by as much as 14% for corn and 35% for alfalfa. These tests were conducted over a range in moisture contents from 30 to 40% W.B. The effect of moisture on dry matter capacity, t/(kW·hr) [T/HP·hr], was minor.

Recutter screens are sometimes fitted beneath chopping cylinders to ensure a small length of cut by not permitting large particles to pass the screen until they have been reduced. Institute tests showed that energy required per tonne of dry matter increased from 9 to 50%, depending on the design of the screen.

The theoretical cut will match the actual cut on corn and sorghum crops that are fed in perpendicular to the cutter, but such is not the case with grass and legume forages that enter at random. The following distribution of grass forage particle sizes was obtained from a 5-cm [2-in.] theoretical cut:

Range in Length, mm	[in.]	% of Material by Weight
0–25	[0–1]	41
25–50	[1–2]	31
50–75	[2–3]	15
75–100	[3–4]	9
over 100	[over 4]	4

Flail-Type Harvesters

Flail-type forage harvesters are machines of simplified design that chop the forage as it stands without requiring preliminary operations. Fig. 11.31 shows a flail machine working in alfalfa. The flailed forage is thrown into an auger, delivered to a blower, and deposited in the trailing wagon.

On some designs the flail mechanism is used as a pickup head. The prechopped material is then delivered to the knife-equipped blower that further reduces the particle size. In general, the particle size or length of cut of the flail-type harvester is longer and more irregular than the conventional shear-bar forage harvester.

Hay Wafering

Hay wafering is another forage processing operation that mechanizes the handling of forage feeds. Hay wafering machines first chop the hay and then

Fig. 11.31. Flail forage harvester.

force the material through converging dies. The hay is compacted into small chunks or cubes as it passes through the dies. The resulting chunks are then handled as a free-flowing material with conveyors and dump wagons. Wafering is commonly done with stationary machines.

Wafering hay has the advantage of creating a dense, free-flowing material from one that was bulky and hard to handle. It has the disadvantages of high power requirements and high investment costs. An operational requirement is that the hay must be very dry for the formation of durable wafers.

Stacking

Hay can be stored in stacks without shelter if care is taken to develop a thatched layer on top to shed rainfall. Machines are available to take the hand labor out of building haystacks. Fig. 11.32 illustrates such a machine depositing a stack. This tractor-pulled trailer is equipped with a flail chopper at the front to pick up windrowed hay. The spout oscillates to spread hay evenly over the trailer bed. Periodically the operator uses hydraulic pressure to compress the stack by lowering the trailer canopy. The completed stack is unloaded to the rear by tipping the trailer and driving slowly forward at the same speed as that of the unloading conveyor. Stacks as long as 6 m [20 ft] and weighing 6 t [6.6 T] are possible with these machines.

Forage Handling Equipment

After forages have been cut, raked, and processed, the forage handling equipment problem arises.

The simplest equipment requirement occurs where hay is stacked. Hay is handled in loose form by large sweep rakes of different types. Sweep rakes may be used to bring the hay to a stacker that lifts the hay onto the stack, or a stacking sweep rake may be used to push or drop the hay directly onto the top of the stack.

Machines have been developed for handling complete haystacks. Fig. 11.33 is a tilt-trailer with a conveyor in its bed. It can be forced under a stack such as that formed in Fig. 11.32. The stack can thus be transported and unloaded as a whole; or the hydraulically driven feeder at the trailer front can chop off slices of the stack and unload them into grinders or other conveyors or onto the ground for feeding range cattle.

After field-baling hay, several handling operations remain. If the bales are scattered over the field they must be collected, loaded, hauled to storage, and unloaded. Bale elevators are commonly used to unload bales into a mow (Fig. 11.34).

A special-purpose, self-propelled wagon has been developed to reduce labor requirements for picking up bales (Fig. 11.35). The bales are elevated, arranged in tiers, and stacked on the tilting bed. Each tier

Fig. 11.32. Mechanical stack maker discharging well-formed stack.

Fig. 11.33. Stack mover and feeder.

Fig. 11.34. Bale elevator.

Fig. 11.35. Self-propelled, single-operator bale harvester.

formed forces the previous one down the slope of the tilting bed. At the storage point the bed tilts and deposits a stack of bales (Fig. 11.36).

Large round bales can be moved with tractor loaders (Fig. 11.26) or with a *bale mover* (Fig. 11.37). The tractor operator can load bales without stopping forward motion. The mover is swung hydraulically to one side of the tractor to allow the front of the mover a clear path to the bale. A two-prong forklift is lowered to the ground and slid under the bale. The forks rotate, plac-

ing the bale on the carrier rails in an upright position. A hydraulically driven lugged chain conveyor moves the bales along the rails to make room for loading the next bale. The conveyor can be reversed to unload the bales into their original orientation—flat (settled) side down—for better weather resistance.

Equipment for handling the material from a forage harvester requires considerable investment since it is practically impossible to move chopped forages effectively by hand. Special forage boxes are manufactured to mount either on trucks or wagon running gears. These boxes can be loaded with minimum effort by controlling the position of the deflector on the forage harvester's discharge pipe. Forage boxes can be unloaded into feed bunks (Fig. 13.5) or into a blower (Fig. 11.38).

Fig. 11.38. Powered unloading wagon emptying into PTO forage blower.

Fig. 11.36 Stacking bales with tractor-trailed bale harvester.

Fig. 11.37. Five-unit bale mover for transporting large round bales.

A forage blower is a part of the chopped harvest system. This machine delivers forage into silos and other storage structures. Compared to mechanical conveyors, blowers are power-consuming; but they have the high capacity needed in a forage harvest system.

Blowers depend more on throwing the material than on blowing it. Close clearance between the blower paddles and the enclosing housing is necessary for efficient operation. Provisions are made for moving the housing closer to the paddles to compensate for wear.

Practice Problems

11.1. Calculate the required forward speed of a PTO powered mower having a 2 to 3 speed increase from the PTO drive shaft to the knife drive. Each stroke of the knife section cuts a 5-cm [2-in.] forward slice. The PTO operates at 540 rpm.

11.2. Determine the forward speed of a 1.8-m [6-ft] single blade rotary mower to give a 2-cm [3/4-in.] forward cut. The 540 rpm PTO drives the blade through a 1:1.5 bevel gear box.

11.3. Calculate the longest distance hay is moved to a windrow if the oblique reel head drive moves the teeth perpendicular to forward travel. The effective width of the rake is 2 m [6.6 ft] and the forward speed is 8 km/h [5 MPH]. The tip of the teeth travel in a 1-m [3.28-ft] diameter circle. The reel is turned by a hydraulic motor at 80 rpm.

11.4. Bales of hay weighing 30 kg [66 lb] each are elevated into a haymow door 10 m [32.8 ft] high at the rate of 4/min. If the elevator has a mechanical efficiency of 15% (ratio of power useful in raising hay to power put in elevator), what is the required power input to the elevator?

11.5. A blower may work well in dry hay or wet ensilage but have poor performance in wilted grass and legume forages (sometimes called haylage) because the moisture content makes the for-

ages limp yet tough. What wear-related adjustment should be checked if an aged blower performs poorly in haylage?

11.6. In double windrowed hay 4-m [13.12-ft] windrow spacing, a baler is reported to have a capacity of 10 t [11T] of hay bales per hour.

a. If the field yields 4.5 t/ha [2 T/a], what forward speed would be required at an 80% field efficiency to attain the reported capacity?

b. If the plunger speed is 1.5 strokes/s, what would be the mass of the average slug?

11.7. Curved bales are sometimes ejected from a rectangular baler because the material on the convex side of the bale is more dense than the material on the concave side. In what way is the baler malfunctioning and what adjustments are indicated?

11.8. A 6-knife forage harvester can provide theoretical cuts of 5 mm [0.2 in.], 10 mm [0.4 in.], 20 mm [0.8 in.], and 40 mm [1.6 in.] by the use of gear sets to give a proper gear ratio between the cylinder and feed roll shafts. The feed rolls are 200 mm [8 in.] in diameter. Assume no slippage as the feed rolls drive the forage into the cylinder. What are the gear ratios that provide the four cuts?

Chapter 12

FARM PROCESSING

Not all crop work is done in the field. Many times various processing operations are required before the crop may be fed or stored. Processing and handling these crop materials are distinct farm mechanization problems.

Processing machinery may be loosely classified as *stationary* and *portable*. The stationary types are designed primarily for large, permanent installations. This type of machine must be purchased with the overall materials handling system well in mind. These machines lend themselves well to electric power. The portable types of machines are designed for more diversified farm operations. In most cases they are driven by a tractor.

This chapter on farm processing machinery is limited to those subjects of most interest: reducing machinery and drying machinery.

Reducing Machinery

Farm crop reducing machines may be classified as

1. *Hammer mills*: Reduction caused by impact.
2. *Attrition mills*: (commonly called burr mills): Reduction caused by twisting pressure.
3. *Roller mills*: Reduction caused by simple normal pressure.
4. *Cutter mills*: Reduction caused by cutting.

Various combinations of these four principal types may be built into one machine. The combination mill is a cutter head and either a burr mill or a hammer mill placed in series. These mills are capable of handling forages as well as grains.

The power requirements for these mills are related to the size of the end product—the finer the particle size, the more power required.

Mechanisms

Hammer mills may use either free-swinging or fixed hammers. The rotor speed must be kept relatively high (2500–3500 rpm) to produce pulverization. A screen that determines particle size is placed below the rotor (Fig. 12.1).

Attrition mills consist of two hard-surfaced circular plates rotating with relative motion. The material is reduced as it passes between the two plates. The common burr mill has one fixed plate and the other rotates (Fig. 12.2).

A roller mill is a very simple reducing machine. Two rollers, spaced with a small clearance, crush the material as it passes between them. The rolls may be grooved; then the machine is called a *crimper-roller* (Fig. 12.3).

The cutter mill mechanism is similar to that in the field harvester. The cutter mill alone will not produce a very small particle but is best used to reduce forages to a size that other reducing mechanisms, such as the combination mill, can handle (Fig. 12.4).

Mills are fed by gravity or by conveyor. Several different ways of taking the processed product away from the machine are drag chain, auger, gravity, or blower. A *cyclone separator* is needed if a blower is used. The blower floats the processed material in an airstream. The cyclone separator is employed to separate the heavier feed particles from the air by centrifugal force. The "cyclone" is formed by the air as it is exhausted from the top of the separator while the feed particles settle to the bottom of the hopper. The principle is illustrated in Fig. 12.5.

Adjustments

In all cases, the particle size adjustment is the most important. In selecting a reducing machine, a buyer should be sure that the reduction mechanisms are easily replaced or sharpened. Grains and especially corn cobs have a definite abrasive effect even on the hardest of chilled cast irons.

The particle size is varied for each particular machine in the following manner.

1. *Hammer mill*: Vary the size of the screen openings.
2. *Burr mill*: Vary spring pressure on the burrs.

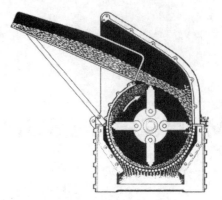

Fig. 12.1. Hammer mill principle.

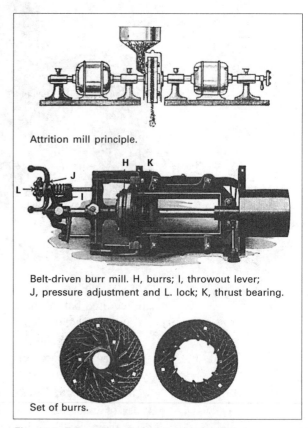

Attrition mill principle.

Belt-driven burr mill. H, burrs; I, throwout lever; J, pressure adjustment and L. lock; K, thrust bearing.

Set of burrs.

Fig. 12.2. Burr mill reducing mechanisms.

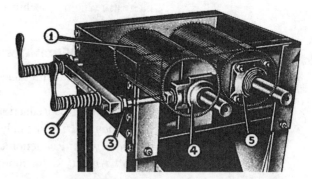

Fig. 12.3. Crimper-roller mill. 1, grooved roller; 2, spring pressure for movable roller; 3, clearance adjustment; 4, sliding bearing mount; 5, stationary bearing mount.

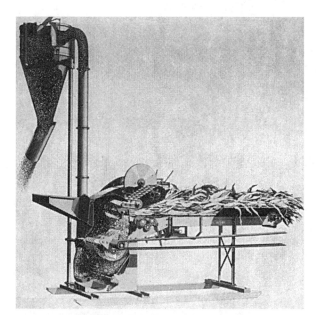

Fig. 12.4. Combination mill.

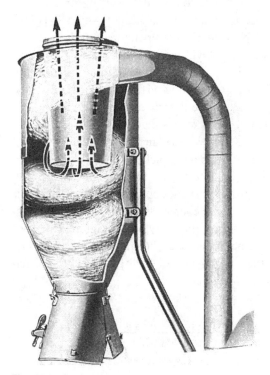

Fig. 12.5. Cyclone separator.

3. *Roller mill*: Vary opening between rolls.
4. *Cutter mill*: Vary the rate of feed of material with respect to the cutter rpm.

Operations

Reduction of grain requires energy and is an expense that must be more than recovered by beneficial results. Increased utilization of the grain (up to a point)

by milk cows was demonstrated by T. E. Heinton (Purdue University).

Degree of grinding of corn and oats	Profit
Whole	$54.23
Cracked and crushed	$65.03
Medium ground	$77.06
Fine ground	$54.00

The required capacity for reduction machinery depends on the design of the system for feeding livestock and on the worth of the operator's labor. Two extreme positions exist. One is that feed grinding can be an unattended low-capacity operation and the time required for grinding is unimportant. The physical size and the capacity of the machinery can be small and inexpensive. The costs are primarily for controls and the permanent installation. The other is that obsolescence should be avoided by keeping the feeding system flexible and nonpermanent; consequently, the reducing machinery should be portable, be tractor operated, and have high capacity to keep the operating time to a minimum. A portable grinder mixer is shown at the beginning of the chapter.

The power requirements for reducing machines depend on the feed rate and the fineness of grind. Table 12.1 lists the energy requirement independent of feed rate for mills. Fine grinding requires more power than coarse grinding; moist grain requires more energy than dry grain. Roller mills require slightly more power than burr mills.

TABLE 12.1. Energy Requirements for Feed Grinding (kW·hr/t [HP·hr/T])

	Hammer Mills	Burr Mills
Shelled corn	6.6 [8] - 7.4 [9]	3 [3.6] - 5.8 [7.1]
Oats	11.5 [14] - 14 [17]	5 [6] - 14 [17]
Barley	9 [11] - 14 [17]	4 [5] - 10 [12]
Ear corn	4.5 [5.5] - 8 [10]	...
Hay	8 [10] - 16 [20]	...
Round bales	3 [3.6] - 9.1 [11](tub grinder)	

Drying Machinery

Stored farm crop materials can be spoiled and consumed by microbiological growth. Such growth can be reduced to insignificance with cool storage temperatures and/or low crop moisture contents. For example, the USDA Grain Storage Research Laboratory at Ames, Iowa, has found the time limits for safe storage of shelled corn (Fig. 12.6) that result in 0.5% dry-matter (d.m.) loss but still produce acceptable grain. For long-term storage, grains need to have less than 13% moisture content, wet basis (W.B.).

The farm manager has the alternative of using natural field drying before harvesting the crop or using drying machinery after the crop has been harvested. The least total cost choice will probably be the most attractive. For artificial or mechanical drying to compete with natural drying, the cost of using the drying machinery must be less than the increased earnings realized from the increased quality and quantity obtained from early harvest.

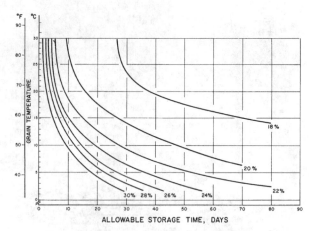

Fig. 12.6. Time limits for stored shelled corn at various moisture contents.

Drying Principles

Stored hay and grain will lose moisture to the surrounding (ambient) air as long as the air is relatively drier. The *equilibrium moisture content* of a crop is that content at which there is no tendency for moisture to leave the crop and enter the surrounding air. A measure of the ability of the air to absorb moisture from the grain is its *relative humidity* (RH), a ratio of its actual water content to the amount it could hold at the same temperature. Relative humidities need to be lower than 70% for air at 21°C [70°F] to be in equilibrium with most grains at 12–13% moisture content.

A *psychrometric chart* (Fig. 12.7) indicates the specific volume and moisture-holding properties of air at various temperatures. These charts were developed especially for crop-drying analysis. Dry bulb temperature, t_{db}, is that sensed by common thermometers. A wet bulb temperature, t_{wb}, is indicated by a thermometer with its mercury bulb covered with a wet cloth that is evaporating moisture into the surrounding air. Both of these thermometers can be mounted on a small panel that is whirled in the air to induce moisture evaporation. This instrument is called a *sling psychrometer*. Heat is required to evaporate water—about 2320 kJ/kg [1000 Btu/lb]. The rate of heat loss from the wet bulb is reflected by that thermometer's lower temperature. If the surrounding air is saturated with moisture, there will be no evaporation and no wet bulb temperature depression from the dry bulb temperature. Saturated air (100% RH) is indicated on the psychrometric chart by the curved line to the left where the t_{wb} and t_{db} lines meet.

The drying process is traced on a psychrometric chart along lines of constant t_{db}. The heat required to evaporate moisture from a crop is provided by a drop in temperature of unsaturated air passing through the

crop. If the path of the airflow is long enough and the grain is wet enough, the air eventually becomes saturated and will do no more drying. For example, if the incoming ambient air has a t_{db} of 25°C [77°F] and an RH of 60%, its temperature will drop and its moisture content will increase along the 19°C [67°F] wet bulb line as the air moves through the drying zone in the grain. If the airflow is not too fast, the drying zone will be narrow and the air will become saturated before it leaves the grain (Fig. 12.8). Each kilogram [pound] of dry air will contain 0.0143 kg [lb] of moisture, having picked up 0.0143 −0.0120 = 0.0023 kg [lb] of moisture from the grain.

Unheated-air drying depends on the ambient air having a moisture content less than saturated air. Using the data from the previous paragraph, the specific volume of the air at the entrance duct would be about 0.86 m³/kg [13.75 ft³/lb] (Fig. 12.7). A fan would need to force

$$374\,\text{m}^3 = \frac{0.86\,\text{m}^3\,\text{air}}{1\,\text{kg air}} \times \frac{1\,\text{kg air}}{0.0023\,\text{kg}\,H_2O}$$

$$\left[5978\,\text{ft}^3 = \frac{13.75\,\text{ft}^3}{1\,\text{lb air}} \times \frac{1\,\text{lb air}}{0.0023\,\text{lb}\,H_2O} \right]$$

of such air through the crop to pick up 1 kg [1 lb] of moisture. Such low removal rates mean that large quantities of air are required, even under good conditions, to dry crop materials fast enough to avoid spoilage. Table 12.2 gives USDA-recommended airflow rates and maximum depths for various crops. Because of the time required for drying, almost all

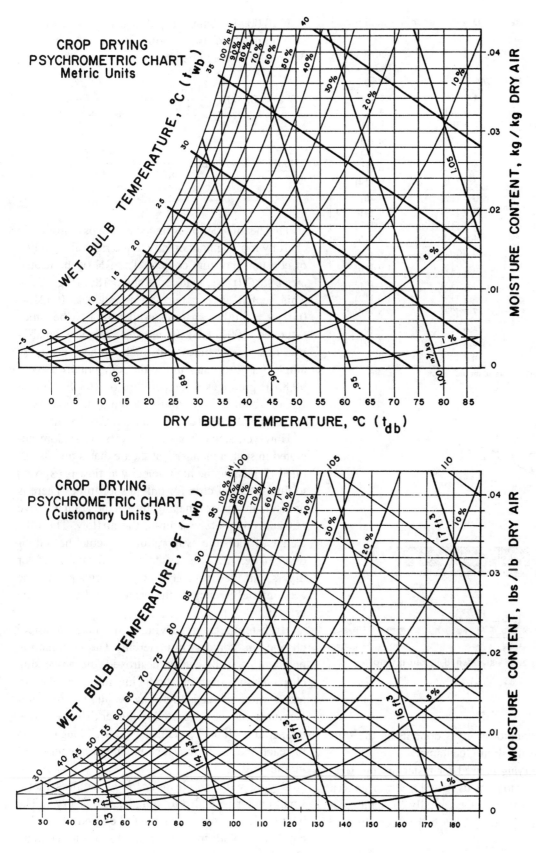

Fig. 12.7. Psychrometric chart.

TABLE 12.2. Unheated-Air Drying Recommendations

Crop	Moisture, % W.B.	Recommended Minimum Airflow, m/min [ft/min]	Recommended Depths, m [ft]
Wheat	20	4.6 [15]	1.8 [6]
	18	4.0 [13]	2.4 [8]
	16	2.4 [8]	3.0 [10]
Oats	25	4.6 [15]	1.8 [6]
	20	4.0 [13]	2.4 [8]
	18	3.6 [12]	3.0 [10]
	16	3.0 [10]	3.6 [12]
Shelled corn	25	6.7 [22]	1.8 [6]
	20	5.8 [19]	2.4 [8]
	18	5.8 [19]	3.6 [12]
	16	4.0 13]	4.9 [16]
Ear corn	30	7.3 [24]	3.6 [12]
	25	5.8 [19]	4.9 [16]
Chopped hay	40	4.6 [15]	2.4 [8]
		5.2 [17]	3.6 [12]
		7.3 [24]	5.5 [18]
Baled hay	35	4.6 [15]	1.8 [6]
		6.7 [22]	2.7 [9]
		9.1 [30]	3.6 [12]

Note: For hay, total airflow should not drop below 13 (m³/min)/t [500 cfm/ton]

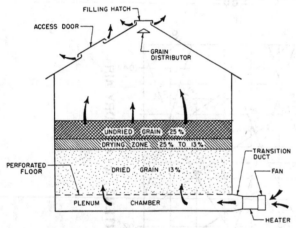

Fig. 12.8. Drying zone established in bin drying.

unheated-air drying of crops is done using the storage structure as the drying container.

Heated-air drying is used for faster drying. Heat is added to the air to raise its t_{db} and to lower its RH. The specific heat for air at essentially constant pressure is 1 kJ/(kg·°C) [0.24 Btu/(lb·°F)]. Temperature rise is given by:

$$\frac{\text{burner output}}{\text{specific heat}} \times \frac{\text{specific volume of air}}{\text{volume of air}} \qquad (12.1)$$

If 10,000 kJ [9480 Btu] of heat were added to the air volume from the previous paragraph, the temperature rise would be

$$\underline{23°C} = 10,000 \text{ kJ} \times \frac{1 \text{ kg} \cdot °C}{1 \text{ kJ}} \times \frac{0.86 \text{ m}^3}{1 \text{ kg}} \times \frac{1}{374 \text{ m}^3}$$

$$\left[\underline{90.85 \text{ °F}} = 9480 \text{ Btu} \times \frac{1 \text{ lb} \cdot °F}{0.24 \text{ Btu}} \times \frac{13.75 \text{ ft}^3}{1 \text{ lb}} \times \frac{1}{5978 \text{ ft}^3} \right]$$

This heating action is traced on the psychrometric chart by moving horizontally to the right along a line of constant moisture content. The RH of the heated air at 48°C (t_{db}) is read to be about 18% (Fig. 12.7). This heated air can pick up as much as 0.0215–0.0120 = 0.0095 kg moisture/kg air if it becomes saturated (100% RH). Its moisture removal capability is over four times that of the unheated air example. In customary units: the temperature rise of 90.85°F produces a temperature of 168°F with an RH of about 5%. Its moisture removal capacity is 0.030–0.012 = 0.018 lb of moisture/lb dry air.

However, actual heated-air dryers are seldom operated in such a manner that their exhaust air is saturated. Typically, the high-velocity airflow is exposed to thin layers of crop for only a short time. There is not enough time for the moisture in the grain to move out and completely saturate the airflow. The efficiency of moisture evaporation for actual heated-air dryers (Fig. 12.17) is less than that for unheated-air dryers, but their capacity is enough greater that the economic efficiency of the heated-air dryer may be greater.

High-temperature drying can cause chemical changes in stored crop materials. The germination potential of seeds can be destroyed, the processing characteristics of grain can be impaired, and the nutritive value of feedstuffs may be altered. Air temperatures may be considerably higher than the kernel temperatures if the kernel is not exposed long enough to reach the maximum recommended temperatures in Table 12.3.

Drying with *supplemental heat* requires that a small burner be added to an unheated-air dryer. The burner operates only when the RH exceeds the RH in equilibrium with the final desired moisture content. Drying temperatures are lower and the efficiency of removing moisture per amount of heat added is much greater than for high-speed, heated-air drying.

TABLE 12.3. Maximum Recommended Drying Temperatures

Kind of Grain	Moisture or Use, % W.B.	Temperature, °C [°F]
Seed corn	over 25% moisture	32-43 [90-110]
Seed corn	under 25% moisture	43-50 [110-120]
Feeding corn	over 25% moisture	50-80 [120-175]
Feeding corn	under 25% moisture	55-82 [130-180]
Commercial corn	for wet milling	55-57 [130-135]
Small grain	seed	50 [120]
Small grain	commercial	82 [180]
Soybeans	seed	43 [110]
Soybeans	commercial	82 [180]
Hay	feeding	66 [150]

Cooling the grain to approximately atmospheric temperature is also a part of heated-air drying. Hot grain placed in storage can deteriorate because of *moisture migration*. Anytime warm grain in the center of a bin is surrounded by cooler grain, a slow current of air circulates up the center and down the sides of the bin. This airflow picks up moisture and releases it when encountering cooler grain. The outside grain becomes drier, but the top surface grain accumulates moisture and may spoil. Fan aeration of stored grain to control moisture is recommended, but aeration does not replace the necessity for cooling hot grain before storage.

USDA researchers at Purdue University have coined the word *dryeration* for the process in which hot grain, dried to within 2–3 moisture percentage points of the desired final moisture content, is removed from the dryer and placed in a holding bin. Hot grain is accumulated and then cooled slowly in 10–12 hr with an airflow of 0.4–0.8 (m³/min of air)/m³ of grain [(ft³/min)/bu]. Additional drying takes place as moisture from within the kernel has time to move to the kernel surface where it can be evaporated. Slow cooling helps reduce corn cracking damage. Dryer capacity is increased to above normal since higher temperatures can be used with wetter corn, cooling time is eliminated, and all the container volume of some dryers can be used for drying.

Mechanisms

Fans are the mechanical hearts of crop dryers, as movement of air through the stored crop is essential. The pressure required to force air through stored crops is not great. The unit most commonly used for conventional drying pressures is the millimetre [inch] of water as measured by a manometer (1 mm of water is equivalent to 9.8 Pa) [1 in. of water is equivalent to 0.036 psi]. Fig. 12.9 indicates typical values expected for farm crops if stored in a clean and

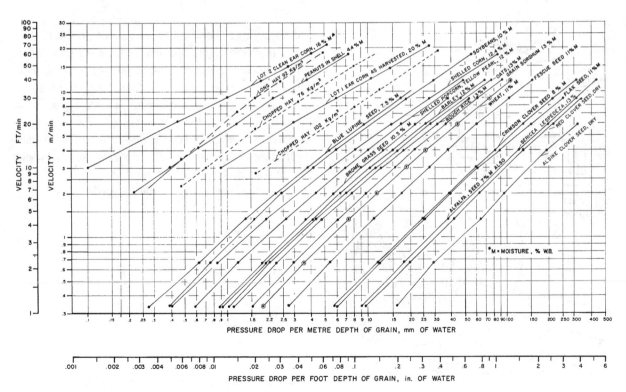

Fig. 12.9. Resistance of clean, stored crops to airflow.

nonpacked condition. If the grains are dirty and packed, resistance values may be 150% of those indicated on the chart. If the moisture content is high, use only 80% of the indicated pressure drop for a given rate of airflow. For uniform drying, equal-length airflow patterns must be maintained during the process. Foreign matter contains unwanted moisture and interferes with the airflow patterns.

Both centrifugal and axial flow fans are used in crop dryers. Fig. 12.10 shows a *backward-curved centrifugal* fan installation. The air is drawn into the center of the fan and then forced into the surrounding scroll by centrifugal force. These fans are relatively quiet, have high efficiency, and do not consume excessive power when delivering against a choked condition.

Axial flow fans pass air parallel to their drive shaft and are typified by the *propeller* fan shown in Fig. 12.11. A *vane-axial* fan, used in the dryer of Fig. 12.15, is a type of propeller fan with fixed vanes that straighten the turbulent air coming from the fan blades. Vane-axial fans are more efficient than regular propeller fans in the pressure ranges at which crop dryers operate. Axial fans as a class are less expensive, more compact, and noisier than centrifugal fans.

Figure 12.12 shows typical pressure (P_x) and efficiency (E_x) curves for the three types of fans used in

Fig. 12.11. Propeller fan.

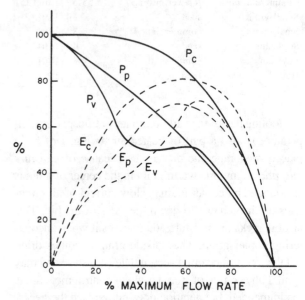

Fig. 12.12. Fan Performance: pressure, P, efficiency, E. Subscripts c for centrifugal, p for propeller, and v for vane-axial fans

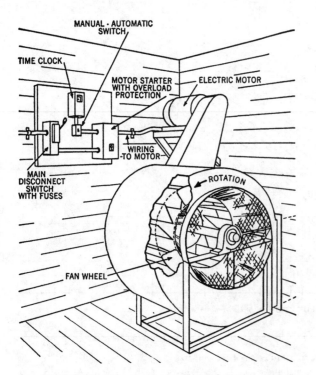

Fig. 12.10. Backward-curved centrifugal fan with controls.

crop drying. The absolute values for pressure, flow, and power required are unique for each size of fan, but any one size tends to be a scale model of all other sizes of its type. The performance values of a type are single curves when reported on a percent-of-maximum basis. The *static efficiency* of a fan is its static power output divided by the power input. These static efficiencies are zero at both choked or zero flow and at maximum flow with no pressure rise.

The power required by fans is given by

$$P = \frac{V\,h}{c\,e} \qquad (12.2)$$

where P = power required, kW [HP]
V = airflow, m³/min [ft³/min]
h = static pressure head, mm [in.] of water
e = static efficiency, decimal
c = constant, 6000 [6350]

Electric power is a popular energy source for smaller dryers and can be used for fans on large stationary dryers if three-phase service is available. The control circuits for the drying process are usually electric too. But for portable dryers and for high-capacity installations when electric service is limited, the PTO drive of a farm tractor is used and the controls may be operated from the tractor battery. The load factor, ratio of maximum actual power required to maximum power available, can be 1 for electric motors but should be no more than 0.8 for tractors, Fig. 2.5. The efficient operating range for dryer fans is 50–80% maximum flow and 50–80% maximum static pressure.

Propane, butane, and natural gas are the common fuels used in crop dryers. The heats of combustion of these fuels, 25 MJ/L [90,000 Btu/gal], 28 MJ/L [100,000 Btu/gal], and 37–41 MJ/m^3 [1000–1100 Btu/ft^3], respectively, are less than for oil or solid fuels, but the clean-burning features of gases permit greater efficiency because the products of combustion can pass through the crop without depositing soot. Dryers using the heavier fuels need a heat exchanger between the burner and the airstream, which reduces the fuel efficiency of the dryer. Gas burners are installed downstream from the fan and are designed to heat the airstream uniformly. A vaporizer is needed for large LP gas burners. Pressure regulators are required to lower the gas supply pressure to burner pressure.

Controls

Safety controls are needed for heated-air dryers. Failsafe solenoid valves are used in the fuel supply lines. These valves close when an electric power failure occurs, a flameout is detected by an electric eye or thermostat, or the maximum temperature thermostat acts. Interlocking controls between the fan and the burner require that the fan be operating to purge the duct system of unwanted fuel before ignition by an electric spark or pilot flame can occur. A 90-s time limit is often designed into ignition attempts. Manual restarting is required after a shutdown for unsafe conditions.

Timers, humidistats, and thermostats are used to control dryers. Fans and burners usually cycle between full-off and full-on. Unheated-air dryers may be shut down by a humidistat if the ambient RH rises too high. Moisture sensors in the crop itself have been used for control in heated-air dryers. Timers are probably the simplest and most reliable controls for both heated- and unheated-air drying. They can turn off an unheated-air fan at night and shut down burners after the estimated time for completion of drying has passed. Thermostats can be used to sense the completion of drying. The exhaust air temperature from wet grain is considerably lower than that of the entrance air if the air picks up a significant amount of moisture. As the grain dries, the temperature depression is less and a thermostat can shut off the burner when the grain moisture drops to the desired level.

Systems

At least four types of crop-drying systems are based on the quantity of crop dried as a unit.

In-storage drying uses storage structures to hold the quantity of crop to be dried. Almost all unheated-air drying is accomplished in this way. Hay dryers normally use this system.

In *batch drying*, only a measured quantity of the crop is processed at a time. If the drying bin is on wheels or skids, it is called a *portable batch dryer* (Fig. 12.13). A batch of grain is loaded, dried, cooled, and then unloaded into trucks, wagons, or storage bins. Dryer bin capacities, called holding capacities, may be as high as 57 m^3 [1600 bu].

Batch-in-bin drying uses a storage structure equipped with proper ductwork (Fig. 12.8). The batches are dried in layers of about 1 m [3 ft]. On completion of drying, the batch is cooled and then augered out to be placed in a regular storage bin. The last batches of grain harvested can be dried and accumulated in the drying bin for storage. This system can be arranged physically as in Fig. 12.14. A centrally located dump, a single portable elevator, and an unloading auger from the drying bin provide the necessary grain handling operations. This system is flexible and easily expanded. Systems with batches as large as 100 m^3 [3000 bu] are used.

A *circulating batch* system is a portable batch dryer in which the grain is constantly moving through the dryer. A more uniform product is obtained, as this mixing action avoids the overdrying that can occur in the other batch systems. Cooling in the batch systems is accomplished by turning off the burner and operating only the fan after drying is completed.

A *continuous flow* system is one in which wet grain enters one end of a drying column, 60 cm [2 ft] thick or less, and comes out the other end both dry and cool (Fig. 12.15). The rate of flow of the grain through the column can be varied to meet the needs of the wet crop. This system lends itself well to automatic operations because the grain can be loaded, dried, cooled, and unloaded without attention as long as wet grain is available in the bin supplying the dryer.

Fig. 12.13. Portable batch dryer.

Fig. 12.14. Batch-in-bin drying and storing system. grain flow through the dryer.

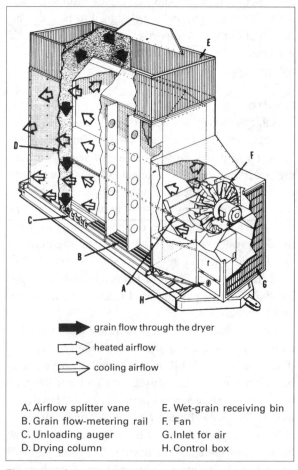

➡ grain flow through the dryer

⇨ heated airflow

⇨ cooling airflow

A. Airflow splitter vane   E. Wet-grain receiving bin
B. Grain flow-metering rail   F. Fan
C. Unloading auger   G. Inlet for air
D. Drying column   H. Control box

Fig. 12.15. Construction of and flow patterns for continuous flow grain dryer.

Fig. 12.16 shows such an installation that can process hundreds of tonnes yearly with little manual labor or operating supervision.

System Performance

Unlike field machines, the performance of dryers is not easily seen. Of most interest is the rate of drying. Three measurements must be taken to estimate the instantaneous drying rate, using Eq. 12.3 below:

$$R = \frac{T\,V}{c\,v} \qquad (12.3)$$

where R = instantaneous rate of moisture removal, kg/ hr [lb/hr]
   T = difference between inlet and exhaust dry bulb temperatures, °C [°F]
   V = volume rate of airflow, m³/min [ft³/min]
   v = specific volume of air moved by fan, m³/kg [ft³/lb]
   c = constant, 38.7 [69.4]

Specifically, the two temperatures and a manometer reading of the static pressure at the entrance to the holding bin are required. The pressure reading indicates the V for Eq. 12.3 if the fan performance curves of the system are known (Fig. 12.12). The specific volume of the air, v, is to be determined from the psychrometric chart at ambient temperatures.

Sample Solution

Assume that a portable batch dryer is measured to have an inlet air temperature of 77°C [170°F] and an exhaust temperature of 10°C [50°F]. The inlet static pressure indicates that the airflow is 565 m³/min [19,950 ft³/min] for the fan. If atmospheric air is 10°C [50°F] and is assumed to be at 100% RH, a psychrometric chart shows the specific volume to be about 0.8 m³/kg [13 ft³/lb]. Then,

$$R = \frac{67 \times 565}{38.7 \times 0.8} = 1222.7 \text{ kg moisture /hr}$$

$$\left[R = \frac{120 \times 19,950}{69.4 \times 13} = 2653.5 \text{ lb moisture /hr} \right]$$

The rate, R, is a momentary rate of drying. As the grain dries, it becomes harder for the air to pick up moisture and the temperature drop lessens. If the measured drop is only half as much near the end of

Fig. 12.16. Farm drying installation.

the drying period, the rate of moisture removal is halved.

The estimated total heat requirements for drying grain are shown in Fig. 12.17. The average rate of drying is thus seen to depend on the initial moisture content of the grain, the ambient air temperature, and the operating temperature of the dryer. If the temperature of the grain and ambient air is less than 15°C [60°F] the amount of heat required to bring their temperature up to 15°C [60°F] must be added to

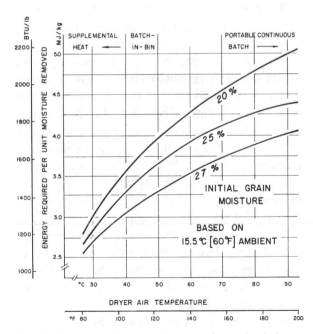

Fig. 12.17. Average heat requirements for grain drying.

the amounts indicated in Fig. 12.17. The average specific heat of all grains can be estimated as 2 kJ/(kg°C) [0.5 Btu/(lb°F)].

The ASAE has standardized the performance rating of heated-air crop dryers. This rating is reported as the amount (wet weight) of clean, yellow, shelled corn dried per hour. The manufacturer reports the drying air temperature and uses either 25.5 – 15.5 or 20.5 – 15.5 reduction in moisture percentage. This rating is based on 10°C [50°F] ambient air and includes the time to cool the dried grain to 42°C [60°F]. Unloading and loading times for batch dryers are not included as part of the drying time. The *holding capacity* of a dryer is defined as the volume of the structure containing the crop when filled to normal operating level.

Operations

Operating a crop-drying system efficiently is a challenge for the machinery manager. Almost all the input elements are variables: the ambient temperature and humidity, the moisture content of the crop as it comes from the harvester, and the airflow resistance of the crop as it is held in the drying bin. The output desired is a uniformly dried crop. Mechanical grain distributors may be required to get a level fill for even airflow when storage bins are used for drying. Uneven airflow results in over-dried or under-dried spots. Over-drying is an expense not recovered by an increase in price for the crop. Under-drying leads to crop spoilage.

The machinery manager must understand the units used in measuring crop quantities and the base for moisture contents to estimate dryer performance. Quantities to be dried are conventionally expressed as wet weight. In customary units the bushel is often used as a quantity term. A bushel of grain is defined as a specified weight of crop at 13% W.B. moisture, except for a shelled corn bushel, which is defined at 15.5% W.B. (The specific weights and densities of crops are given in Table 13.1.) The bushel is an inconvenient unit to use for drying calculations. Only the mass of crop material, as measured by its weight, will be used in this presentation.

Two means of expressing moisture content are used. The wet basis (W.B.) moisture percentage relates the mass of water per unit to the total mass of the unit. The dry basis (D.B.) percentage relates mass of water to the dry matter (d.m.) content of the crop unit. These relationships are shown in.

$$\% \text{ W.B.} = \frac{\text{mass } H_2O}{\text{mass d.m.} + \text{mass } H_2O} \times 100 \qquad (12.4)$$

$$\%\text{D.B.} = \frac{\text{mass } H_2O}{\text{mass d.m.}} \times 100 \qquad (12.5)$$

Use of the dry basis percentage is preferred since the base for the percentage does not change. However the wet basis percentage is more commonly used and will be used in this book.

A reliable moisture meter is necessary for the machinery manager who dries large quantities of grain. Samples can be oven-dried overnight and weighed to determine the moisture content, but most managers want faster answers. Fig. 12.18 illustrates a meter that gives rapid indications of grain moistures.

Calculating the amount of water removed using a wet basis percentage is facilitated by constructing and filling in the *moisture calculation table* below. For example, to calculate moisture removal for drying 1 unit of hay at 40% W.B. to 20% W.B. the table would appear as:

	%	Units H_2O +	Units d.m. =	Units total
Initial	40	0.4	0.6	1
Final	20	0.2x	0.6	x

If at the start 40% of the unit is moisture, the remaining 60% or 0.6 units must be dry matter that remains constant throughout the drying process. The unknown final mass is assigned the value x. The lower line of the table then expresses the mass relationships in terms of x.

$$0.2x + 0.6 = x$$
$$x = 0.6/0.8 = 0.75 \text{ units}$$

If the final mass of the crop is 0.75 units, then 0.25 units of water must have been removed. This answer may be verified by substituting 20% of x or 0.15 units into the final units-of-water blank and subtracting from the initial 0.40 units to get 0.25 units of water removed.

Sample Estimation Procedure

The operation of a dryer can be estimated using the relationships discussed previously. An example of the estimation procedure should be helpful in understanding dryer operations.

Fig. 12.18. Moisture meter.

A heated-air continuous dryer is used to reduce the moisture content of 500 t [550 T] of shelled corn from 25% to 15% W.B. The burner rating is 3000 MJ/hr [2,844,141 Btu/hr]. Propane fuel is available at 10¢/L [37.85¢/gal] and has an energy (heat) content of 25 MJ/L [90,000 Btu/gal]. The vane-axial fan characteristics are 700 m³/min [24,720 ft³/min] against zero pressure rise and zero flow against 150 mm [6 in.] of water. The dryer manufacturer reports that 50 mm [2 in.] is the expected operating static pressure for this dryer when filled. The grain and ambient air is at 10°C [50°F]. The air is assumed to be at 70% RH. Find:

1. Required amount of moisture to be removed
2. Airflow rate
3. Power required
4. Operating temperature
5. Fuel consumed
6. Time required for drying
7. Average drying rate

1. A moisture calculation table

	%	Units H²O	+ Units d.m.	= Units total
Initial	25	0.25	0.75	1
Final	15	0.13	0.75	0.88
		0.12		0.12

gives the moisture to be removed as 500 × 0.12 = 60 t [550 × 0.12 = 66 T].

2. From Fig. 12.12, 33% maximum P_v (for a vane-axial fan) indicates 80% maximum flow rate or 560 m³ [19,776 ft³]/min.

3. Figure 12.12 shows the fan efficiency to be about 60% at 80% flow. The fan power can be found from Eq. 12.2.

$$P = \frac{V\,h}{c\,e} = \frac{560 \times 50}{6000 \times 0.6} = 7.77 \text{ kW}$$

$$\left[P = \frac{V\,h}{c\,e} = \frac{19,776 \times 2}{6350 \times 0.6} = 10.38 \text{ HP} \right]$$

The power used for unloading augers and controls could add 7.4 kW [10 HP] more to give a total power requirement of 15.2 kW [20.4 HP].

4. At 10°F the specific volume of air going through the fan would be about 0.8 m³/kg [13 ft³/lb]. The operating temperature is estimated from the airflow, the specific heat and the specific volume of the air, and the burner output, Eq. 12.1.

$$\underline{71.4°C} = \frac{3000 \text{ MJ}}{\text{hr}} \times \frac{\text{kg} \cdot °C}{1 \text{ kJ}} \times \frac{0.8 \text{ m}^3}{1 \text{ kg}} \times \frac{\text{min}}{560 \text{ m}^3}$$

$$\times \frac{1000 \text{ kJ}}{1 \text{ MJ}} \times \frac{\text{hr}}{60 \text{ min}}$$

$$\left[\underline{129°F} = \frac{2,844,141 \text{ Btu}}{1 \text{ hr}} \times \frac{\text{lb} \cdot °F}{0.24 \text{ Btu}} \times \frac{13 \text{ ft}^3}{1 \text{ lb}} \right.$$

$$\left. \times \frac{\text{min}}{19,776 \text{ ft}^3} \times \frac{1 \text{ hr}}{60 \text{ min}} \right]$$

The dryer operating temperature would be 10 + 71.4 = 81.4°C [50 + 129 = 179°F].

5. From Fig. 12.17, a dryer temperature of 81.4°C [179°F] indicates an evaporative efficiency of 4300 kJ/kg [1850 Btu/lb] of moisture removed. The heat required would be

$$258{,}000 \, \text{MJ} = \underline{60 \, \text{t}} \times \frac{1000 \, \text{kg}}{\text{t}} \times \frac{4300 \, \text{kJ}}{\text{kg}} \times \frac{\text{MJ}}{1000 \, \text{kJ}}$$

$$\left[244{,}200{,}000 \, \text{Btu} = \underline{66 \, \text{T}} \times \frac{2000 \, \text{lb}}{\text{T}} \times \frac{1850 \, \text{Btu}}{\text{lb}} \right]$$

Since the data in Fig. 12.17 are based on 15.5°F ambient air, the heat required to raise the freshly harvested grain from 10°C to 15.5°C [50°F to 60°F] must be added to the evaporative heat. Using the specific heat for shelled corn as 1.7 kJ/(kg°C) [0.4 Btu/(lb°F)];

$$4675 \, \text{MJ} = \underline{500 \, \text{t}} \times \frac{1000 \, \text{lb}}{\text{T}} \times \frac{1.7 \, \text{kJ}}{\text{kg} \cdot {}^\circ \text{C}} \times \underline{5.5 \, \text{C}}$$

$$\left[4{,}400{,}000 \, \text{Btu} = \underline{500 \, \text{T}} \times \frac{2000 \, \text{lb}}{\text{T}} \times \frac{0.4 \, \text{Btu}}{\text{lb} \cdot {}^\circ \text{F}} \times 10^\circ \text{F} \right]$$

The total heat requirement would be the sum or 262,675 MJ [248,600,000 Btu].

The propane consumption would be

$$10{,}507 \, \text{L} = 262{,}675 \, \text{MJ} \times \frac{\text{L}}{25 \, \text{MJ}}$$

$$\left[2762 \, \text{gal} = 248{,}600{,}000 \, \text{Btu} \times \frac{\text{gal}}{90{,}000 \, \text{Btu}} \right]$$

6. The time required for drying is 262,675/3000 [248,600,000/2,844,141] or 87.5 hr.
7. The ASAE drying rate would be 500/87.5 or 5.7 t/hr [6.3 T/hr]. The harvesting rate can be no greater than 11.4 t/hr [12.6 T/hr] for a 12-hr day assuming the dryer is operated 24 hr/day.

Unheated-air drying could be used. Table 12.2 shows that a depth of 1.8 m [5.9 ft] is recommended for the drying mass. The cross-sectional area of airflow can be computed knowing the density of the grain, 719 kg/m³ [44.8 lb/ft³], as given in Table 13.1. But such densities are for dry grain–15.5% W.B. for shelled corn. The density for the wet grain can be approximated from a moisture calculation table for a cubic metre [cubic foot] of wet grain.

	%	kg [lb] H₂0	+ kg [lb] d.m.	= kg [lb] total
Dry	15.5	111 [6.94]	608 [37.86]	719 [44.8]
Moist	25.0	203 [12.62]	608 [37.86]	811 [50.48]

The density of the wet grain is approximately (ignoring shrinkage) 811 kg/m³ [50.48 lb/ft³]. Dividing the mass by the density times the depth gives the cross-sectional area required for airflow.

$$342.5 \, \text{m}^2 = \underline{500 \, \text{t}} \times \frac{1000 \, \text{kg}}{\text{t}} \times \frac{\text{m}^3}{811 \, \text{kg}} \times \frac{1}{1.8 \, \text{m}}$$

$$\left[3693.4 \, \text{ft}^2 = \underline{550 \, \text{T}} \times \frac{2000 \, \text{lb}}{\text{T}} \times \frac{\text{ft}^3}{50.48 \, \text{lb}} \times \frac{1}{5.9 \, \text{ft}} \right]$$

Three round bins of the type in Fig. 12.8 with diameters of 12 m [39.6 ft] could provide adequate capacity.

The maximum amount of moisture that could be picked up would be (Fig. 12.7)

0.0065 − 0.0055 = 0.0010 kg/kg [lb/lb] air

To remove 60 t [66 T] of moisture would require 60,000 t [66,000 T] of air or 48.59 Mm³ [1,716,000,000 ft³] at a specific volume of 0.8 m³/kg [13 ft³/lb].

Fig. 12.6 shows that the time for drying with 10°C [50°F] air can be no longer than 18 days or 432 hr. The required air velocity over the drying area must be

$$5.46 \frac{\text{m}}{\text{min}} = \frac{48 \, \text{Mm}^3}{432 \, \text{hr}} \times \frac{1 \, \text{hr}}{60 \, \text{min}} \times \frac{1}{3 \, \text{bins}}$$

$$\times \frac{\text{bin}}{113 \, \text{m}^2} \times \frac{1{,}000{,}000 \, \text{m}^3}{1 \, \text{Mm}^3}$$

$$\left[17.9 \frac{\text{ft}}{\text{min}} = \frac{1{,}716{,}000{,}000 \, \text{ft}^3}{432 \, \text{hr}} \times \frac{1 \, \text{hr}}{60 \, \text{min}} \right.$$

$$\left. \times \frac{1}{\text{bins}} \times \frac{1 \, \text{bin}}{1232 \, \text{ft}^2} \right]$$

At such velocities the pressure drop for the typical packed wet and dirty grain from the field is 13.3 ×

$1.5 \times 0.8 \times 1.8 = 28.7$ mm of water [$0.16 \times 1.5 \times 0.8 \times 6 = 1.15$ in of water] (from Fig. 12.9).

The air power required for each bin (Eq. 12.2) would be

$$\underline{2.91} = kW \frac{113\,m^2}{6000} \times \frac{5.37\,m}{min} \times \frac{28.7\,mm}{m}$$

$$\left[\underline{3.93}\,HP = \frac{1232\,ft^2}{6350} \times \frac{17.6\,ft}{min} \times \frac{1.15\,in.}{ft} \right]$$

At a fan efficiency of 60%, the installed motor power rating should be 4.85 kW [6.56 HP] or more on each bin.

The energy required at an 80% motor efficiency would be

$$\underline{7857}\,kW \cdot hr = 3\,motors \times \frac{4.85\,kW}{motor} \times 432\,hr \times \frac{1}{0.80}$$

$$\left[\underline{7916}\,kW \cdot hr = 3\,motors \times \frac{6.56\,HP}{motor} \times 432\,hr \right.$$

$$\left. \times \frac{1}{0.80} \times \frac{0.746\,kw}{1\,HP} \right]$$

Drying with a batch-in-bin system requires close coordination with the harvesting operation. The drying bin must accumulate the day's harvest, dry it, cool it, and empty it before receiving the first truckload from the next day's harvest. A batch-in-bin system is flexible in that the elements can be sized to meet specific needs. For example, suppose the machinery manager wishes to have a 40 t [44 T]/day capacity. Batch-in-bin systems are limited to about a 1-m [3.28-ft] depth of grain in the bin; thus the drying area required would be (for shelled corn):

$$\underline{55.6}\,m^2 = 40\,t \times \frac{1000\,kg}{t} \times \frac{m^3}{719\,kg} \times \frac{1}{1\,m}$$

$$\left[\underline{598.9}\,ft^2 = 44\,T \times \frac{2000\,lb}{T} \times \frac{ft^3}{44.8\,lb} \times \frac{1}{3.28\,ft} \right]$$

If a round bin were used, its diameter would be about 8.4 m [27.6 ft].

Allowing 8 hr/day for drying at the same moisture and weather conditions as above, this system would need to evaporate $40 \times 0.12 = 4.8$ t [$44 \times 0.12 = 5.28$ T]/day of water. (The 0.12 units of moisture per unit of wet weight is taken from the first moisture calculation table.) On an hourly basis, the evaporation rate would be 600 kg [1320 lb] of water.

Typical values from Fig. 12.17 show a 45°C [113°F] drying temperature and an efficiency of 3.5 MJ/kg [1500 Btu/lb]. The required size of the burner must be

$$\underline{2100}\frac{MJ}{hr} = \frac{600\,kg\,water}{hr} \times \frac{3.5\,MJ}{kg}$$

$$\underline{1,980,000}\frac{Btu}{hr} = \frac{1320\,lb}{hr} \times \frac{1500\,Btu}{lb}$$

The fan size required would be

$$\underline{622}\frac{m^3}{min} = \frac{2100\,MJ}{hr} \times \frac{kg \cdot °C}{1kJ} \times \frac{1000\,kg}{1MJ} \times \frac{0.8\,m^3}{kg}$$

$$\times \frac{1\,hr}{60\,min} \times \frac{1}{45°C}$$

$$\left[\underline{21,536}\frac{ft^3}{min} = \frac{1,980,000\,Btu}{hr} \times \frac{lb \cdot °F}{0.24\,Btu} \times \frac{13\,ft^3}{lb} \right.$$

$$\left. \times \frac{1\,hr}{60\,min} \times \frac{1}{83°F} \right]$$

Air velocities would be $(622\,m^3/min)/55.6\,m^2 = 11.18$ m/min [$(21,536\,ft^3/min)/598.9\,ft^2 = 35.96$ ft/min]. Static pressures expected would be 43 mm [1.7 in.] of water (see discussion of Fig. 12.9 for wet and dirty grain).

The power required at 60% fan efficiency would be

$$P = \frac{V\,h}{c\,e} = \frac{622 \times 43}{6000 \times 0.6} = 7.43\,kW$$

$$\left[P = \frac{V\,h}{c\,e} = \frac{21,536 \times 1.7}{6350 \times 0.60} = 9.6\,HP \right]$$

Total drying time would be 500/40 = 8 days or 64 hr.

Shrewd operators can increase the average efficiency of drying several ways. Taking advantage of favorable ambient conditions (high temperatures, low RH) will improve drying efficiency and consequently drying capacity. Early-season drying benefits from higher temperatures and lower humidities, but the crop moistures are highest at this time also. The benefits and disadvantages are balanced to get the optimum dryer capacity. Harvesting operations are usually limited to the daylight hours, but a dryer can operate continually. The harvesting and drying capacities per day should be matched so that each is working in an efficient manner. In some systems overdrying is unavoidable, but the shrewd operator will blend the overdried grain with underdried grain to save on dryer operation. Cooling can be done more quickly with low nighttime temperatures, although some rewetting may occur. The dryeration scheme mentioned earlier is a good example of an increase in drying capacity by varying the mode of operation.

Management

Drying machinery management problems are similar to field machinery management problems. Estimates of costs are used to measure the efficiency of operation and to make management decisions. As with field machines, drying machines have fixed costs and variable costs (see Chapter 4). Their fixed-cost percentage would be about 13% of the purchase price annually if a 15-year life can be realized. Repair and lubrication can be estimated as 2% P. Labor costs may be quite small if the dryer is automatic. The primary variables of operating costs for a dryer are power and fuel. If tractor PTO drives are used, the tractor power required can be determined by using the methods in the Power Performance section. Electric power consumption is measured by the power company's meter. Costs will depend on the rate structure of the local company and will be assumed at 4¢/(kW· hr).

Costs are reported as $/t [$/T] of wet crop dried. Drying costs are difficult to compare because the entering and exit moisture contents are so variable. A cost per unit of wet mass per moisture percentage point removed gives a fair comparison of drying costs. As with field machinery, increased usage reduces unit costs.

The three systems analyzed in the section on Operations can be used to illustrate cost calculations.

Assume a purchase price of $20,000 for a continuous dryer. The fixed costs plus repair and lubrication costs per unit dried per year would be (assume 15% P):

$$6.00 \frac{\$}{t} = 0.15 \times \$20,000 \times \frac{yr}{500\,t}$$

$$\left[5.45 \frac{\$}{T} = 0.15 \times \$20,000 \times \frac{yr}{550\,T} \right]$$

Fuel costs would be

$$4.20 \frac{\$}{t} = 10,507 L \times \frac{20¢}{L} \times \frac{1}{500\,t} \times \frac{\$1}{100¢}$$

$$\left[3.80 \frac{\$}{T} = 2762\,gal \times \frac{75.7¢}{gal} \times \frac{1}{550\,T} \times \frac{\$1}{100¢} \right]$$

If a $10,000 tractor used 400 hr/yr for other operations consumes $4.50 of fuel and oil for each hour of dryer operation, the tractor cost would be (FC% = 17)

$$0.195 \times 10,000 \times 87.5/487.5 = \$350$$

Repair and maintenance
$$0.000,12 \times 10,000 \times 87.5 = \$105$$

Fuel and oil
$$\$4.50/hr \times 87.5 = \$393.75$$

Total tractor costs would be $848.75 or $1.70/t [$1.54/T]. The total drying costs would be $11.90/t [$10.81/T].

The costs for unheated-air drying are separated from the bin costs since the bin would be purchased for storage anyway. Unheated-air drying equipment could be expected to last 20 yr and have a 13% fixed cost percentage. If three-phase electric power is available, the investment in three fans would be about $3500. Ductwork in the bins would be another $2000. The total fixed cost charge would be about $1.43/t [$1.30/T].

Electric power costs at 4¢/(kW·hr) would be

$$0.63\frac{\$}{t} = 7857\,kW\cdot hr \times \frac{4¢}{kW\cdot hr} \times \frac{1}{500\,t} \times \frac{\$1}{100¢}$$

$$\left[0.58\frac{\$}{T} = 7916\,kw\cdot hr \times \frac{4¢}{kW\cdot hr} \times \frac{1}{550\,T} \times \frac{\$1}{100¢}\right]$$

The total costs for the unheated-air example would be $2.06/t [$1.88/T].

The batch-in-bin drying costs are complicated by the potential use of the dryer bin as a storage bin also. If the bin is so used, the costs of the basic bin are assigned to the storage function, but the investment costs of all installed equipment necessary for drying (ductwork, grain distributor, loading and unloading augers, and the dryer itself) should be charged against the drying function. The expected life is 20 yr, and a 13% annual fixed cost charge is appropriate. Investment prices for the batch-in-bin example system are:

Dryer and controls, three-phase power	$3400
Floor ductwork	$2000
Grain distributor	$500
Unloading augers	$2000

Annual fixed costs would be 0.13 × $7900 or $1027. Fixed costs would be $2.05/t [$1.86/T]. The power cost for 80 hr would be (80% motor efficiency)

$$0.0476\frac{\$}{t} = \frac{7.43\,kW}{0.80} \times \frac{4¢}{kW\cdot hr} \times 64\,hr \times \frac{1}{500t} \times \frac{\$1}{100¢}$$

$$\left[0.0416\frac{\$}{T} = \frac{9.6\,HP}{0.80} \times \frac{0.7457\,kW}{1\,HP} \times \frac{4¢}{kW\cdot hr} \times 64\,hr\right.$$

$$\left.\times \frac{1}{550\,T} \times \frac{\$1}{100¢}\right]$$

Fuel costs would be

$$2.15\frac{\$}{t} = \frac{2100\,MJ}{hr} \times \frac{L}{25\,MJ} \times \frac{20¢}{L} \times \frac{\$1}{100¢} \times \frac{64\,hr}{500\,t}$$

$$\left[1.94\frac{\$}{T} = \frac{1,980,000}{hr} \times \frac{1\,gal}{90,000\,Btu} \times \frac{75.70¢}{gal}\right.$$

$$\left.\times \frac{64\,hr}{550\,T} \times \frac{\$1}{100¢}\right]$$

The total cost for the batch-in-bin drying system would be $4.25/t [$3.84/T].

The speed of optimum crop drying is an important decision for a machinery manager. As shown above, high-speed drying is less efficient and more expensive. But the speed of harvest with heated-air systems depends on the speed of drying. The cost of the combined harvest and drying system is most important. Timely, fast harvest of crops avoids field losses and crop deterioration. The increased costs for fast drying can be offset by the decrease in loss due to fast harvest. An economic evaluation of the costs associated with timeliness are in Chapter 16.

Practice Problems

12.1. Estimate the average power requirement for a hammer mill used to grind oats at a rate of 5 t [5.5 T]/hr.

12.2. A stack of hay made by the machine in Fig. 11.32 has a mass of 5 t [5.5 T] at 25% W.B. moisture.

    a. How much dry matter is there in the stack?

    b. What is the percent moisture on a dry basis?

    c. How much moisture evaporates if the stack dries to 20% W.B.?

12.3. What is the momentary drying rate of a dryer if the temperature drop of a 600 m³/min [21,200 ft³/min] airflow through the crop is 20°C [36°F]? Entering air is at 80% RH and 10°C [50°F].

12.4. Sixty tonnes [66 T] of clean wheat (wet weight) are to be dried from 20% W.B. to 13% W.B. with unheated air. Use the minimum recommendations of Table 12.2 and densities from Table 13.1 to determine:

    a. Circular bin diameter and fan volume flow

    b. Power required for a propeller fan whose pressure at zero flow is 70 mm [2.76 in.] of water and which is to be operated at maximum efficiency.

    c. Estimated drying time if the average ambient air temperature is 20°C [68°F], 80% RH, and the exhaust air is saturated. The fan runs continuously.

12.5. A portable batch dryer has a 50-cm [19.7-in.] drying depth, a holding capacity of 10 m³ [353 ft³], a 3000-MJ/hr [2,844,141-Btu/hr] heater, and a vane-axial fan performance of 1000 m³/min [35,315 ft³/min] at 100 mm [3.94 in.] of water static pressure. Estimate the drying capacity for shelled corn being dried from 20% to 15.5% W.B. and cooled. The ambient air temperature is 10°C [50°F] and its RH is 100%. The grain before drying is at ambient air temperature. Assume the cooling requires 10 min/batch.

MATERIALS HANDLING

The whole area of materials handling machinery depends greatly on the use of conveying equipment. Feedstuffs need to be moved into and from storage, to and from processing operations, and to the feed bunk. Waste materials and bulk fertilizers are also conveyed mechanically from storage to eventual destination.

Conveying machinery as used on the farm may be classified as

1. *Flight conveyors*: Inexpensive, versatile, noisy, slow.
2. *Augers*: Inexpensive, simple, compact, for short transfers.
3. *Belt conveyors*: Quiet, fast, expensive.
4. *Blowers*: High capacity, high power requirement, short distance transfer.
5. *Wagons*: Versatile, high capacity.

Flight Conveyors

This type of conveyor may have chain-carried scrapers for flat conveying or formed buckets for vertical work. In many farm implements single-chain flight conveyors are used, but for heavy-duty conveyor work at least two chains are required. A single chain conveyor is shown in Fig. 11.34.

Capacity

The capacity, CAPΔ of a flight conveyor can be estimated quite closely by knowing dimensions of flights, chain speeds, flight spacing, and the angle of the conveyor with the horizontal. Refer to Fig. 13.1 for a definition of flight conveyor variables

where w = width of flight, cm [in.]

s = center-to-center spacing of flights, cm [in.]

h = height of flight, cm [in.]

t = thickness of flight, cm [in.]

Ø = angle of conveyor, rad [deg]

S = speed, m/s [ft/s]

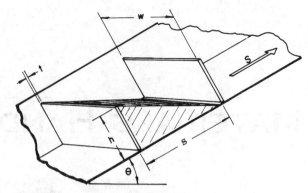

Fig. 13.1. Variables affecting flight conveyor capacity.

The cross-hatched area represents the level-fill volume per cell. If this area is triangular in cross section, the capacity for the flight conveyor is given as

$$CAP\Delta = \frac{wh^2 S}{cs \tan \emptyset} \frac{m^3}{min}\left[\frac{ft^3}{min}\right]$$

(13.1)

where c = 333.3 [4.8]

If the cross-hatched area is trapezoidal the capacity is given by

$$CAP = \frac{2wSh(s-t) - wS(s-t)^2 \tan \emptyset}{cs} \frac{m^3}{min}\left[\frac{ft^3}{min}\right]$$

(13.2)

The vertical distance that the grain is elevated, y, may be expressed in terms of the length of the conveyor, L, and the angle, Ø, as in

$$y = L \sin \emptyset \; m \; [ft]$$

(13.3)

When the techniques described in Appendix B are used, the product of capacity and height of elevation are maximized at approximately the critical angle,

$$critical \; angle = arc \; tan \; h/(s-t)$$

(13.4)

The critical angle describes the condition of the cell bottom just covered by grain but the fill volume not trapezoidal. At such a condition the maximum performance of the flight conveyor as expressed by

the product of capacity and height, CAPΔ, is given by

$$CAP\Delta = \frac{wh(s-t)S}{cs} \frac{m^3}{min}\left[\frac{ft^3}{min}\right]$$

(13.5)

Long elevators placed at shallow angles are required to satisfy the optimum capacity relationship in Eq. 13.4 and 13.5. The purchase price is almost directly proportional to length; therefore, if minimum cost is more desirable than maximum capacity, the machinery manager will probably purchase a shorter elevator and operate it at a steeper angle. The value of time should be carefully considered when making this decision.

Flight conveyor capacities are also functions of flight speed. Large, permanently installed conveyors have flight speeds of 0.5 m/s [1.64 ft/s] or less to keep wear to a minimum. Small drag-chain conveyors used on field machinery may run as high as 2 m/s [6.5 ft/s] or more, as wear is not critical for machines with only a few hundred hours of life. Portable farm elevators usually operate in the speed range of 1–1.25 m/s [3.3–4 ft/s].

Power Requirements

The power required to operate farm elevators is not very great if speeds are kept low. For estimation purposes one can consider that only 30% of the power input to this type of conveyor actually goes into lifting the grain to the height desired. For example, calculate the power required to elevate 509 kg [20 bu]/min of flaxseed 4.5 m [14.76 ft]. Refer to Table 13.1 for data.

$$1.87\,\text{kW} = \frac{509\,\text{kg}}{\text{min}} \times \frac{9.807\,\text{N}}{1\,\text{kg}} \times \frac{4.5\,\text{m}}{1} \times \frac{1}{0.20}$$

$$\times \frac{1\,\text{W}\cdot\text{s}}{1\,\text{N}\cdot\text{m}} \times \frac{1\,\text{min}}{60\,\text{s}} \times \frac{1\,\text{kW}}{1000\,\text{W}}$$

$$\left[\, 2.5\,\text{HP} = \frac{20\,\text{bu}}{\text{min}} \times 14.76\,\text{ft} \times \frac{56\,\text{lb}}{\text{bu}} \right.$$

$$\left. \times \frac{1\,\text{HP}\cdot\text{min}}{33,000\,\text{ft}\cdot\text{lb}} \times \frac{1}{0.20} \right]$$

Flight conveyors do lend themselves to more detailed analysis if the coefficients of friction are available as in Table 13.2. The total horsepower is computed as the sum of the individual power requirements:

1. Power to lift the grain
2. Power to overcome sliding friction of the grain

TABLE 13.1. Common Seed Densities (Dried)

Crop	kg/m³	Density [lb/bu or	lb/ft³]
Barley	616	[48]	[38.4]
Bluegrass	180-385	[14-30]	[11.2-24]
Corn, popcorn			
ear	900	[70 (approx.)]	[56]
shelled	719	[56]	[44.8]
Cottonseed	410	[32]	[25.6]
Flaxseed	719	[56]	[44.8]
Grain sorghum	719-642	[56 and 50]	[44.8]
Oats	410	[32]	[25.6]
Orchard grass	180	[14]	[11.2]
Rice (rough)	577	[45]	[36]
Rye	719	[56]	[44.8]
Timothy	577	[45]	[36]
Wheat	770	[60]	[48]
Legumes (field beans, soybeans, cowpeas, alfalfa, clovers, vetch)	770	60	48
	M³/t	ft³/T	
Long, loose hay	12.5	400	
Baled hay	6.0	192	
Chopped hay	11.0	362	

TABLE 13.2. Coefficients of Sliding Friction

Metal on wood	0.50
Cast iron on steel	0.23
Steel on steel	0.57
Grain on wood	0.30
Grain on metal	0.35
Straw, hay on metal	0.60
Straw, hay on wood	0.50

3. Power to overcome sliding friction of the metal parts

Note that no power is consumed in lifting the weight of flights and chain as an equal weight is to be lowered simultaneously, but friction requirements apply both up and down. Shaft bearing friction is minimal if the bearings are kept lubricated.

Augers

Augers, used as components in many implements, are the simplest of conveyors. Large augers are used to transport small grain from combine grain tanks to wagons or trucks. These augers have very high capacities and may operate at an incline of more than 1.2 rad [70°].

Portable augers like the one shown in Fig. 13.2 may be used in place of flight conveyors for elevating grain. Quite commonly augers are used as feed conveyors to cattle (Fig. 13.3).

Capacity

The capacity of an auger may be limited by its inlet restrictions; but if the inlet opening is large, the potential capacity of an auger per revolution is the circular area between the tube and auger shaft multiplied by the axial spacing of the flighting. This circular area is seldom completely filled and depends on the rotational speed and inclination of the auger. The percent fill of the circular area and the power factor given in Table 13.3 were developed from data for augering barley reported by W. M. Regan and S. M. Henderson (Agricultural Engineering Department, University of California, Davis). Maximum capacity seems to occur at 700–800 rpm for 15 cm [6 in.] augers and 1000 rpm for 10 cm [4 in.] augers. Such speeds are common in farm-implement unloading and conveying augers, but large, heavy-duty augers for

Fig. 13.2. Auger conveyor.

Fig. 13.3. Auger used as feed conveyor.

TABLE 13.3. Auger Performance

Incline rad [deg]	% Fills at given rpm				Power Factor
	100	300	500	700	
0 [0]	85	84	69	54	1500
0.35 [20]	82	70	54	40	1085
0.70 [40]	74	59	43	33	845
1.00 [57.3]	65	49	36	28	630
1.22 [70]	47	36	29	22	612

permanent installations are run 150 rpm or less. The flight spacing in a standard pitch auger is commonly made equal to the nominal tube diameter.

Standard augers as well as flight conveyors seem to have an optimum inclination. The product of the quantity moved and the height lifted seems to be maximized at inclination of about 1 rad [57°] from the horizontal regardless of the speed of operation.

Power Requirements

Almost all the power required by augers is used in overcoming friction and energy losses between the flighting and the tube. Only a very small portion (less than 1%) of the power is actually used for lifting the weight of the grain. Horsepower for small-diameter, fully loaded farm augers can be estimated by

$$P = W \frac{x}{f\,c} \qquad (13.6)$$

where P = total pewer required to operate at full load, kW[HP]

W = mass rate of flow of material, kg/min[lb/min]

x = effective length of auger, m[ft]

f = power factor from Table 13.3

c = 1[5.39]

Belt Conveyors

Belt conveyors, not generally used on modern farm field implements, are used in stationary processing. Belts may have very high capacities, will run very quietly, but are expensive to purchase.

Belt conveyors are of two types, flat and troughed. In either case the incline for grain is limited to about 0.26 rad [15°]

Capacity

Most farm grains will pile up at about a 0.35 rad [20°] angle on either a flat belt or troughed belt. The effective area is equal to about three-fourths the resulting triangle for flat belts and equal to the triangle for troughed belts. The upper limit in belt speed is 4 m/s [13.12 ft/s].

Blowers

Blowers are used for conveying grain and forage either vertically or horizontally. As used on farms, blowers have higher capacities than other farm conveyors and require more power per unit of weight handled. (Fig. 11.38).

High-velocity air is required to float grain and forage particles in air. Minimum velocities for some farm crops are given in Table 13.4.

In general, it requires about 2–3 m³ [40–50 ft³] of air to float 1 kg [1 lb] of grain. The energy required ranges from 0.33 to 0.4 kW·hr/t [0.4 to 0.5 HP·hr/T].

TABLE 13.4. Minimum Air Velocities for Flotation

Material	m/s	ft/min
Beans	30	6000
Shelled corn	29	5800
Oats	22	4500
Wheat	30	6000
Chopped hay	20	4000
Ensilage	30	6000

The most efficient blowing practice is to feed material directly into the fan where some initial impetus is given to the grain particles. The grain may be damaged by this practice; therefore, some blowers are designed to introduce the material into the airstream after the fan. A feeder should be designed so that it will permit no air loss from the duct.

Wagons

A very important materials handling problem on farms is the transport of seed, fertilizer, manure, and pesticides to the fields; crop yields from the field to the farmstead; and feeds about the farmstead. Trucks and/or wagons are used as batch transporters for these materials.

Whether a truck or a wagon is used depends on several factors. Trucks must usually be used for highway transport, too, to achieve unit costs low enough to be economic for field use. On large farms where large quantities of material are hauled considerable distances, a truck can be economic for field transport alone. Such trucks may need to be modified for field use with special wheel tread to fit crop row spacings, extra low transmission gear ratios to keep pace with slow-moving harvesters, and flotation tires to work over soft soils.

The wagon is a versatile farm transport vehicle and has a much lower investment cost than a truck. It can be pulled by any tractor with adequate power. Several wagons can be hitched in series to form a train with a large transport capacity. Wagons can serve as temporary storage for materials without obligating a tractor or a truck.

The important part of a wagon is its running gear. Running gears may be purchased with different wheelbases, wheel treads, load-carrying capacities, and axle designs. Fig. 13.4 shows a running gear with oscillating tandem rear wheels and a rocking rear bolster that combine to ease heavy loads over rough ground with less strain on the box and less motion to the load.

Two types of steering mechanisms are used for wagons. *Pivoting spindle* steering, used by the wagons in Fig. 13.4 and 13.6, has a fixed front axle and the wheel support for the load is always at the ends of the axle. *Fifth wheel* or pivoting front axle steering, shown in Fig. 13.12, provides superior trailing characteristics at high speeds and may permit shorter turns than the pivoting spindle design.

Two-wheeled wagons or trailers are used for transporting light loads and for maneuverability around

and in buildings. Wagons used for filling feed bunks are often two wheeled (Fig. 13.5 and 13.7).

Some farm transports are dedicated to a single use. The wagon body in Fig. 13.5 serves as a bunk feeder but can be used to transport chopped forage from the field to the farmstead. The gravity unloading wagon (Fig. 13.6) is designed specifically for transporting free-flowing grain. The auger trailer in Fig. 13.7 is for delivering feed, but if the auger delivery spout is turned back toward the box, it also serves as a portable feed mixer.

Fig. 13.4. Wagon running gear.

Fig. 13.5. Cross-conveyor wagon used for bunk feeding.

Fig. 13.6. Gravity unloading wagon with telescoping tongue and flotation tires.

Fig. 13.7. Wagon unloaded by PTO-driven auger.

Fig. 13.9. Tank spreader transports liquid and solid wastes.

Fig. 13.8. Heavy-duty manure spreader.

Fig. 13.10. Manures are spread from side of tank spreader.

The manure spreader (Fig. 13.8) is a specialized transport but has a powered mill at the rear to spread the manure uniformly over the ground. A manure spreader of simple design is shown in Fig. 13.9. Essentially it is a cylindrical tank unloaded by flail action (Fig. 13.10). Liquid manure transport is accomplished with the tank trailer shown in Fig. 13.11. Fig. 13.12 is a transport for liquified petroleum gases (LPG) and for anhydrous ammonia. As an ammonia carrier, this wagon may be towed behind and serve the field applicator. Fig. 13.13 is a tank trailer for the application of low-pressure liquid fertilizer. The tandem wheel design offers less rolling resistance than duals on soft soils; the front wheel compacts the soil into a track that provides little rolling resistance for the second wheel. Such a design does give more scuffing on turns and has faster tire wear for turns on concrete surfaces.

The machinery manager must consider several factors in selecting wagons. The load capacity is listed by the manufacturer and includes the limits on wheels and tires. These limits must be respected for safety

Fig. 13.11. Liquid manure transport-spreader with filling pump.

and long wagon life. The tread width (center-to-center spacing of wheels on an axle) should be compatible with the various row spacings on the farm. Wide tread gives greater stability but overall width should be kept under 2.4 m [8 ft] for highway travel. The wheelbase should be kept short for maneuverability but be 60% of the box length for optimum sill strength. If much highway travel is anticipated, particularly with wagon trains, wagon brakes should be considered for safety reasons. A safety chain paralleling the hitch is required by OSHA.

Fig. 13.12. Liquified gas transport with pivoting front axle steering.

Fig. 13.13. Tandem-wheel trailer for applying liquid fertilizer.

Tire equipment is an important selection as tire performance and life need to be high. Oversize tires permit operation on soft soils and also reduce damage to crops and soil. M. M. Boyd (University of Massachusetts) measured a definite decrease in the damage by a truck to a grass and legume crop due to the replacement of the standard 8.25–20 tires with 15.5–20 tires. However, oversize tires are expensive and may limit permissible row curvatures in the field. (Fig. 1.4).

Operations

The number of field transports needed can be determined with the cycle diagrams presented in Chapter 1. The whole system—wagons, harvesters, and unloaders—needs to be considered when determining the economic optimum number of transports.

Maximizing the performance of transports is a goal of the machinery manager. Faster road speeds, larger boxes, shorter travel distances, faster unloading, and less waste time all contribute to fewer numbers of wagons required. Faster transport speeds are limited by available tractor speeds and rough ground. Efforts toward keeping field surfaces level to aid transportation should be taken during tillage and cultivation operations.

Rubber tire deterioration is a cost of transport equipment that the machinery manager should minimize. Observing the limits of tire loading as related to inflation pressure should lengthen tire life (Table 13.5).

TABLE 13.5. Implement Tire Static Load Limits, kg[lb], at Various Inflation Pressures, kPA[psi]

| Pressure | Diagonal (bias) ply tires. Speeds limited to 40 km/hr[25 MPH] | | | | | | |
	170 [24]	190 [28]	220 [32]	250 [36]	280 [40]	300 [44]	Maximum
5.00-15 SL	325 [715]	365 [805]	400 [880]	437 [965](**4**)			
6.40-15 SL	437 [965]	487 [1070]	530 [1170](**4**)	580 [1280]	615 [1360]	650 [1430]	690 [1520]@330 kPa[48 psi](**6**)
6.70-15 SL	487 [1070]	530 [1170]	580 [1280](**4**)	630 [1390]	690 [1520]	730 [1610](**6**)	
7.60-15 SL	560 [1230]	630 [1390](**4**)	690 [1520]	750 [1650]	800 [1760](**6**)	850 [1870]	950 [2090]@360 kPa[52 psi](**8**)
10.0-15 SL	1030 [2270]	1150 [2540]	1250 [2760]	1360 [3000]	1450 [3200](**8**)	1550 [3420]	1800 [3960]@390 kPa[56psi](**10**)
5.50-16 SL	387 [855]	425 [935]	475 [1050]	515 [1140](**4**)			
6.50-16 SL	515 [1140]	580 [1280]	630 [1390]	690 [1520]	730 [1610]	775 [1710](**6**)	
7.50-16 SL	670 [1480]	750 [1650](**4**)	825 [1820]	875 [1930]	950 [2090]	1000 [2200]	1215 [2680]@410 kPa[60psi](**10**)
9.00-16 SL	900 [1980]	1000 [2200]	1090 [2400]	1180 [2600]	1250 [2760]	1320 [2910](**8**)	1500 [3300]@360 kPa[52psi](**10**)
13.5-16.1 SL	1600 [3520](**6**)	1750 [3860]	1950 [4300](**8**)	2120 [4680]	2240 [4940](**10**)	2360 [5200](**12**)	
7.50-20 SL	730 [1610]	800 [1760](**4**)	875 [1930]	950 [2096](**6**)			
7.50-24 SL	750 [1650]	850 [1870](**4**)					
9.00-24 SL	1150 [2540]	1285 [2830](**6**)	1400 [3080]	1500 [3300]	1600 [3520](**8**)		
11.25-24 SL	1500 [3300]	1650 [3640]	1850 [4080](**8**)				
11.25-28 SL	1550 [3420]	1750 [3860]	1900 [4180]	2060 [4540]	2300 [5080]	2430 [5360](**12**)	

Notes: Bold figures in parentheses denote ply rating for which the loads and inflations are maximum.
The loads indicated here may be increased 15% for speeds not exceeding 16 km/hr[10 MPH].

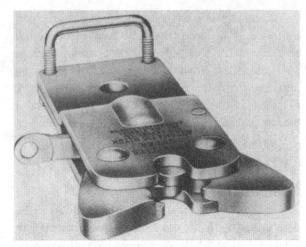

Fig. 13.14. Automatic hitch for farm wagons.

Considerable time is wasted in hitching and un-hitching wagons and tractors. The machinery manager should investigate the use of automatic hitches of the type shown in Fig. 13.14. Engagement is made by backing into the clevis pin on the wagon tongue. The hitch is released by the pull of a rope.

Telescoping wagon tongues (Fig. 13.6) are especially helpful when hitching loaded wagons in soft fields. By releasing a hand latch, the wagon tongue may be extended to make the hitch with a tractor, another wagon, or a harvester. On completion of the hitch, the whole outfit is backed until the tongue of the wagon recouples at its short length.

Loaders and Forklifts

Hydraulically operated loaders and forklifts mounted on tractors are popular and economic materials handling vehicles in fields and around farmsteads. Fruit and vegetable harvests placed on pallets or in bins are effectively handled by forklifts. Loaders are used for manure handling (Fig. 13.15) and earth moving.

The skid-steer loader shown in Fig. 13.15 is an especially maneuverable vehicle. An operator can reverse the rotation of the drive wheels on either side to cause this loader to pivot within its own wheelbase. Such maneuverability and its small size make this machine especially effective in cramped quarters inside buildings.

Loader performance is measured several ways. The *lifting capacity* is the mass in kilograms [pounds] that can be raised to full height using the lift cylinders only. The *breakout force* is the maximum force that

can be developed at the forward lip of a bucket or at the tip of manure fork tines when at ground level by use of the lift cylinders only. Bucket volume (cubic metres [feet or yards]) is a meaningful rating for low density materials and soil handling. The ultimate measure is the quantity of material per hour moved to its destination.

Operations

The machinery manager should observe several precautions when operating materials handling vehicles. A proper amount of counterweighting is needed to balance the moment of force produced on the vehicle by the lifting action. An equilibrium equation defines the mass, M, that can be lifted.

$$M = \frac{(W\,c + C\,a - R\,b)}{d} \qquad (13.7)$$

where forces R and F and forces due to masses M, C, and W are given in newtons [pounds] and distances a, b, c, and d are horizontal distances in m [ft] related to point O in Fig. 13.16. In the SI system masses are converted to newtons of force by multiplying the mass in kilograms by the acceleration of gravity, 9.807 m/s^2.

The achievable maximum, M, is the point at which the rear wheel soil force, R, goes to zero. Should M be too heavy or the breakout force too great, the hydraulic lift cylinders merely jack the rear wheels off the ground. Adding more counterweights will improve vehicle stability.

Fig. 13.15. Skid-steer loader.

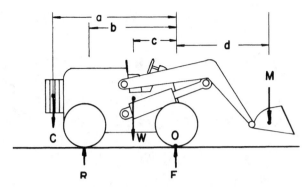

Fig. 13.16. Forces acting on loader.

Materials handling vehicles sustain heavy loads on their tires. Overloading is not as critical at low speeds as it is for high speeds.

Loaders and forklifts should operate at slow forward speeds, particularly on rough ground and with a raised load. In the raised position the mass of the load raises the center of gravity of the forklift or loader, making it less stable and more susceptible to sidewise overturning.

Practice Problems

13.1. A horizontal, troughed belt conveyor has an effective conveying cross-sectional area of 80 cm^2 [12.4 in^2] and travels 2 m/s [6.6 ft]/s. What is the capacity in tonnes [tons] per hour for wheat?

13.2. A horizontal 10-m [32.8-ft] auger having a single-thread screw, a 10-cm [4-in.] diameter tube, and a 2.5-cm [1-in.] shaft turns at 400 rpm. What is the estimated capacity for shelled corn in kg [bu] per hour?

13.3. Barley is conveyed by a flight conveyor inclined at 1 rad [57.3°]? The trough is 250 mm [10 in.] wide and 100 mm [4 in.] in height. The flights are 100 mm [4 in.] in height, 3 mm [0.12 in.] thick, are spaced at 250 mm [10 in.], and move at a speed of 1 m/s [3.28 ft]/s.

   a. What is the critical angle for this elevator?

   b. What is the estimated capacity in tonnes [tons] per hour?

   c. What is the power requirement if the barley

is elevated 4 m [13.12 ft]? The conveyor has chains and flighting weighing 2kg/m [1.34lb/ft] that rub steel sheeting.

13.4. Assume the load is carried evenly on the 9.00-16 tires of the running gear in Fig. 13.4. What is the maximum load in tonnes [tons], including box weight, this wagon can carry:

   a. At 40 km/hr [25 MPH]

   b. At 8 km/hr [5 MPH]

13.5. The weight of a 4000-kg [8800-lb] tractor is carried 75% on its rear wheels and 25% on its front wheels when equipped with a front-mounted loader. The wheelbase of the tractor is 2 m [78.74 in.] and the center of the loader bucket, when at ground level, is 1 m [39.37 in.] in front of the front axle. What is the maximum breakout force this loader can develop without tipping forward if a 500kg [1100 lb] counter weight is attached 0.5m [19.9 in.] behind the rear wheel.

Chapter 14

SPECIAL CROP MACHINES

Many special machines are available for crops other than grain and forages. In many cases machines used for grains and forages can be adapted with few changes for use on special crops—particularly tillage, seeding, and cultivation equipment. Such adaptations can be quite economical, as the manufacturing volume of grain and forage equipment permits the economies associated with large-scale industry.

But the harvesting of special crops often requires a machine completely different from anything even remotely connected to grain or forage harvesting. Such machines may be quite expensive because they either involve many special mechanisms and/or because the volume of sales is so low that the construction of such machines is on a machine-shop basis.

Several special machines are used in such quantities that they are regular production items for some companies (Fig. 14.1–14.4). Machines falling in this category are bare root transplanters; planters for potatoes, beets and beans; cultivators for vegetables; and harvesters for cranberries, cotton, green beans, peanuts, sugar beets, sugar cane, and sweet corn.

Cotton harvesters are good examples of machines built to accommodate a special crop. Unlike other special crop machinery, the value of cotton pickers and strippers sold each year is significant in the farm equipment industry even though the numbers sold are small.

Cotton harvesters are different from most other harvesters in that they are designed for use with specific crop varieties. The *cotton stripper* is used with varieties whose bolls open slowly if at all. This machine is for once-over operation since it harvests open and closed bolls indiscriminately. The separation of the fiber from

Fig. 14.1. Two-row sugar beet harvester.

Fig. 14.2. Four-row potato planter.

the boll is left to the ginning operation. The *spindle picker* pictured at the beginning of this chapter is used on varieties whose bolls open fully. It is a selective harvester in that its rotating spindles pluck only the fiber from an opened boll. Fig. 14.5 shows the action of a spindle picker where rotating spindles enter the row, wind up the fiber, and withdraw from the row. The cotton is subsequently stripped from the spindles and carried away by an airstream. The spindle machine may be used twice or more on the same row as the bolls open progressively.

Special wagons are used to transport harvested seed cotton. Because the density of cotton is so low, the reach of a 4.5t [5 T] running gear is lengthened to produce a wheelbase of as much as 5.75 m [19 ft] (Fig. 14.6). The box dimensions can be as long as

Fig. 14.3. Sugarcane harvester.

Fig. 14.4. Peanut harvester.

Fig. 14.5. Spindle picker mechanism: 20 spindles/ bar, 16 bar front, 12 bar rear, moistener pads, doffers, suction delivery to basket, automatic height sensor.

Fig. 14.6. Cotton box.

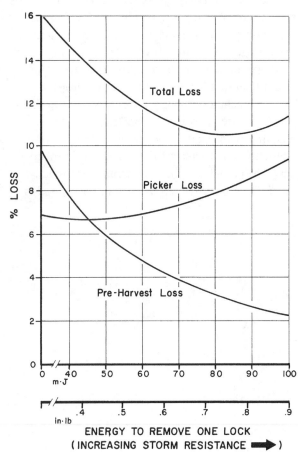

Fig. 14.7. Picker efficiency as function of plant characteristic.

7.3 m [24 ft], as wide as 2.4 m [8 ft], and as high as 1.8 m [6 ft].

Plant characteristics can have a significant influence on the operation of cotton pickers. Fig. 14.7, developed by T. E. Corley (Agricultural Engineering Department, Auburn University) shows the losses associated with varieties having various degrees of resistance to fiber removal. The farm manager of special crop machines must be aware of the limitations imposed by the crop on these machines.

The effect of stripper roll spacing on the harvest of storm-resistant cotton was investigated by D. F. Wanjura and A.D. Brashears (USDA, Lubbock, TX.). Stripper roll clearances were spaced from 0 to 2.2 cm [0.87 in.] on a machine equipped with rolls containing alternate rows of brushes and rubber strips. The more aggressive settings (closer spacings) resulted in the least harvest loss but increased the foreign matter (primarily plant stems) in the yield. High plant-moisture content at harvest time reduced the foreign matter in the yield. Taller plants contributed more foreign matter to the yield than did smaller plants.

Probably every crop, be it fruit, nut, berry, root, fiber, or leaf, can be mechanized from seeding to harvest. Current commercial machinery is available for at least partially mechanizing the production.

The extent to which mechanization is possible depends not so much on engineering technology as on the economics of applying the machinery. Each year, however, the cost of labor rises and its availability decreases, thus bringing ever closer the time when all crops will be mechanized.

Management of special crop machines does not differ in principle from that of grain and forage production machines but it may be much more important; costs are likely to be higher, thus the money lost by poor management can be greater. The timeliness for the operation may be somewhat more critical, as special crops tend to be of higher value and more susceptible to loss and quality reduction than grains and forages. Superior preventive maintenance and field operation practices will probably give a greater return than that for the more common crops.

The machinery manager for special crop enterprises is more likely to compare the costs of a special machine with hand labor than with other machine alternatives, particularly if the machine is relatively new. The problem is one of determining a *break-even point*, the point at which the size of the operation is such that machinery costs exactly equal the cost of doing the job with hand labor.

At the break-even point, (annual machine fixed costs) + (annual machine operating costs) + (reduced value of crop due to reduction in quality, if any) = (annual hand labor costs).

The symbols for the approximate cost analysis (Eq. 4.8) may be used and rearranged to express the break-even area, A, in

$$A = \frac{(FC\%)\,(P)}{\left(\dfrac{nL_h}{C_H} - \dfrac{c}{Swe}\left[(R\,\&\,M)P + nL_m + O + F + T\right] - D\right)100}$$

(14.1)

where the new symbols mean

n = number of laborers required

L_H = hand labor rate, $\dfrac{\$}{hr}$

L_M = machine labor rate, $\dfrac{\$}{hr}$

C_H = capacity of total hand labor force, $\dfrac{ha}{hr}\left[\dfrac{a}{hr}\right]$

D = value for quality discount, $\dfrac{\$}{ha}\left[\dfrac{\$}{a}\right]$

For many harvesting jobs, area is not the most appropriate unit, as hand labor rates may be more often quoted on the pound, barrel, or crate basis.

Part IV

POWER

Driving that tractor Sank didn't look like a humble and degraded tiller of the soil. He didn't look like a hay-chewing rube with chin whiskers, or a dunghill yokel, or a peasant without thought or hope. The tractor had done that. It had changed a farmer from a clod into an operator; from a dumb brute into a mechanic, all over the world. The tractor had done more than anything that had ever happened in the whole history of agriculture since the invention of the wheel.

—MORROW MAYO

"The Man with a Tractor," *Harper's Magazine*, 1938

THE FARM TRACTOR

Management of the power source has not been considered in the previous discussions. Although the field machine requires power to do its work, the source, the type, and the mechanism of the power were immaterial as long as they didn't interfere with the functioning of the machine. But the cost of power for the machine approaches the cost of use of the machine for many operations; thus efficient management of power is fully as important as efficient management of machinery.

The farm power manager's major goal is to minimize the costs of power while maximizing the returns from the farm enterprises. To do so, an understanding of fundamentals, economics, details of operation, and energy sources of power plants is needed.

The power plant for field work today must be compact and light as it is a moving power plant. Field work has been accomplished with semistationary power plants in the past (cable plowing, for example) and might be again in the future. It is almost inconceivable that it will ever be practical to bring the field to the power plant rather than take the power plant to the field; yet the highest efficiencies obtainable are with large, stationary power plants.

Three major shifts in sources of field power have been humans to animals, animals to external combustion engines (steam), and external combustion to internal combustion engines. Doubtless within the lifetime of many presently living, a fourth shift will occur. Very likely the future power shafts will be turned by electricity but the source for the electric energy may be gas turbines, fuel cells, solar cells, atomic energy, or some source unknown at the present.

Today, practically all field power comes from internal combustion engines, and most of these engines are mounted in tractors. This section covers the portions of development, operation, performance, mechanisms, repair, and maintenance that are of value to the farm power manager.

The application of power other than human to agriculture has been a matter of great concern through the ages. Animal power was used well before recorded history to supplement human labor. With the advent of steam power in the early nineteenth century, efforts were made to utilize steam force for agriculture. After the Civil War and until 1920, steam power was commonly used for plowing and belt work. However, the steam power unit was bulky and heavy, and although the fuel was cheap and water was usually free, considerable labor was required to operate and fuel the engine. When the engine was placed on a carriage and the rear wheels were powered, the resulting combination, though forceful, was cumbersome and not applicable to many of the jobs found on farms.

The first traction engines using an internal combustion engine appeared about 1890 and had the same disadvantages as the steam engines (Fig. 15.1). However, as internal combustion engines were improved, the horsepower-to-weight ratio surpassed that of the steam engine and the era of the modern lightweight, powerful, adaptable tractor arrived. The name *tractor* was coined about 1906 by a farm machinery salesman.

Early tractor engines were designed to burn kerosene (Fig. 15.2), for kerosene was plentiful and inexpensive during the first third of the twentieth century.

Tractor design has varied considerably in the past and may be expected to change even more radically in the future. The first tractors were merely self-propelled power units whose belt pulley power was used for threshing and other stationary jobs. Early plowing tractors were able to exert considerable draft but were slow of speed and therefore had rather low power ratings.

Manufacturers soon saw that a tractor that could replace the horse in all work would be desirable. Consequently, several two-wheeled front drive tractors were produced that could attach directly to horse-drawn implements (Fig. 15.3). These tractors were rather difficult to operate; but with power steering and the specifically designed implements common today, this type of tractor design might have persisted.

A ready farm market during World War I plus the economies of mass production jumped tractor numbers and tractor popularity (Fig. 15.4). Prosperity

Fig. 15.1. Otto traction engine. Power, 15 kW [20 HP]; mass, 4500 kg [10,000 lb].

Fig. 15.2. Early kerosene-burning tractor, 1914.

Fig. 15.3. Two-wheeled front drive tractor, 1918.

also induced many small manufacturers to produce tractors. By 1920 there were an estimated 100 tractor manufacturers.

The Nebraska Tractor Tests, established at this time, were an attempt to evaluate the tractors sold in Nebraska. These tests and the agricultural price recession in the early 1920s eliminated from the scene many smaller-financed, inefficient tractor manufacturers. Only the more strongly financed companies and those whose tractors had a good record in the field endured.

Fig. 15.4. One of first mass-produced tractors. 1917.

High clearance was the next step in making tractors useful for all farm jobs. A tractor-mounted cultivator released the horse from its last farm job. After some experimentation, manufacturers settled on a dual front wheel arrangement to give tractors necessary row crop clearance (Fig. 15.5).

A crawler tractor (also called a *track-layer*) was developed in 1904 by Benjamin Holt as a solution to the problem of keeping farm tractors from miring in the rich, spongy soil along the San Joaquin River in California. The tractor's rear drive wheel was replaced by a large, toothed sprocket that drove an endless chain, called *track*, carrying cross-mounted wooden slats around a smooth front wheel. The weight of the tractor was also carried by a series of idler wheels spaced between the front and rear wheels that rolled on the back side of the track. This design produced a large bearing area on the soil over which to distribute the tractor's weight. With very wide tracks, the pressure produced by the tractor weight can be less than that created by human foot pressure. In securing increased flotation, the tractive effort was also increased. In the words of the developer, "A 40 HP tractor (equipped with tracks) could pull four gangs of plows 2 inches deeper than a 60 HP wheel model pulled three similar gangs."

Fig. 15.5. One of the first high-clearance tractors, 1924.

The track design has persisted through the years for conditions demanding high flotation, large drawbar pulls, or both. The design was adapted for World War I tanks to provide cross-country travel. By 1928 earth-moving tractors utilized the track design almost exclusively (Fig. 15.6).

Since 1920 the history of farm tractors is one of gradual increases in power and efficiency and sudden adoptions of new mechanisms that greatly improve the tractor's utility. With the advent of the power take-off shaft in the 1920s, field machines no longer

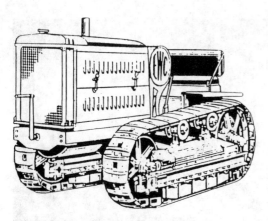

Fig. 15.6. Track-laying tractor, 1928.

needed auxiliary engines. Tractor power could be applied directly to the functional needs of harvesters, tillers, etc.

The adoption of pneumatic tires in the mid-1930s decreased tractor rolling resistance and permitted higher field speeds. High compression engines for tractors were initiated at this time to take advantage of the increases in the octane levels of gasolines. Cranking motors, lights, and hydraulic implement lifts appeared on many 1940 models.

After World War II the price structure made other fuels competitive in cost with gasoline. Some manufacturers provided a choice of gasoline, diesel, or LP gas engines for many tractor models. Mounted implements and the three-point hitch were developed before the war but were generally adopted during the 1950s. Concurrently, most manufacturers incorporated some type of traction-assist with their mounted hitch mechanism. During this period the number of transmission gear ratios available on tractors was expanded greatly and several shift-on-the-go features were introduced.

Increased use of hydraulics was typical of 1960 tractor models. The farmer could control implements, steer, brake, and change gear ratios quickly and easily. Turbochargers were adopted during this period also.

Concern for the operator's safety, comfort, and convenience was reflected in the tractor designs for the 1970s. Tractor cabs strong enough to withstand overturning and equipped with heaters and air conditioners are popular options available to farmers. Concern for noise damage to the ears of both the operator and the bystander led to changes in cab and engine design.

The decade of the 80s included several develop-

ments. Diesel engine use was extended even to small tractors and for some pickup trucks. Rapid advances were made in the use of electronics for monitoring and in the use of microprocessors for control. The development of a rubber track for a crawler tractor advanced the use of that form of tractor for agriculture.

Because today's society has a wide range of tractor power needs, manufacturers provide many different forms of tractors. The garden tractor (Fig. 15.7) meets the needs of many suburban home owners, gardeners, and farmers. By contrast, the 4-wheel-drive tractor meets the needs of farmers for powerful wheel tractors (Fig. 15.8). The utility tractor, as its name implies, is useful for field work and for the many transport, materials handling, and processing operations on farms (Fig. 15.9). It is small, highly maneuverable, and easy for an operator to mount and dismount; it furnishes hydraulic and PTO power, accepts three-point hitch-mounted implements, and has traction-assist. The construction and manufacturing industries use it as a general power source.

It is economically sound to have the power source as a separate machine (a tractor). The idea of the tractor is to get the greatest use of an expensive engine and power train by coupling them with many different

Fig. 15.7. Small tractor for garden work.

Fig. 15.8. Four-wheel-drive tractor.

Fig. 15.9. Utility tractor.

implements. An alternative is for all implements to be self-propelled with their own dedicated engine and gear train as are some combines, windrowers, and sprayers. The power unit on self-propelled machines is necessarily limited to use for that machine. The power cost for self-propelled machines may be very high unless they have considerable annual use.

A self-propelled design has economic advantages, especially in harvesting machines, in which absence of excessive headland travel and land opening losses reduces costs.

The following sections discuss details of engines and power trains of use to machinery managers.

Engine Principles and Mechanisms

The basic mechanism of the internal combustion engine is the linkage illustrated in Fig. 15.10. High pressures due to the combustion of fuel are

brought to bear on the piston head. The connecting rod transmits this force into a torque on the crankshaft.

Engine position is denoted by the distance from top dead center (TDC) or bottom dead center (BDC), expressed as an angle of the circle described by the connecting rod-crankshaft bearing. The position shown is 1 rad [57°] after top dead center (ATDC) if a clockwise rotation is assumed.

The size of an engine is given by its total piston head displacement volume, a cylindrical volume in which the stroke is the height of the volume and the bore is the diameter. By convention, cylinder specifications are reported as diameter first and then length of stroke as in 114mm × 135mm [4.49in × 5.32in.].

Compression ratio is the ratio of combustion chamber volume at BDC to the volume at TDC. The volume remaining above the piston head at TDC is referred to as the clearance volume. Compression ratio, CR, can be expressed in terms of the clearance volume and the displacement volume as in

$$CR = \frac{\text{displacement vol} + \text{clearance vol}}{\text{clearance vol}} \qquad (15.1)$$

An analysis of the theoretical cycles for the two types of engines, spark ignition (SI) and compression ignition (CI), produce efficiency equations based on compression ratio. For the

$$CI \text{ engine : eff} = 1 - \frac{1}{CR^{k}-1}\left[\frac{R^{k}-1}{k(R-1)}\right]$$

$$SI \text{ engine : eff} = 1 - \frac{1}{CR^{k-1}} \qquad (15.2)$$

where k is 1.2, a modified ratio of the specific heats of air at constant pressure to that of constant volume.

R, called the cutoff ratio, is the ratio of the volume when injection stops to the volume at TDC.

In numbers, the efficiency of a CI engine having a CR of 16 and a 1.1 cutoff ratio would be 42%. At a similar compression ratio, the SI engine efficiency would be 42.5%, but engine knock limits the CR for

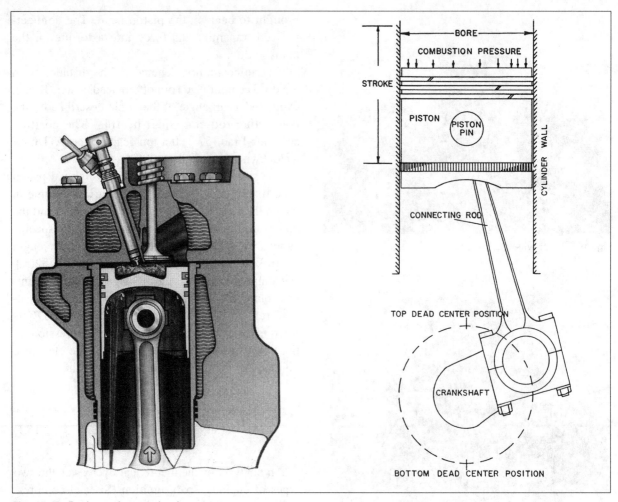

Fig. 15.10. Basic engine mechanism.

SI engines to be no more than 9, which produces an efficiency of only 35.5%.

Cycles

Farm tractor engines follow a 4-stroke cycle. A cycle refers to the engine movement required before the events repeat. The necessary events in any internal combustion engine are (1) the intake of air, (2) the compression of the air, (3) combustion and expansion, and (4) exhausting the burned products. Fuel may be inducted with the air in carbureted engines or it may be injected during the combustion and expansion process.

In the 4-stroke cycle, approximately 1 stroke is required for each event, or 2 complete revolutions are required for a cycle to be completed. In the 2-stroke cycle the exhaust and intake events occur simultaneously—the other events are shortened so that a complete cycle may occur in 1 revolution of the engine crankshaft.

Two methods are used for allowing a 2-stroke cycle engine to intake and exhaust simultaneously.

In one method a separate blower or high-capacity fan powered from the crankshaft maintains an air pressure in the intake manifold. When the intake and exhaust valves are both open, the high-pressure air blows out the exhaust gases and charges the combustion chamber with fresh air. The second method uses crankcase compression to charge the combustion chamber with air and fuel. By using check valves in the sealed crankcase the downward motion of the piston forces air into the combustion chamber. The fuel mixture used in this type of engine must be viscous enough to provide crankcase and cylinder wall lubrication since no crankcase oil is employed.

Two-stroke cycle engines are lightweight for the power produced. Using crankcase compression and a non-float carburetor, they are able to operate in any position, are simply constructed, and are inexpensive power for such tools as portable chain saws, weed trimmers, and leaf blowers.

Four-stroke cycle engines have greater fuel efficiency at the high engine speeds most commonly used by farm tractors.

Cylinder Numbers

The number of cylinders used in an engine is a compromise between expense and smoothness of operation. Farm tractors are built with engines having 3 to 8 cylinders. Engines with 6 cylinders or more have overlapping power strokes. Then the flywheel function of providing power between engine power strokes is unnecessary but the flywheel is used to store energy that is given up in the event of sudden overloads.

Valves and Valve Timing

Good valve action is essential to proper engine operation. Valves must seat tightly to seal off the combustion chamber during the power stroke. The exhaust valve must be able to seal effectively while exposed to gas at very high temperatures. The tractor engine in Fig. 15.11 has four valves per cylinder for improved gaseous transfer.

Today's tractors use an overhead valve placement design. The valve-actuating linkage begins at the camshaft with the cam follower or tappet. The tappet lifts the rod, which operates the rocker arm that bears on the valve stem. As the engine temperature rises, these linkage parts expand and would hold the valves

open except for the planned gap or *valve train clearance* between the rocker arm and the valve stem. The clearance values for farm tractors range from 0.2 mm to 0.5 mm [0.008 to 0.020 in.] with the exhaust valve clearance often greater than the intake. The manufacturer may recommend the clearance on either a hot or a cold engine temperature basis (Fig. 15.12).

In engines with hydraulic valve lifters, the valve train clearance volume is taken up with trapped oil. As the cam lifts, the oil transmits the action to the push rods. As heat expands the parts, less oil is needed in the clearance volume. This arrangement reduces noise and provides a gentle, full valve range regardless of the engine temperature.

Valves may fail from

1. Sticking open because of overlubrication or excessive engine head temperature
2. Burning because of insufficient tappet clearance, lean fuel-air mixture, warped valves or valve seats, or deposits on the valve seat or valve face
3. Breaking because of poor valve materials or excessive tappet clearance

Fig. 15.11. Six-cylinder, diesel tractor engine with electronic fuel injection.

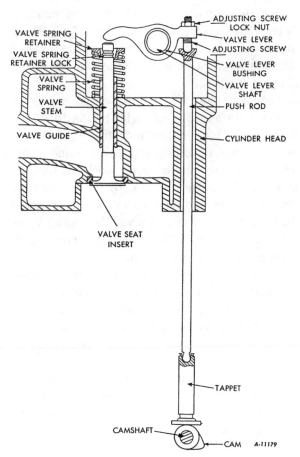

Fig. 15.12. Engine valve linkage.

Valve rotators are used on the exhaust valves of many farm tractors. Rotating the valve every time it is lifted reduces valve stem sticking and valve seat and valve face deposits that lead to valve burning (Fig. 15.13).

For prolonged high-temperature operation, hollow, sodium-filled, exhaust valves are used. The sodium promotes rapid heat transfer from the head of the valve through the stem and valve guide to the surrounding cooling water.

The valve timing spiral in Fig. 15.14 represents a typical valve timing for farm tractors. The valve notations illustrated are points where the valves *begin* to open and *completely* close. Because of the shape of the cam, the valve is completely open only at the midpoint between the opening and closing points.

Valve range refers to the number of degrees of crankshaft rotation from the time of valve opening to closing. The valve range for high-speed engines is greater than that for low-speed engines. Such increased range is necessary for the valves to be open long enough to do an adequate job of filling and emptying the cylinders.

The position of the crankshaft and piston mechanism is shown externally by a pointer and markings on the flywheel or fan drive pulley (Fig. 15.15). These markings are for cylinder number 1 and are revealed with a stroboscopic flash unit while the engine is running.

The engine block is a cast-iron structure that serves as a base for all the other engine mechanisms. The tractor engine block differs from other engine blocks in that it may serve as a part of the frame of the tractor. Furthermore, the tractor block has replaceable cylinder liners (Fig. 15.16). These liners or sleeves may be a dry type or a wet type as shown in the photo for Problem 15.5 where the cooling water comes in contact with the sleeve. Wet sleeves require sealing at both top and bottom.

A tractor engine may be completely renewed upon a deterioration in performance because of wear. The piston-connecting rod mechanism (Fig. 15.17) is made up of replaceable parts. The compression rings, the oil control ring, the piston, the piston pin, and the connecting rod all may be replaced. These replacements along with new sleeves, valves, and valve guides may in effect produce a new engine. Occasionally the crankshaft, camshaft, and their bearings must be replaced, too.

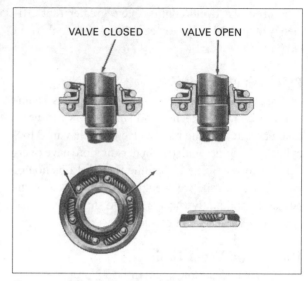

Fig. 15.13. Valve rotator mechanism.

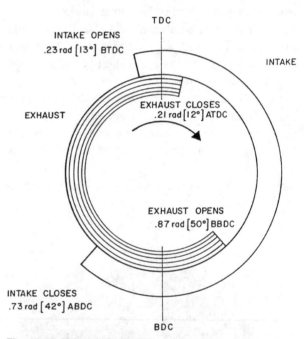

Fig. 15.14. Valve timing spiral.

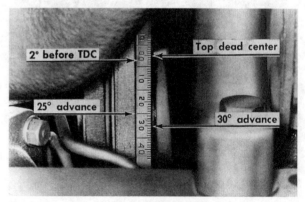

Fig. 15.15. Timing marks on fan drive pulley.

Fig. **15.16.** Engine block with removable cylinder liners.

Fig. **15.17.** Piston-connecting rod assembly.

Practice Problems
(Engine Principles and Mechanisms)

15.1. Given: A 4-cylinder, 100 mm × 100 mm [4 in × 4 in.] engine turning at 1600 rpm.

a. What is the total engine displacement?

b. If the instantaneous piston head pressure is 690 kPa [100 psi] at 1.57 rad [90°] ATDC, what is the torque on the crankshaft?

c. If the mean effective pressure per power stroke is 550 kPa [80 psi], what is the indicated power of the engine?

d. If the individual cylinder's clearance volume is 130 cm³ [8 in³], what is the engine's compression ratio?

15.2. A certain tractor engine has a 1-5-3-6-2-4 firing order. When the timing marks show TDC, a valve clearance adjustment may be made on which cylinders?

15.3. A 4-stroke cycle engine is running at 2400 rpm. If the intake valve range is 3.84 rad [220°], for what fraction of a second is the intake valve open?

15.4. Cylinder wear has been related to the average piston speed in metres per second [feet per minute]. What is the average piston speed for a 100 mm × 120 mm [4 in × 5 in.] engine running at 2400 rpm?

15.5. A left-side sectioned view and a rear sectioned view of the same spark ignition engine are shown in the figures on the next page. Identify the parts listed below with arrows and appropriate key numbers. Have the instructor check for accuracy. Practice identification of parts by studying the completed illustrations without referring to the list.

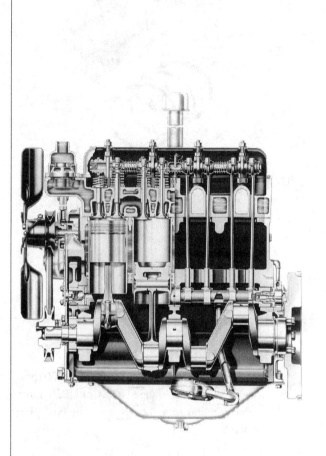

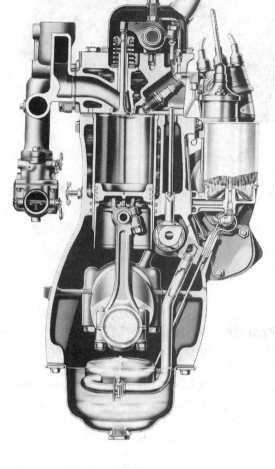

Left-side view of spark ignition engine
(Problem 15.5).

Rear view of spark ignition engine
(Problem 15.5).

1. Fan	16. Camshaft	1. Exhaust manifold	16. Push rod
2. Fan belt	17. Cam follower	2. Intake manifold	17. Oil dipstick
3. Thermostat	18. Cam lobe	3. Updraft carburetor	18. Oil filter
4. Water pump	19. Oil intake	4. Block coolant drain	19. Coil
5. Camshaft gear	20. Flywheel	5. Cylinder liner	20. Distributor
6. Oil pressure control	21. Oil pump	6. Piston	21. Spark plug
7. Cylinder liner	22. Push rod	7. Compression rings	22. Rocker arm
8. Piston	23. Oil to rocker arm	8. Oil control ring	23. Oil to rocker
9. Oil control ring	24. Rocker arm shaft	9. Piston pin	24. Valve clearance adjustment
10. Piston pin	25. Rocker arm	10. Connecting rod	25. Valve rotator
11. Connecting rod	26. Valve spring	11. Crankshaft	27. Valve spring
12. Connecting rod bearing	27. Exhaust valve rotator	12. Oil intake	27. Valve guide
13. Crankshaft	28. Exhaust valve	13. Camshaft	28. Valve
14. Center main bearing	29. Intake valve	14. Cam lobe	29. Coolant
15. Center camshaft bearing	30. Valve guide	15. Cam follower	30. Breather

Fuels and Combustion

Engines create power from combustion of fuels. The availability, energy value, combustion characteristics, and economy of the fuel are of prime importance to the machinery manager. Table 15.1 lists the properties of some tractor fuels.

Burning of fuels is a chemical oxidation of carbon, C, and hydrogen, H. The reaction equation for gasoline is

$$2\, C_8H_{18} + 25\, O_2 \rightarrow 16\, CO_2 + 18\, H_2O$$

The masses involved are

$2(96 + 18) + 25(32) \rightarrow 16(12 + 32) + 18(18)$
228 units C_8H_{18} + 800 units $O_2 \rightarrow$
704 units CO_2 + 324 units H_2O_2

Since air is about 23.2% oxygen by mass, the correct air-fuel (A-F) ratio is 3448 units air to 228 units gasoline or 15.1:1. Other elements and compounds enter into the actual combustion equation; large quantities of inert nitrogen are carried through and elemental carbon, carbon monoxide, and hydrogen may be found in the exhaust gases.

For spark ignition engines, *volatility*, the ability to vaporize, is an important fuel property because liquids will not burn. Since refined petroleum fuels are not homogeneous, the percent of the fuel vaporized at any time depends on the temperature of the fuel at that time. The distillation curves (Fig. 15.18) show the boiling points of several fuels. The highly volatile liquified petroleum (LP) gases, propane and butane, are not shown on the chart; these fuels are more homogeneous because of their simpler molecular structure and they tend to have constant boiling points that depend only on pressure. At atmospheric pressure propane boils at $-42°C$ [$-44°F$] and butane at $0°C$ [$32°F$].

The low volatility fuels cannot be used conveniently in a spark ignition (SI) engine, but they are commonly used in compression ignition (CI) engines. Fuels with low initial vaporization temperatures exhibit easy cold weather starting. A high end point temperature indicates crankcase oil dilution tendencies due to incomplete combustion.

Marketers of gasoline consider seasonal temperatures when blending fuels. The cracking towers of petroleum refineries boil crude oil and then cool the vapors. The heavy oils condense first. The lighter fuels condense last and are the most volatile liquids. (See Table 15.1 to compare the mass per volume of volatile gasoline with low-volatility diesel fuel.) Liquids must vaporize before they will burn. Cold starting of SI

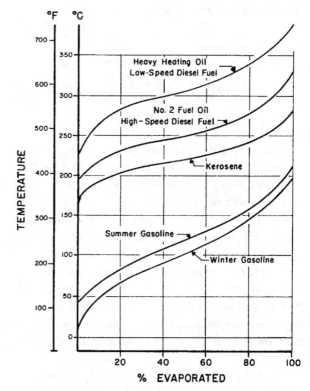

Fig. 15.18. Distillation curves.

TABLE 15.1. Typical Properties of Fuels.

Fuel	API Gravity, deg	Mass per Volume, kg/L [lb/gal]	Energy per Mass, MJ/kg [Btu/lb]	Energy per Volume, MJ/m³ [Btu/gal]	Octane Number, avg.	Practical Compression Ratio
Propane	146	0.51 [4.25]	50.31 [21,680]	25,719 [92,300]	100	10.0:1
Premium gasoline	62	0.73 [6.06]	47.15 [20,320]	34,330 [123,200]	91	8.5:1
Regular gasoline	60	0.74 [6.13]	47.06 [20,280]	34,636 [124,300]	87	8.0:1
Diesel fuel	35	0.85 [7.08]	45.46 [19,590]	38,676 [138,800]	40[a]	20.0:1
Methanol	46	0.80 [6.63]	19.95 [8,600]	15,890 [57,000]	119	12.5:1
Ethanol	46	0.79 [6.61]	26.69 [11,500]	21,180 [76,000]	115	12.0:1

[a]Cetane rating.

engines requires fuels that will vaporize at low temperatures. Warm temperature starting can use the lower-cost, less-volatile fractions of gasoline. Gasolines contain many different additives. The antiknock additive is referred to in the discussion on octane. Other additives counter combustion chamber deposits, corrosion, rust, oxidation during storage, emulsion with water and fuel line freezing.

Octane rating is the second important fuel property for spark ignition engines and is related to the ability of a fuel to burn in an engine. Mechanical damage to the engine can result if severe detonation (knocking, pinging, autoignition) occurs. Detonation is detected as a noise arising from combustion pressure vibrations. These vibrations occur when the unburned fuel and air experience spontaneous combustion. Spontaneous combustion temperatures may be reached as a result of

1. Engine deposits
2. Spark occurrence advanced too far
3. Overloading and overheating
4. Engine compression ratio too high for the fuel being used

Fuels vary in their detonation characteristics. These variations are expressed as an octane rating for that fuel. This rating is determined by comparative engine tests with two base fuels, normal heptane and 2, 2, 4 trimethyl pentane (isooctane). These two fuels are rated as 0 and 100 octane, respectively, in their ability to burn in an engine without detonating. If a specific fuel detonates in an engine identically as does a mixture of 60% isooctane and 40% heptane, the fuel is rated as 60 octane. Since some fuels are even better in their antiknock characteristics than isooctane, these fuels are said to have octane ratings of over 100.

The LP gas fuels are naturally highly resistant to detonation, while the heavy fuels are very likely to detonate. Natural gasoline has a moderate octane rating that may be raised greatly with additions of alcohols or other organic and inorganic compounds.

Diesel engines are subject to detonation also, but in a different way than SI engines. Diesel fuels detonate because they do not burn fast enough. An ignition delay occurs on injection because the fuel needs to absorb heat and mix with oxygen. The accumulated fuel explodes (detonates) rather than burns as it is injected. This effect is most noticeable at idle as the effect is nearly masked out when the engine is under load. The *cetane rating* is the rating given to diesel fuels for their ability to burn without detonating.

Table 15.2 lists the grades of diesel fuel. A low flash point indicates an easy starting fuel. A high 90% point indicates possible incomplete combustion and smoke. A high viscosity reading defines a fuel that may not flow to the pump in cold weather. The 1-D fuel tends to burn cleaner but has less heat energy per volume than does 2-D. The lowest operating temperature for 2-D is about −7°C [20°F] to get adequate fuel flow. At lower temperatures the machinery manager will dilute 2-D with 1-D or kerosene, or use straight 1-D. Additives for diesel fuel include cetane improvers, oxidation inhibitors, biocides for bacteria and fungi, rust preventers, metal deactivators, pour-point depressants, demulsifiers, and de-hazers (haze is a suspension of very small water droplets in the fuel), smoke depressants, detergent-dispersants, conductivity improvers (dissipates electrostatic charge), deicers, and dyes.

Fuel Impurities

Sulfur is the most serious fuel impurity. When sulfur is burned, SO_2 is formed. When combined with water and oxygen, SO_2 can form a powerful acid. The sulfur content of gasoline is usually held to less than 0.25%, while for fuel oils it may reach 2%. The sulfur problem can be minimized by oil additives and by operating the engine at a high temperature.

Gum formation in the fuel occurs because of slow oxidation of the fuel. The gum formed can clog carburetor passages, fuel injection nozzles, and the intake manifold; can lacquer the pistons, rings, and

TABLE 15.2. Properties of Diesel Fuels

Grade	Flash Point °C [°F]	90% Distillation Point, °C [°F] Min.	90% Distillation Point, °C [°F] Max.	Viscosity, cST at 40°C Min.	Viscosity, cST at 40°C Max.	Cetane No.	Sulfur, % by Weight
1-D	38 [100]	•••	288 [550]	1.3	2.4	40	0.5
2-D	52 [125]	282 [540]	338 [640]	1.9	4.1	40	0.5
4-D	55 [130]	•••	•••	5.5	24.0	30	2.0

cylinder walls; and can stick the valves. Using fresh fuel minimizes gum troubles.

Water and dirt impurities can enter during fuel storage and transfer. In addition, water vapor can condense in vented fuel tanks because of atmospheric temperature changes. SI engines are tolerant of small amounts of these impurities, but proper fuel storage and handling facilities plus correct engine operating temperatures should be maintained at all times. If the water vapor can be exhausted before cooling and condensing, there should be no engine damage.

Fuel Storage

Proper fuel storage facilities are a part of good power management. Fig. 15.19 shows a design for an aboveground fuel tank that can be emptied by gravity or a pump. Gasoline fuel tanks are subject to vapor loss. Pressurized fuel caps and cool storage temperatures provided by shaded or underground tanks reduce this loss. An aboveground tank, 1135 L [300 gal], painted a dark color, vented, and exposed to sunshine can lose 36 L/mo [9.5 gal/mo] during normal use.

Carburetion

A carburetor is a device that meters and mixes fuel and air into a combustible mixture. The mixture is pulled into the combustion chamber by the intake (suction) stroke of the piston. Carburetors are inexpensive and are now used almost exclusively on small, single cylinder engines.

While the air-fuel ratio of 15.1:1 is theoretically correct for gasoline, actually the carburetor should deliver different air-fuel (A-F) ratios for different engine speeds as shown in Fig. 15.20. When the engine is idling, the A-F mixture should be richer since the valve timing values are unsuited to slow speeds. The other speed ranges are economy and power; the former requires as lean an A-F ratio as permissible, while the latter sacrifices economy for maximum power performance.

The simplest up-draft carburetor appears in Fig. 15.21. The float and the float needle valve maintain a definite fuel level. As the engine pistons go down on the intake stroke, air is sucked up through the carburetor. When air passes the main tube, the venturi restriction causes the air velocity to momentarily increase. As the velocity increases, its static pressure decreases and the difference between that pressure and the float chamber pressure forces fuel out the tube and into the airstream.

On some small, single cylinder, spark ignition en-

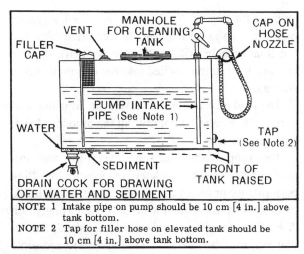

| NOTE 1 | Intake pipe on pump should be 10 cm [4 in.] above tank bottom. |
| NOTE 2 | Tap for filler hose on elevated tank should be 10 cm [4 in.] above tank bottom. |

Fig. 15.19. Good fuel storage facility.

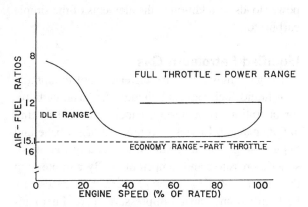

Fig. 15.20. Air-fuel ratio requirements.

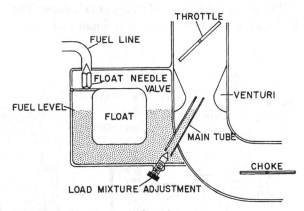

Fig. 15.21. Simple up-draft carburetor.

gines the carburetor sits atop the fuel tank. Engine suction lifts fuel from the tank into a constant level chamber which supplies the tube in the venturi section. A float is not needed to control the fuel level in the chamber as a constant level is maintained by an overflow design.

The throttle controls the amount of air and fuel

entering the engine. As the choke is closed, a greater pressure drop exists between the venturi section and the float chamber (if the throttle is open); thus an extra-rich mixture is pulled into the engine to aid in starting.

The load mixture needle valve is an adjustment to vary the A-F ratio by controlling the amount of fuel entering the tube.

This simple carburetor works well for a constant-speed engine but is completely inadequate for idling performance or for a variable-speed engine. A simple carburetor has a tendency to produce increasingly rich A-F mixtures with increasing engine speed. An actual engine carburetor has an idling system and some type of compensating system for leaning the mixture at economy operation, an enriching system for full power loads, in addition to the elements of the simple carburetor.

Liquified Petroleum Gas

LP gas engines require different fuel equipment than liquid fuel engines although the basic carburetor operations of metering, vaporizing, and mixing are unchanged. LP gas is composed of propane and butane, singly or in mixtures, and is a gas at atmospheric pressures and temperatures. By compressing the gas and lowering the temperature, these gases may be liquified and carried in pressure tanks. Fuel handling must be accomplished with high pressure hoses, pumps, and other fittings. (Fig. 13.12).

Fig. 15.22 indicates the fuel system elements. The fuel leaves the tractor tank as a liquid and passes

through the filter and a high-pressure regulator that lowers the gage pressure to about 40 kPa [6 psi]. Boiling takes place in the vaporizer coils where circulating hot water from the engine's cooling system furnishes the heat of vaporization. The low-pressure regulator reduces the pressure to atmospheric. As the engine operates, a vacuum occurs in the venturi section as in a gasoline carburetor and the fuel vapors are drawn into the airstream.

Fuel Injection

Fuel injection systems have replaced carburetors for most automotive, industrial, and large farm engines. The carburetion fuel system has deficiencies in precise control of the A-F mixture, in complete fuel vaporization, and in uniform delivery of fuel to each cylinder. Fuel injection improves on these deficiencies to give increased engine performance and fuel efficiency. Intake manifold heating is no longer needed and there is less restriction in intake airflow, acceleration is quicker, and the system is not as sensitive to altitude nor to sloping ground. Most important, injection systems lend themselves to microprocessor control.

There are significant differences between the injection system for volatile fuels such as gasoline and those for the nonvolatile fuels used in diesel engines. The diesel engine requires that fuel be injected directly into the combustion chamber and ignition is automatic when the air-fuel mixture is compressed and heated above the autoignition temperature. Gasoline injection systems place the nozzle in the intake

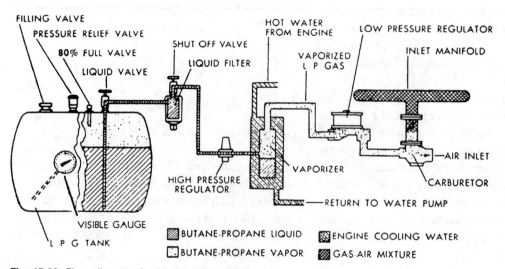

Fig. 15.22. Flow diagram for LP gas fuel system.

manifold near the intake valve to the cylinder. An electric spark is needed for ignition. Fig. 15.23 shows schematically the fuel injection system for the compression ignition engine and Fig. 15.24 shows that for the spark ignition engine.

Fuel temperatures for self-ignition at 1380 kPa [200 psi] compression pressures are

LP gas	360°C [680°F]
Gasoline	290°C [560°F]
Fuel oils	230°C [450°F]

The low self-ignition temperature for fuel oils make them an ideal CI engine fuel. Their relatively high viscosity provides good lubrication for the close fitting parts of the high-pressure injection pump and of the injectors themselves.

A diesel engine is a high compression ratio engine (15:1 or more). The intake air is compressed to a pressure of 3000–5000 kPa [435–725 psi] and temperatures as high as 700°C [1300°F] may be realized. Liquid fuel is sprayed into the hot air and after a short delay (0.001 s) begins to burn. The expanding hot gases force the piston down. Combustion proceeds as long as injection continues. The quantity of fuel injected per stroke determines the time for combustion to continue. The last fuel injected provides combustion pressure at a crank angle several degrees past TDC. This portion of the combustion contributes to the high-torque characteristic of the diesel engine.

Fig. 15.25 shows the left or fuel injection side of an engine installed in a tractor. Fig. 15.26 shows the right or exhaust side of the same engine. Air enters through the air cleaner (1, Fig. 15.25), is compressed in the compressor end of the turbocharger (3, Fig. 15.26), and is then piped to the intake manifold on the fuel injection side of the engine. The tractor is fueled through the cap on the fuel tank (2, Fig. 15.25). Fuel from the tank goes through the primary filter (4), through the transfer pump on the end of the injection pump (7), through the final filter (5), and to the injection pump. This pump raises the fuel pressure to 17 MPa [2500 psi] or higher, and then delivers it to the six lines leading to the individual injectors.

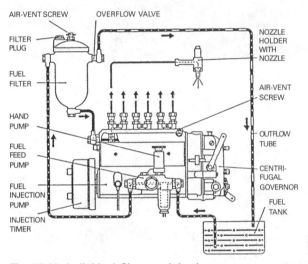

Fig. 15.23. Individual CI pump injection system.

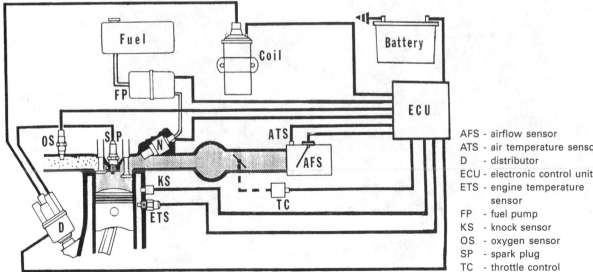

AFS - airflow sensor
ATS - air temperature sensor
D   - distributor
ECU - electronic control unit
ETS - engine temperature
      sensor
FP - fuel pump
KS - knock sensor
OS - oxygen sensor
SP - spark plug
TC - throttle control

Fig. 15.24. Microprocessor-controlled SI engine.

1. Air cleaner
2. Fuel filler cap
3. Ether container
4. Primary fuel filter
5. Final fuel filter
6. Oil filler level gage
7. Fuel injection pump
8. Hand priming pump
9. Fuel filter vent valve
10. Crankcase coolant drain plug
11. Radiator drain cock
12. Engine coolant heater

Fig. 15.25. Diesel fuel injection system.

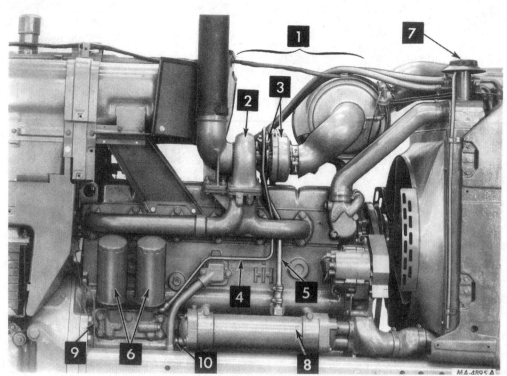

1. Turbocharger
2. Turbine housing
3. Bearing housing and compressor
4. Oil inlet tube
5. Oil return tube
6. Engine oil filters
7. Radiator filler cap
8. Oil cooler
9. Rule drain
10. Oil cooler drain plug

Fig. 15.26. Turbocharger and exhaust system.

Fig. 15.25 also shows some auxiliary equipment. The ether injection system (3) is a source of fuel with a low self-ignition temperature. For cold weather starting, the operator can inject ether into the intake manifold and achieve immediate combustion. A connection to an electric heater in the block (12) permits the warming of the coolant and the engine before attempting a start. The hand pump (8) aids in priming the fuel supply system to the injection pump. The bleed valve (9) permits the escape of entrapped air when the system is being purged.

The pump and nozzle mechanisms are the principal parts of the diesel injection system. A *unit injector* has the overhead rocker arm-actuated pump contained in the nozzle and is controlled by an external governor or, as in Fig 15.27, the delivery may be controlled electronically. Other injection systems have a separate pump and integral governor unit driven by the timing gears, with high-pressure tubing leading to the nozzles located in the engine head. These systems may have an *individual pump* system (Fig. 15.23) with a separate pump for each nozzle or *distributor pumps* that serve several nozzles through a fluid distributor system as in Figs. 15.25.

Metering may be accomplished in several ways. One method is with a rotatable scroll on the plunger as illustrated in Fig. 15.28. The delivery is determined by a rack position. The rack rotates the plunger independent of its in-and-out motion, and while the stroke of the plunger remains constant, the effective delivery is varied. Injection continues until the scroll portion of the plunger uncovers the relief hole whereupon the fuel trapped by the plunger face is now free to flow into the cavity beneath the scroll and through the hole in the stem back to the fuel tank. Different amounts of fuel are injected when the scrolled plunger is partially rotated by the control rack. Lubrication for the injection pump is provided by the fuel itself.

A pressure-time injection nozzle is shown in Fig. 15.29. The amount of fuel injected is changed by varying the pressure on the fluid system as it is forced through a metering orifice. The engine governor operates a pressure regulator, which establishes the rather low pressure flow to the injector. The fuel flows down the side of the injector past the feed passage valve (FV), which is a necked or reduced section of the rocker arm-actuated, spring-loaded plunger (P). After injection, the plunger lifts and uncovers the metering orifice (MO), which allows the cup (C) to fill with the fuel amount as determined by the pressure of the system. Excess fuel flows through the bypass

Fig. 15.27. Electronically controlled unit injector.

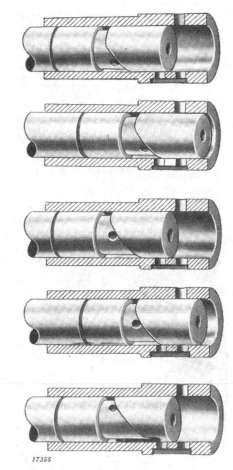

Fig. 15.28. Rotating scroll method of metering. Top to bottom plunger position: retracted, full delivery, rotated for half delivery, half delivery, no delivery.

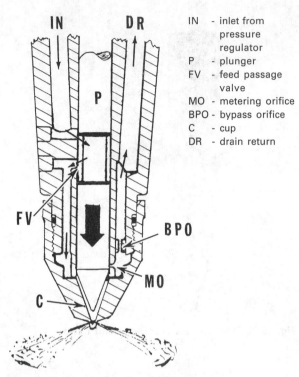

IN	-	inlet from pressure regulator
P	-	plunger
FV	-	feed passage valve
MO	-	metering orifice
BPO	-	bypass orifice
C	-	cup
DR	-	drain return

Fig. 15.29. A pressure-time fuel injector.

orifice (BPO) and returns to the system (DR) as drainage. The plunger descends past the metering orifice, traps the metered fuel in the cup, and forces it out at very high pressure through the small holes in the nozzle tip. This injection system has no external high-pressure tubing.

Fuels for diesel engines must be kept clean. Several filters and a sediment bowl-water trap are usually placed in the fuel supply line to protect the injection pump and the nozzles from stoppages and damage due to dirt.

In the event that the engine runs out of fuel, bleeding the lines of air is necessary before the engine will run again. Notice the bleed provisions in the system illustrated in Fig. 15.23.

The hole nozzle is used on engines with direct or open chamber combustion. A tip beneath the needle seat is drilled with several small holes (as small as 0.15 mm [0.006 in.]) that spray the fuel to all parts of the combustion chamber. (Fig. 15.10 and 15.29).

The nozzle orifices can become plugged and they can erode with use. Periodic inspection and clean fuel should keep the engine running at high performance rates for long periods of time.

Supercharging

Supercharging is the name applied to the process in which intake air is compressed before being admitted to an engine's intake manifold. The compressed air contains more oxygen and consequently can burn more fuel per power stroke. Supercharging for spark ignition engines is limited by the octane rating of the fuel. In effect, supercharging raises the compression ratio of the engine, which may be desirable at high altitudes. Usually it means increased and undesirable detonation for SI engines, but no such problem exists with compression ignition engines and supercharging can produce significant increases in engine performance.

Supercharging is accomplished in two ways. Large industrial engines have used gear driven blowers to increase the charge of air. The diesel tractor uses a *turbocharger*. As shown in Fig. 15.30, an exhaust driven turbine operates a centrifugal air compressor to compress the inlet air charge.

The potential for turbocharging has allowed tractor engine manufacturers increased options. Three power output levels can be obtained from the same basic engine. The comparative performances of a bare engine and those equipped with turbochargers are

Normally aspirated engine	100%
Same engine with turbocharger	120-130%
With turbocharger and aftercooler	150-170%

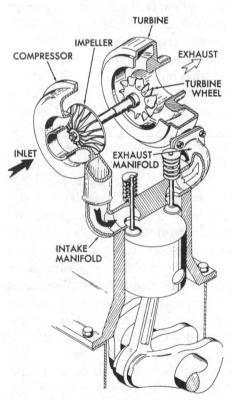

Fig. 15.30. Turbocharging action.

The *aftercooler*, also called *intercooler* (Fig. 15.31) is a heat exchanger that uses airflow to the engine's radiator to lower the temperature of the compressed air coming from the compressor side of the turbocharger. The lower temperature air has greater density and contains more oxygen per volume than air coming straight from the compressor. Delivery from the fuel injector can be increased and more power obtained than possible without an aftercooler.

Turbocharging is of interest to both manufacturers and farm machinery managers. Up to 50% more power can be obtained from an engine with very little increase in weight. The price of turbocharging is less per power gained than the price of comparable power gained by an increase in engine displacement.

Injection timing for CI engines is established by the manufacturer and should not be changed. P. H. Schweitzer (Pennsylvania State University) found that for optimum timing the smoke density was least (Fig. 15.32). The least smoke indicates the efficient use of fuel.

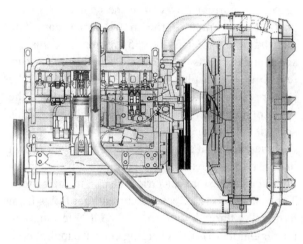

Fig. 15.31. Air-to-air intercooler.

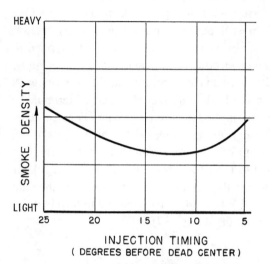

Fig. 15.32. Effect of ignition timing on CI smoke production.

Practice Problems
(Fuels and Combustion)

15.6. For each litre [gallon] of regular gasoline burned in an engine, how many cubic metres [feet] of air are required?

15.7. What are the symptoms of a carburetor float level that is too high?

15.8. How many cubic millimetres [inches] per stroke of diesel fuel must an injector deliver for a 100 ×127 mm [4 × 5 in.] turbocharged engine running with 120% theoretical air? The chemically correct air-fuel ratio is 15:1.

15.9. What is the effect on engine performance if the spring on a pintle nozzle is weakened?

Ignition and Electrical Circuits

One of the earliest uses for electricity was to ignite the fuel-air mixtures in internal combustion engines. Modern engines use electricity for cranking and for several additional functions.

Spark Ignition

A spark ignition system and its accompanying electrical circuits are shown in Fig. 15.33. Several points in the circuits are energized with voltage at all times whether the engine is running or not. The battery (8) serves as storage for electrical energy. One battery post is attached to the frame of the tractor, which serves as a return for the current supplied to the various electrical components. The tractor frame thus serves as an electrical ground for all circuits. A heavy cable leads from the other battery terminal to a terminal on the solenoid (16). Wire (1) carries current through the ammeter (19), which will show a discharging condition as the current flows out through wire (3) to the battery terminal (B) of the regulator (15). Internally, terminal (B) is connected to the load terminal (L). Wire (4) carries current to terminal (B) of the ignition switch (17).

The ignition system is energized when the ignition switch (17) is in the "on" position. Current leaves from terminal (I) and is conducted by wire (2) to the coil (10). Surging through the primary winding, the current stores energy in the core of the coil. Emerging from the coil, the current is carried to the breaker points and condenser in the distributor (11). See Fig. 15.34 and 15.35. When the engine runs or is being cranked, the breaker points open and close, creating a pulsating primary current. The resulting buildup and collapse of the magnetic field in the core of the coil induces a very high voltage (24,000 V or more) in the secondary circuit of the coil. This pulsating secondary current flows from the center tap of the coil to the center of the distributor's cap. The distributor's rotor (Fig. 15.35) rotates at camshaft speed and connects (or distributes) the secondary current to the spark plug (9) in the cylinder when the compression stroke nears TDC. A spark jumps the spark plug gap inside the combustion chamber. This spark is intense enough to start the burning process.

Other electrical circuits used on tractors are shown in Fig. 15.33. One of the most important is the charging circuit that stores electrical charge in the battery. The generator (12) creates a current that flows from terminal A through wire (6) to the GEN terminal on the regulator (15). This charging current supplies both terminals (L) and (B) of the regulator. When current flows from terminal (B), it is conveyed backward

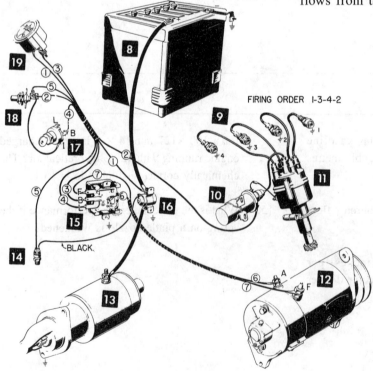

FIRING ORDER 1-3-4-2

BLACK

8. Battery
9. Spark plugs
10. Coil
11. Distributor
12. Generator
13. Cranking motor
14. Safety start switch
15. Charging regulator
16. Cranking motor solenoid
17. Ignition switch
18. Starting button

Fig. 15.33. Spark ignition tractor's electrical systems.

SPRING ATTACHING SCREW

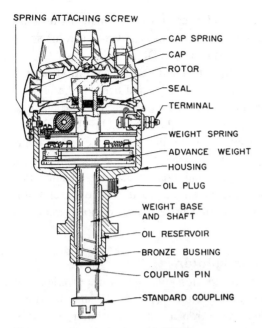

- CAP SPRING
- CAP
- ROTOR
- SEAL
- TERMINAL
- WEIGHT SPRING
- ADVANCE WEIGHT
- HOUSING
- OIL PLUG
- WEIGHT BASE AND SHAFT
- OIL RESERVOIR
- BRONZE BUSHING
- COUPLING PIN
- STANDARD COUPLING

Fig. 15.34. Cross section of distributor.

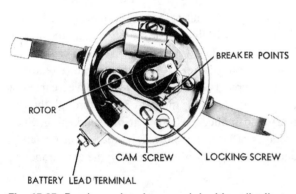

BREAKER POINTS

ROTOR

CAM SCREW — LOCKING SCREW

BATTERY LEAD TERMINAL

Fig. 15.35. Breaker points in a spark ignition distributor.

through the ammeter (19), which shows a charging condition, to the battery. Forcing current backward through the battery increases the charge held on its plates. The charging rate is varied by the flow of current from terminal (F) on the regulator through wire (7) to terminal (F) on the generator. The L terminal of the regulator supplies current to lights and other electrical loads.

Alternators, which generate alternating current and then use diodes to convert to direct current, have replaced the older, direct current generators with their brush and commutator problems. The wiring diagrams are similar. (Fig. 15.44).

The cranking circuit starts at the starter switch (18). When its button is depressed, current flows through wire (5) to the safety start switch (14). This switch is open unless the clutch pedal is depressed or the trans-

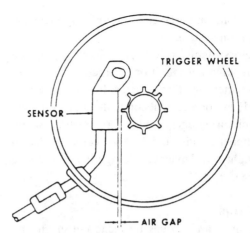

TRIGGER WHEEL

SENSOR

AIR GAP

Fig. 15.36. Electronic "breaker points" in a distributor.

mission is in neutral. From the safety switch the current travels to the cranking motor solenoid (16). Such flow through the solenoid to the ground closes the heavy-duty switch that connects the battery directly to the cranking motor (13).

A schematic diagram of a diesel tractor's electrical system is shown in Fig. 15.37. Besides the absence of an ignition circuit, the diesel's electrical system differs in that it provides current for a solenoid-operated electric heater in the intake manifold. This heater warms incoming air and is used only while cranking for cold weather starting.

Magneto Ignition

A magneto is an electric generator with permanent magnetic fields. Magnetos have been specially developed and widely used for internal combustion engine ignition. As with battery ignition, both primary and secondary circuits are involved and identi-

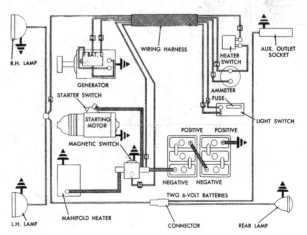

Fig. 15.37. Diesel tractor's electrical system.

cal mechanisms are used in the remaining elements of the ignition system.

Magneto systems are compact, lightweight, and have more favorable ignition characteristics under load than battery ignition systems; but since magnetos provide ignition currents only, the more versatile battery ignition system is used on farm tractors. Small, single cylinder engines and 2-cycle engines use compact and lightweight magnetos for ignition power (see Fig. 15.38).

Spark Plugs

Spark plugs are devices that lead current to the spark gap in the combustion chamber. Fouling of the plugs may cause a short circuit of the plug gap, with no ignition spark resulting. Fouling may be reduced by using the proper heat-range plugs. The temperature should be maintained high enough to burn adequately the fuel at the tip of the plug. Any less heat will cause deposits—any more may cause preignition.

Spark Timing

The occurrence of the spark in an engine's cycle is referred to as ignition timing. At slow speeds and for cranking, the spark should occur near the TDC position. At fast engine speeds the spark needs to occur as much as 0.61 rad [35°] BTDC, which is referred to as an advanced spark. *Spark advance* is the number of degrees between the retarded spark position (slow speed) and the advanced position. Small engine magnetos generally have no spark advance and the spark occurrence is at or near TDC.

Engines equipped with battery ignition have a flyball weight governor located beneath the breaker point assembly (Fig. 15.34). This governor changes the relative position of the cam and breaker points as engine speed varies. At high speeds the spark occurrence is advanced; as the speed decreases, the spark occurrence is gradually retarded. Spark advance may be controlled by engine load. The breaker point mounting is shifted in response to intake manifold vacuum.

Electronic Ignition

Electronic ignition overcomes the problems of breaker point adjustment (Fig. 15.35) and deterioration. This system replaces the distributor cam, rubbing block, and points with a rotating timer (Fig. 15.36) having the same number of projections as cylinders in the engine. The projections rotate past a magnetic pickup and an induced current is sent through the primary winding of the coil. Note that the air gap between the projections and sensor is a critical adjustment, but once set it should not require readjustment. A typical electronic system will have a capacitor that can discharge as much as 300 V through the primary and induce 30,000 V at the spark plug. Such voltages offer easier starting, increased performance, and extended plug life. Physically, the appearance of

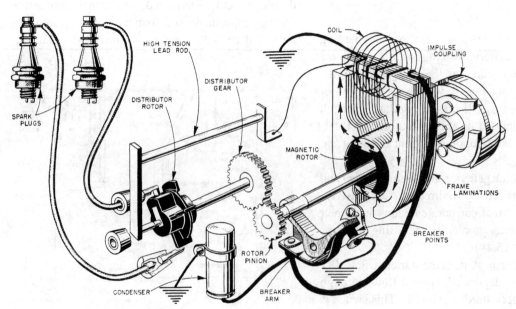

Fig. 15.38. Magneto ignition system.

the elements of the electronic system are similar to the mechanical one with the exception of the capacitor discharge element.

Microprocessor control of the ignition and fuel system can lead to increased engine performance and efficiency. Fig. 15.24 is a schematic of a system of sensors that monitor the engine state, and an electronic control unit (ECU). Measurements of engine speed, air temperature and flow, engine temperature, knock occurrence, and the amount of excess air in the exhaust as indicated by the amount of oxygen are input to the ECU, which then controls the throttle, the ignition timing, and the air-fuel ratio to give the optimum adjustment for the performance desired by the operator.

Batteries

The battery is the central point for all the electrical systems on a tractor. The commonly used lead-acid storage battery consists of lead and lead oxide plates suspended in a solution of sulfuric acid and water. Each such cell can develop 2 V; therefore, 3 cells are needed for a 6-V battery and 6 cells are needed to produce 12 V.

The chemical reaction of charging and discharging is diagrammed in Fig. 15.39. Note that water is formed in the solution as discharge occurs; sulfuric acid is formed when charging. The charge of a battery is determined by using a hydrometer to find the specific gravity of the solution (1.29 is fully charged, 1.08 or less is fully discharged).

The size or capacity of a battery is determined by its plate area: more plates, more potential output. Ratings are usually given in ampere-hours. If a battery is rated as 100 A·hr, it could deliver 5 A for 20 hr without the voltage dropping below 1.75 V/cell.

Use caution when near a charging battery since highly explosive hydrogen gas may escape through the battery vents. When working with tractor electrical circuits, disconnect the battery ground cable to eliminate spark flash accidents. Electrical systems may be either positively (+) or negatively (−) grounded.

Cranking Motors

Cranking a tractor engine requires a heavy-duty electric motor that may draw a hundred amperes of current. At such loads the motor should be operated for less than a minute; otherwise overheating and motor deterioration will result.

Two methods of engaging the cranking motor to the engine flywheel ring gear are used in farm trac-

tors. The *Bendix drive* is an automatic engaging and disengaging device that requires only a heavy-duty electrical switch to actuate it (Fig. 15.40). An *overrunning clutch drive* has a manual shift lever to engage the pinion with the flywheel ring gear (Fig. 15.41). When the engine starts, an overrunning clutch protects the cranking motor from damage. The motor

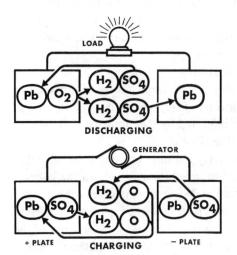

Fig. 15.39. Chemical action in lead-acid battery.

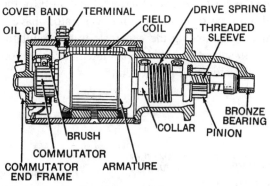

Fig. 15.40. Bendix drive.

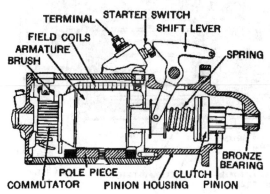

Fig. 15.41. Overrunning clutch drive.

switch is a part of the shift lever device to ensure that the pinion is engaged before power is put to the cranking motor.

Solenoids are often used in cranking circuits for convenience and to reduce the length of cable and consequent power loss from the battery to the cranking motor. For a Bendix drive, the solenoid operates only as a remote electric switch; for the overrunning clutch drive, the solenoid must first slide the pinion into engagement and afterward actuate the switch.

Charging Circuits

The charging circuit originates in the armature of the generator (Fig. 15.42), passes through the cutout relay, and is indicated by the ammeter as it goes to the battery (Fig. 15.43).

The cutout relay is a one-way, electrical valve that allows current to travel through only in the battery-charging direction. When the engine and the generator stop, the cutout relay points spring open and prevent the reversal of current flow and the discharge of the battery through the grounded brush in the generator. The cutout relay is designed to open any time the generated voltage is less than normal battery voltage.

Tractor generator terminals are usually stamped with an A for the armature terminal and an F for the field terminal connection. Some generators have a third wire grounding the generator to the regulator frame. In such a case there is no internal connection of the armature and field circuits.

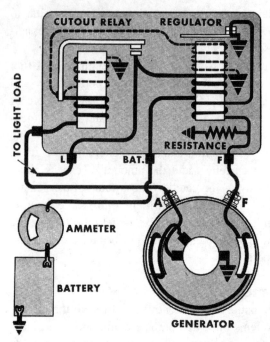

Fig. 15.43. Third brush generator and combined current-voltage regulator.

Regulators

Because the generator produces voltages that vary with engine speed and because the battery charge condition varies, some type of charge controlling or generator regulator is required. The basic principle is to vary the resistance in the field circuit of the generator that affects the armature output.

The simple regulator pictured in Fig. 15.43 is sensitive to charging voltages and charging currents and is called a combined current-voltage regulator. Regulation is accomplished by varying the resistance of the generator field circuit. This particular method uses vibrating regulator points to permit adjustment to small changes in the battery demands on the generator. The notation shown for the regulator connections is standard. The L (load) terminal should be connected to lights, ignition, etc.

The charging rate of the DC generator is often inadequate at low engine speeds. That problem has been solved by using an AC generator that has good output at slow speeds. The AC generator, or *alternator,* is less costly, does not have commutator problems, and is used on most large engines.

Fig. 15.44 diagrams the circuits for an alternator. Enclosed in the alternator housing are current generation, rectification from AC to DC, and voltage regulation functions. This design uses slip rings to pass current into the rotating field (F) which is turned by engine power. Current is generated in the three-

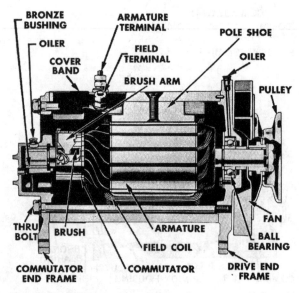

Fig. 15.42. DC generator cross section.

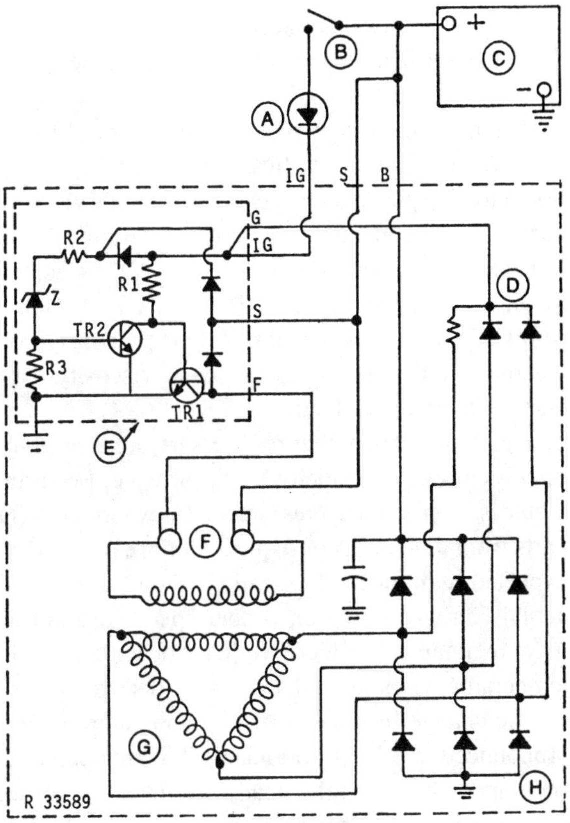

A—Diode
B—Switch
C—Battery
D—Diode/Resistor Trio
E—Regulator
F—Rotor (Field)
G—Stator
H—Rectifier Bridge

Fig. 15.44. Alternator circuitry includes a rectifier and solid-state voltage regulation.

winding stator (G), is converted to DC by the 6 diodes in the rectifier, H, and sent to the regulator (E). This current enters at G and flows through resistance R1 to transistor TR1. When TR1 is switched ON, the regulating current flows to ground, the rotating field is energized, and battery charging current is being generated. When the voltage output rises well above 12 volts, the zener diode Z senses this increased voltage and allows the regulating current to flow through R2 to TR2 and then directly to ground. TR1 is thus turned OFF, no current flows through the rotating field, and no charging current is being generated. This alternating process repeats many times per second and is designed to maintain the system voltage at a proper level. (Note the similarity of action between the vibrating mechanical points in the old regulators (Fig.15.43) to the transistor switches. This circuitry is sealed and nonadjustable. Test equipment reveals failures and indicates appropriate replacement of the elements.

Accessories

Electrical accessories have increased the load on tractor electrical systems. In addition to the instrumentation and control needs for electrical current, there are current draws for turn signals, wiper blades, heating and air conditioning, radio, tape players, communication equipment and many lighting circuits. Circuit breakers are used to protect these circuits and the machinery manager needs to know the location of the circuit breaker panel. For trailing implements, a two-wire system should be employed as the ground connection through the hitch may not be adequate.

Practice Problems
(Ignition and Electrical Circuits)

15.10. A coil theoretically transforms voltage in the same ratio as the ratio of secondary turns to primary turns. What turn ratio is required in a 12-V coil to produce 25,000 V? Note: Primary circuit voltages may reach instantaneous voltages of 250 V.

15.11. In a 6-cylinder, 4-stroke cycle engine using a 6-sided breaker point cam, the points remain closed for 0.5 rad [30°] between openings. If the engine runs at 1800 rpm, what is the time in seconds for the primary current to flow?

15.12. Why do low-charged batteries freeze in winter while highly charged batteries do not?

15.13. How does a regulator sense a fully charged battery and reduce the charging rate?

15.14. Explain how the lighting load in Fig. 15.43 can cause the ammeter to show discharge.

Engine Lubrication

Internal combustion engines have many close-fitting, rapidly moving parts that would wear quickly if not lubricated. A typical lubrication system, Fig. 15.45, shows the flow of oil through an engine. Oil flows from the oil pump, located in the oil pan, through the oil cooler, A. If the resistance through the cooler becomes too great, valve B causes the flow to bypass the cooler and go directly to the oil pressure regulating valve, E. Spring force holds the valve against the flow to create a system pressure level. Flow through the filter, C, may be bypassed by the spring-loaded valve, D, should the filter become plugged. Notice the oil supply line to the turbocharger. Oil then enters the block-length main gallery which supplies the seven main bearings, F, and the four camshaft bearings, H. The crankshaft is drilled to allow oil flow from the main bearing to the connecting rod bearing, G. Oil is also supplied to nozzle I which sprays the cylinder walls and the underside of the piston, (Fig. 15.10). Oil enters the piston pin bearing through a hole on the top of the connecting rod. The rocker arm shaft is lubricated from a passage connected to the rear camshaft bearing. Valve stem lubrication results from oil escaping the rocker arms.

Engine crankcase oil has multiple functions:

1. Reduces wear and friction by separating rubbing parts
2. Acts as a coolant by carrying heat away from high-temperature spots
3. Absorbs shocks that occur on bearings
4. Acts as a seal for the combustion chamber

The most important property of a lubricant is its *viscosity* or resistance to flow. An oil's ability to support a load depends on its viscosity. An oil's viscosity varies with temperature as shown by the curves in Fig. 15.46. This variation is a basis for classification of oil as shown in Table 15.3. The SAE oil number followed by a W is an oil developed for easy cold weather starting but with sufficient viscosity for a warm engine under load.

Fig. 15.47 shows that oil pressure areas are built up with a relative velocity between two close-fitting surfaces. Note that the pressure area prevents the metal parts from touching; consequently no wear should be expected and perfect lubrication is said to exist. In actual engines, however, proper speeds are not always maintained; sudden overloads occur; and as the temperature rises, the oil's viscosity decreases.

The change in viscosity with temperature is very pronounced in oils. Unfortunately, high temperatures accompany heavy engine loads and the ability of the oil to carry the loads is reduced because of the thinning of the oil. High-viscosity oils would help to solve this problem, but considerable energy would be required to overcome the oil's resistance to flow. Furthermore, at cold starting temperatures, the oil pressure area may not form immediately because of the poor flow of cold, high-viscosity oil.

Contamination and dilution of crankcase oil inter-

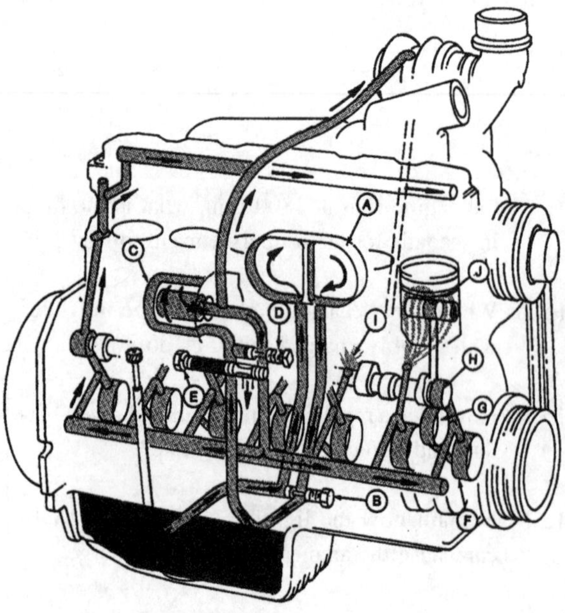

Fig. 15.45. Engine lubrication system.

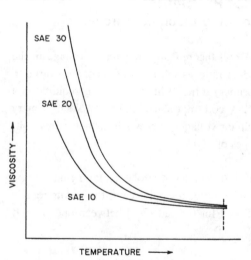

Fig. 15.46. Change in oil viscosity with temperature.

TABLE 15.3. SAE Viscosity Grades for Engine Oils

SAE Viscosity Grade	Viscosity (cp at Max. Temp. °C)		Viscosity (cSt at 100°C)	
	Cranking	Pumpability	Minimum	Maximum
0W	3250 @ -30°	30,000 @ -35°	3.8	•••
5W	3500 @ -25°	30,000 @ -30°	3.8	•••
10W	3500 @ -20°	30,000 @ -25°	4.1	•••
15W	3500 @ -15°	30,000 @ -20°	5.6	•••
20	•••	•••	5.6	<9.3
20W	4500 @ -10°	30,000 @-15°	9.3	•••
25W	6000 @ -5°	30,000 @ -10°	9.3	•••
30	•••	•••	9.3	<12.5
40	•••	•••	12.5	<16.3
50	•••	•••	16.3	<21.9

Notes: 1 centipoise (cp) = 1 mPa · s
1 centiStoke (cSt) = 1 mm²/s

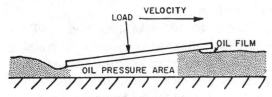

Slider-bearing lubrication.

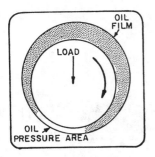

Journal-bearing lubrication.

Fig. 15.47. Speed and close clearances produce load-carrying oil pressures.

feres with good lubrication. In addition to the unburned fuel washed down from the cylinder walls, lead salts, dirt, metal particles, water, products of combustion, and acids will accumulate in an engine's crankcase. High crankcase temperatures and crankcase ventilation will do much to reduce contamination although high temperatures also tend to burn and break down the oil.

Oil filters, used to screen out the larger particles, are not a substitute for the periodic oil change. The full-flow filter must be equipped with a bypass valve that will open if the filter becomes plugged.

A provision for varying the oil pressure is provided in some tractors although the gage reading of oil pressure does not in itself guarantee adequate engine lubrication.

Various additives are included in engine oils to improve their performance. To aid the machinery manager in selecting properly compounded oils, API (American Petroleum Institute), SAE (Society of Automotive Engineers), and ASTM (American Society for Testing Materials) have devised service classifications for oils according to light-duty car and pickup truck use (S) or heavy-duty engine use (C). (Table 15.4).

TABLE 15.4. Service Classification of Oils

API Engine Service Classification	Engine and Operating Characteristics
Service Category SJ	For all automotive engines currently in use.
Commercial Category CH-4	For high-speed, four-stroke engines designed to meet 1988 exhaust emission standards. Compounded for use with diesel fuels ranging in sulfur content up to 0.5% weight. Can be used in place of CD,CE,CF-4, and CG-4 oils
CG-4	For severe duty, high-speed, four-stroke engins using fuel with less than 0.5% weight sulfur. Required for engines meeting 1994 emissions standards. Can be used in place of CD, CE, and CF-4 oils.
CF-4	For high-speed, four-stroke, naturally aspirated and turbocharged engines. Can be used in place of CE oils.
CF-2	For severe duty, two-stroke-cycle engines. Can be used in place of CD-II oils.
CF	For off-road, indirect-injected and other diesel engines including those using fuel with over 0.5% weight sulfur. Can be used in place of CD oils.

Over the years the classification of oils has changed to meet the needs of newly developed engines. The first classification, SA, was a straight run mineral oil with no additives and was recommended only for engines operating under mild conditions. New classifications can be expected as engine developments require the increased protection from new oil additives and for reduced emissions.

Other Engine Systems

Governors

Unlike many other automotive engines, a tractor engine is a governed engine, since many farm operations require a constant speed at variable loads.

In carbureted engines the governor controls the throttle plate. The diesel engine governor controls the fuel-metering device. Tractor governors are variable-speed governors; the spring force counteracting the flyball weights is adjustable through a speed control lever.

A perfectly governed engine would maintain a constant speed regardless of the load, but actual governors permit a small decrease in speed as the load increases. This speed decrease from no load to rated load is called *speed droop* and should not exceed 10% of the no-lead speed.

Governor hunting occurs due to worn or sticky linkages and to incorrect A-F ratios.

Air Cleaners

A tractor often operates in very dusty conditions and requires a heavy-duty air cleaner if rapid engine wear is to be avoided. A good air cleaner must take out all the dirt in the air while presenting the least obstruction to airflow.

Various types of air cleaners have been developed. One type is the oil bath, self-washing element air cleaner. The air picks up oil and passes through a metal mesh element. Dust and dirt are trapped in the oil film on the mesh. The airborne oil impinges on the mesh, replenishes the film, and washes the trapped dirt down into the outer portion of the oil cup.

The oil viscosity, required for the air cleaner is the same as for the engine. Either overfilling or underfilling the oil cup is to be avoided. The oil in the cup should be changed as soon as the cleaner can no longer perform its function adequately. If the time interval recommended for changing the oil in the cup is followed, no exceptional wear should be observed.

Most farm tractors are equipped with dry air cleaners (Fig. 15.25). Air travels down the intake stack and is started spinning by inclined vanes, Fig. 15.48, whereupon the heavy dust particles are thrown out of the airstream. These dust particles are trapped and may be unloaded by hand or automatically when the engine stops. The partially cleaned air then passes through the paper filter element that removes most of the remaining particles. The cleaned air inside the center of the filter than travels out the end of the cleaner and to the engine. The filter element may be renewed by removing and gently tapping it to dislodge the trapped dust particles, or it may be washed in a detergent solution. Occasionally the element should be replaced.

A *precleaner* on the intake stack (Fig. 15.49) reduces the cleaning load on air cleaners by screening out very large particles and centrifuging out heavy dust particles. The dust collects in the transparent bowl and is unloaded by the tractor operator as a maintenance duty.

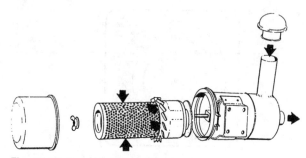

Fig. 15.48. Dry air cleaner for tractors.

Fig. 15.49. Engine air precleaner.

Cooling Systems

A more accurate term for the cooling system would be the *temperature control system*, since the objective is not to run an engine as cool as possible but at the most efficient temperature for power, economy, and wearing rate. Table 15.5 lists the effects of coolant temperature on a gasoline engine's power, fuel consumption, and wear as engine temperature is reduced below the recommended 80°C [180°F]. Note particularly the tremendous increase in cylinder wear as the temperature lowers.

An accounting of the heat energy liberated by burning fuel in a loaded engine reveals that approximately one-quarter of the heat goes to power, one-third is wasted in exhaust gas heat, one-third is given up to the cooling water, and the rest is lost by friction and radiation.

Since the specific heat of air is 1 kJ/(kg · °C) [0.24 Btu/(Ib · °F)], it may be seen that large quantities of air are needed for engine cooling, whether the heat is dissipated in the air directly as in air-cooled engines or the air is used to cool a water radiator as in the liquid cooling system. (Fig. 15.50 and 15.51). For the greatest heat transfer to the surrounding air, large, clean surfaces are needed. On air-cooled or liquid-cooled engines, dirt and trash must be kept out of the cooling fins or radiator.

For liquid cooled engines, thermostats are temperature actuated valves used to limit coolant flow through the radiator until proper operating temperatures are reached. Bypass thermostats permit circulation through the block when the thermostat is closed. (F and D, Fig. 15.51).

Pressurized radiator caps are used to seal the cooling system to prevent evaporation losses. The cap includes two spring actuated valves. One relieves excess pressure and the other prevents a vacuum in the system when the engine is not operating.

Fig. 15.51 depicts a system that has an engine oil cooler, J, and an intercooler H, for lowering the tem-

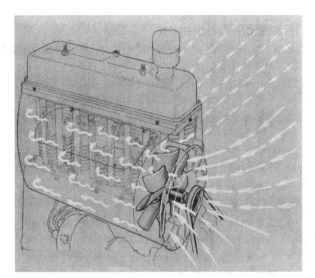

Fig. 15.50. Airflow through air-cooled engine.

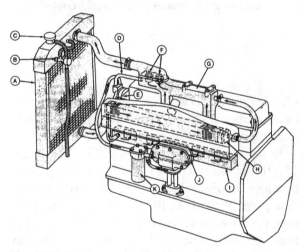

Fig. 15.51. Tractor engine liquid cooling system.

perature of the turbocharged incoming air. A filter, K, adds anti-corrosives to the coolant while screening-out large particles.

Power Transmission

Clutches

The clutch, which is in the power train immediately after the engine flywheel, is called the master or main clutch, as some tractors have separate clutches for the PTO shaft.

Clutches used on tractors may operate dry (Fig. 15.52) or run in oil (wet) (Fig. 15.55). The wet clutch is often multiple plate; that is, it has more than one driving disk. Wet, multiple-plate clutches are designed for smooth engagement, high-slip applications.

TABLE 15.5. Effect of Coolant Temperatures

Coolant Temp., °C [°F]	Power, % of Max	Fuel Consumption, % of Min.	Cylinder Wear, % of Min.
80 [180]	100	100	100
70 [160]	98	103	166
60 [140]	96	114	334
40 [100]	92	125	668
4 [40]	88	136	2670

Practice Problems
(Engine Lubrication and Other Engine Systems)

15.15. If an engine oil pump circulates 7.57 L/min [2 gal/min] at 207 kPa [30 psi], how much power is required to operate the pump?

15.16. What oil classification would one purchase for a diesel engine used continuously for irrigation pumping?

15.17. If an engine runs at a governed 1800 rpm, what would the no-load speed be if the governor has a 5% speed droop?

15.18. What flow rate of water is required to maintain a constant temperature in a liquid-cooled en-

gine operating under full load and burning 15 L [4 gal]/hr of diesel fuel? Assume a block-radiator temperature differential of 20°C [36°F]. The specific heat of water is 4.18 kJ/(kg·°C) [1 Btu/(lb·°F)] and water density is 1000 kg/m³ [8.33 lb/gal].

15.19. What flow of 11°C [60°F] air having a specific volume of 0.82 m³/kg [13.2 ft³/lb] is necessary to maintain a constant temperature in the radiator of problem 15.18 if the air leaving the radiator is at 60°C [140°F]?

Clutches may be held in engagement by springs or by an overcenter linkage. Hydraulic clutches are held in engagement by fluid pressure.

In the dry, single-plate clutch (Fig. 15.52 and 15.53), the clutch is released when the throw-out bearing is brought to bear on the forks pivoted on the base plate. These forks lift the pressure plate away from the driven disk. As the pressure is removed between the driven disk and the flywheel, the ability to transmit torque drops.

Excessive slippage results in rapid wear. As wear occurs, the forks will begin to ride against the throw-out bearing when in the engaged position. The mechanical stop for the clutch pedal keeps the forks from

completely disengaging. Absence of clearance, or play, in the engagement linkage means that the pressure plate cannot exert its full force on the clutch plate in the engaged position. The clutch plate then slips under high-torque loads, wear occurs, and even more slippage results. Clearance in the foot linkage must be maintained at all times to prevent the throw-out bearing from riding on the clutch forks.

New materials are being used in tractor clutches. One is a ceramic material reported to have high torque transmission capability and tolerance to high temperatures. These clutches do not use the conventional driven disk; instead, the ceramic material is formed into small plates that contact the flywheel.

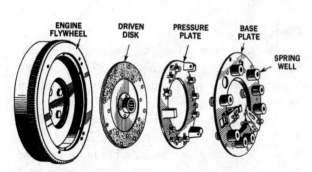

Fig. 15.52. Dry, single-plate, spring-loaded main clutch.

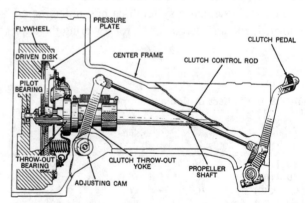

Fig. 15.53. Foot-operated clutch disengagement linkage.

Clutch maintenance chores are lubrication of the throw-out and pilot bearings. On most tractors this lubrication is provided from the transmission lubrication system.

The inner construction of a hydraulically actuated tractor clutch is shown in Fig. 15.54. Hydraulic lines lead from the foot-pedal-operated valve to a hydraulic cylinder mounted in place of a throw-out bearing for actuating the clutch forks. When hydraulic pressure is removed, the forks release all the load on the pressure plates regardless of the amount of clutch plate wear. This clutch has disks fastened to the flywheel alternating with plates splined to the output shaft. This clutch is oil cooled and provides smooth, long-wearing service. Fig. 15.55 is a clutch where the force is made hydraulically instead of by springs. The clutch is engaged by applying hydraulic pressure directly onto a pressure plate.

Transmissions

The transmission section of a tractor is contained within the box-shaped casting between the engine clutch housing and the final drive casting that includes the axles. It provides the operator with control over engine power to the rear axles. Because of the many speed and load requirements for field operations, tractor manufacturers have designed transmissions specially for the wide range of drive-wheel torques and typical field speeds needed.

Manually selected gear transmissions are the simplest. One of a series of gear pairs is engaged to produce a desired gear ratio. Typical values of gear ratios range from 1:1 to 15:1 speed reduction.

Gear ratio changes in *sliding gear* transmissions involve moving a gear axially along a splined shaft to engage a matched gear on a countershaft. Forward travel must be stopped to accommodate the shift and engagements are sometimes difficult when the gears must be rotated slightly to permit the gear teeth to mesh.

The ability to change gear ratios without stopping is provided by an *underdrive* section of the transmission. Two constant engagement gear pairs provide the tractor operator with a choice of direct drive through the section or of a speed reduction. This alternative doubles the possible ratios available from the transmission. The underdrive typically gives a 20–25% speed reduction and thus a 20–25% torque increase. As illustrated in Fig. 15.56, the gear box (B) turns with the drive shaft (C) when the clutch (A) is engaged in direct drive. The gear cluster (E-D) is locked and does not turn. The drive shaft (F)

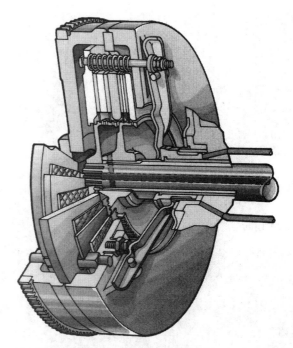

Fig. 15.54. Hydraulically actuated tractor main clutch.

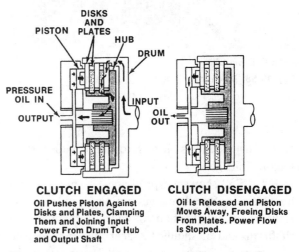

CLUTCH ENGAGED
Oil Pushes Piston Against Disks and Plates, Clamping Them and Joining Input Power From Drum To Hub and Output Shaft

CLUTCH DISENGAGED
Oil Is Released and Piston Moves Away, Freeing Disks From Plates. Power Flow Is Stopped.

Fig. 15.55. Hydraulic clutch for underdrive.

then turns at the same speed as the gear box. Note that the overrunning clutch (G) is inactive. To actuate the underdrive, the clutch is released, the reaction to torque causes the overrunning clutch (G) to lock the gear case, the gear cluster (E-D) now rotates on its shaft, and the output shaft (F) then rotates at a slower speed than (C). Underdrives can be actuated hydraulically rather than mechanically.

Some tractors have an *overdrive* as well as an underdrive. These options plus direct drive give three

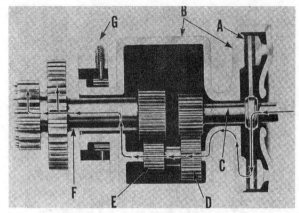

DIRECT DRIVE.

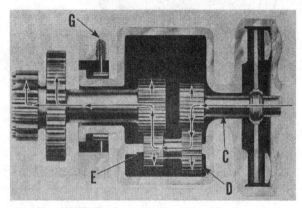

REDUCED SPEED.

Fig. 15.56. Mechanically actuated underdrive.

speeds that are accessible to the operator without stopping to shift gears.

Fig. 15.57 shows a typical constant mesh tractor transmission. Speed changes are made with shifting forks that couple the hubs of engaged gears to their shafts instead of sliding the gear into engagement. This transmission has three parallel gear sets in the selective gear section. The output from this section goes through the speed range section where a gear set, in series with the previous section, doubles the possible gear ratios as well as provides for three reverse speeds.

A second type of transmission employs a planetary gear mechanism (Fig. 15.58). In this arrangement the input shaft turns the planetary carrier. If the brake band is applied to the drum that is fastened to the sun gear, the planetary gears will turn and cause the ring gear and the output shaft to turn somewhat faster than the input shaft. When the brake band is released the sun gear can spin instead of the loaded ring gear; thus no power is transmitted.

If a one-way or overrunning clutch (Fig. 15.59) is placed between the carrier and the output shaft, a 1:1 gear ratio will exist through the unit while the sun gear spins or freewheels. With such an arrangement a gear ratio change can be made without ever actually engaging or disengaging gears.

A hydraulically shifted transmission is shown in Fig. 15.60. A series of multiple plate wet clutches provides a selection of 18 forward gear engagements.

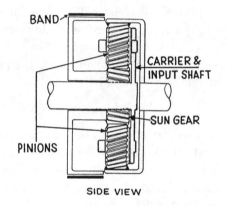

SIDE VIEW

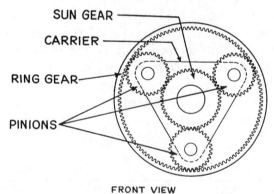

FRONT VIEW

Fig. 15.58. Planetary gear system.

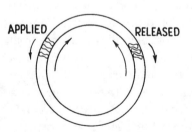

Fig. 15.59. One-way clutch.

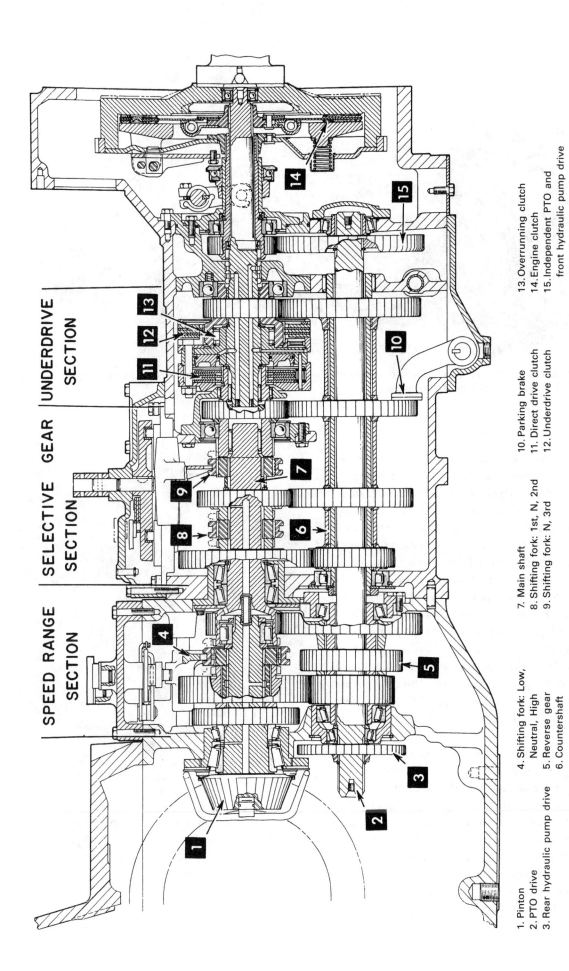

1. Pinton
2. PTO drive
3. Rear hydraulic pump drive
4. Shifting fork: Low, Neutral, High
5. Reverse gear
6. Countershaft
7. Main shaft
8. Shifting fork: 1st, N, 2nd
9. Shifting fork: N, 3rd
10. Parking brake
11. Direct drive clutch
12. Underdrive clutch
13. Overrunning clutch
14. Engine clutch
15. Independent PTO and front hydraulic pump drive

Fig. 15.57. Typical tractor transmission with twelve forward speeds, six reverse speeds, underdrive, and independent PTO.

Fig. 15.60. Tractor transmission composed of series of planetary gear sets. Electronically actuated hydraulic clutches used to actuate the gear sets.

A third type of transmission is called a hydrostatic drive (Fig. 15.61). The engine drives a hydraulic pump (6) whose output operates a hydraulic motor (9) that rotates the final drive gears. The tractor operator varies forward speed by moving control (3), which causes servo cylinder (4) to change the angle of the swash plate (5), which affects the displacement of the individual plungers (2) in the pump. The swash plate for the motor (12) is at its maximum angle when the pump output is zero. This condition provides a high starting torque. As the pump output increases, the angle of the motor's swash plate automatically decreases. The motor and the tractor's final drive then run at a faster speed. This transmission is very popular for self-propelled field machines where incremental control of forward speed is desired.

There is an optimum range of speed and load combinations for peak efficiency of this transmission. The tractor operator should adjust the speed and load combination (which is really the power requirement of the attached load) so that the engine operates at its rated speed, or a little lower, at full-open governor setting. Two gear sets are provided in the transmission to assist in optimizing the speed-load combination. The peak efficiency of this transmission is still somewhat less than for an all-gear transmission, but the greater control over ground speed and the conve-

Fig. 15.61. Hydrostatic transmission.

nience of operation tend to counterbalance the factors of greater investment and lower efficiency.

Lubrication of tractor transmissions is not as demanding as for the engine crankcase. A high viscosity oil gives greater gear tooth protection, but the heat generation and dissipation aspects are not as good as for a lower viscosity oil. Oil is necessary for the rolling contact bearings in the transmission to keep the bearings cool as well as to provide protection against wear and rust. New type transmissions may require specially formulated oils. One should follow closely the manufacturer's recommendations as to type of oil and frequency of oil changes.

Final Drive

The final drive for a farm tractor is built heavier than for automobiles or trucks because the gear reduction and the torque transmitted are greater.

Following along the sketch in Fig. 15.62, the output shaft from the transmission terminates in a pinion that drives the ring gear. Inside the ring gear is a cluster of gears called the *differential*. The differential permits the torque of the drive shaft to be divided equally between the two drive wheels. This arrangement aids in turning corners with the tractor, as it allows one wheel to slow down and the other wheel to speed up.

Many tractor manufacturers provide a *differential lock* that can be activated and deactivated by the tractor operator at will. In a forward travel situation where variable tractive surfaces are encountered, canceling the action of a differential will improve tractor performance. Locking the differential gear case to the axles reduces the power loss that results when a drive wheel on a poor tractive surface spins while the other drive wheel, though on a better tractive surface, is ineffective because a differential gear set is directing half the torque to a wheel incapable of using it.

Fig. 15.63 shows the operation of a tractor differential lock. Normally, the ring gear, O, and its attached

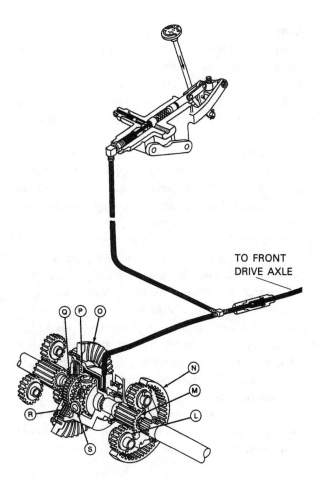

Fig. 15.63. Hydraulically actuated differential lock.

housing drive the axles through the differential pinion, S. If one drive wheel is slowed by turning or braking, the pinion gears rotate causing the other drive wheel to increase speed. When the tractor operator presses a foot pedal, a piston in a small hydraulic cylinder causes a pressure spike that is carried into the differential through a sliding connection. This pressure acts on the piston, P, which forces the clutch plates, R, against the ring gear, O. As the clutch plates are splined to the differential lock gear, Q, the whole differential assembly turns as one unit and the drive wheels are locked together.

The planetary gear final drive design of Fig. 15.63 has replaced the bull gear design of Fig. 15.62. The design is more compact and multiple gear teeth carry the torque instead of a single bull gear tooth. The ring gear, N, is held stationary in the axle housing. The axle is driven by the carrier holding the planet gears, L. The sun gear, M, is driven from the differential. If the planet gears are the same size as the sun gear, the speed reduction and the torque multiplication is 4:1.

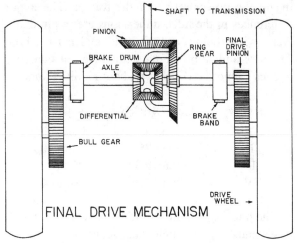

Fig. 15.62. Final drive for tractor.

Brakes

Brakes are used to slow the tractor, to hold it stationary when stopped, and to aid in turning it at the ends of a field. The power-absorbing capacity of brakes should be greater than the power output of the tractor engine. This absorbed power is converted into heat.

Individual drive-wheel brakes are called differential brakes, since if only one is applied, the differential speeds up the other wheel and assists in turning corners. If both brakes are applied together, forward motion is slowed or stopped.

A mechanical differential tractor brake is shown in Fig. 15.64. This brake has self-actuating features. As the load on the brake increases, the balls wedge even tighter and the holding force of the brake increases.

Large tractors require brakes capable of absorbing more power than their engines can produce; hydraulically actuated brakes have greater capacity than most mechanical brakes. Hydraulic pressure is used to exert large clamping forces on the disks that rotate with the final drive. The multi-plate, hydraulic actuated brakes are often wet with oil and have the same long periods of use without the need for adjustment as do multi-plate wet clutches.

Hitches

A typical tractor hitch is shown in Fig. 15.65. This mechanism accommodates the three classes of implement hitches. Hitches are classified by the way they restrain implements. A trailing or towed implement is attached to the tractor drawbar with a single pin and clevis. This hitch is classified as a *one-point hitch* (Fig. 14.2). The implement is free to pivot horizontally about the hitch point, usually with enough flexibility for considerable vertical pivoting as well. The swinging drawbar in Fig. 15.65 can be used for this class of hitch.

A semimounted implement is an example of a *two-point hitch*. Transverse implement pivoting is restrained, but hinging in the vertical plane about the axis through the two hitch points is not. In Fig. 15.65, the hitch is made through the two eyes at the ends of the draft arms. Fig. 5.15 is an example of a semimounted, two-point implement hitch.

A mounted implement hitch is one in which implement action is restrained in the vertical and both horizontal directions. Front-mounted and midmounted implements may have many attachment points. Rear-mounted implements are attached with a *three-point hitch*. The top link (third point) in Fig. 15.65 is used.

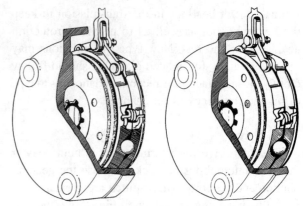

Fig. 15.64. Mechanical disk brakes.

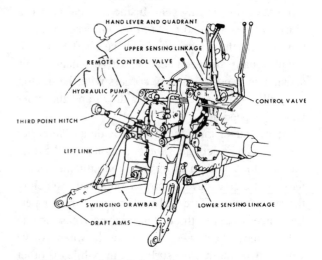

Fig. 15.65. Rear hitch linkage for tractors.

Example three-point hitches are shown in Fig. 11.1 and 8.3. The row cultivator in Fig. 7.3 operates through a three-point hitch but is towed in the transport mode with a two-point hitch.

ASAE Standard S217.11 has established four categories of dimensions for three-point hitches according to the power rating of the tractor. The major differences in the hitch dimensions are in pin diameters, width between draft arms, and mast height. Tractors and implements with different category hitches cannot be coupled without modification.

Category	Max. DBP kW	[HP]
I	15 to 35	[20 to 45]
II	30 to 75	[40 to 100]
III	60 to 168	[80 to 225]
IV	135 to 300	[180 to 400]

The hitch system in Fig. 15.65 is very versatile. It raises and lowers mounted and semimounted implements. It restrains implement action yet provides flexibility when needed through the free-swinging drawbar and attachment arms. It can sense the draft load on the tractor and send a signal to the traction-assist system.

A *fast hitch* attachment was developed to make the hitching of mounted implements to tractors easier and safer. The apparatus shown in Fig. 15.66 attaches to the conventional three-point hitch. The implement is fitted with pins for the lower attachment points and with an eyehole at the top of the mast. The tractor operator is able to back the tractor to the implement with the hitch lowered, align the hitch points, raise the hitch, and drive away with the secured implement without leaving the tractor seat.

Power Take-Off Shafts

The PTO shaft (Fig. 15.67) is an important method of tractor power transmission. Most PTO drive systems are independent; the main clutch is bypassed by the PTO drive. The shaft is driven directly from the engine and has its own separate clutch.

Two ASAE standards have been set up for farm tractor PTO power transmission. The 540 rpm standard provides

1. Speeds of the shaft when operating under rated load shall be 540 ± 10 rpm in a clockwise direction when facing in the direction of the tractor's forward travel.
2. The shaft shall be 35 mm [1$\frac{3}{8}$ in.] nominal diameter with a 6-tooth spline.
3. The shaft shall be located within 76 mm [3 in.] of the tractor centerline; the spline end shall be 356 mm [14 in.] horizontally from the drawbar hitch point and within a range of 152–381 mm [6–15 in.] vertically above the drawbar hitch point.

The 1000 rpm standard provides

1. The speed of the shaft when operating under load shall be 1000 ± 25 rpm clockwise.
2. The shaft shall be 35 mm [1$\frac{3}{8}$ in.] nominal diameter with a 21-tooth spline.
3. The shaft shall be located within 76 mm [3 in.] of the tractor centerline; the spline end shall be 356 mm [14 in.] horizontally from the hitch point and within a range of 152–305 mm [6–12 in.] vertically above the drawbar hitch point.

Both standards provide for a 381 ± 51 mm [15 ± 2 in.] drawbar height above the ground surface when

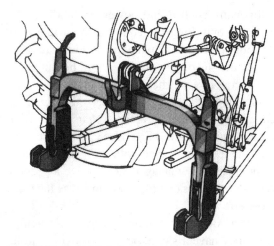

Fig. 15.66. Fast hitch for tractors.

using the PTO. Additional standards listed are for machining dimensions, safety shielding, and alternative equipment.

A PTO drive to a trailing implement uses universal joints in its drive line to permit power transmission while making turns. A pair of universal joints will give constant velocity transmission only if they both have the same degree of angularity and if the forks of the center shaft are in line. The center of the middle shaft should be vertically above the hitch point of a trailing implement, but such is not always accomplished.

Hydraulic Systems

A tractor's hydraulic system has became one of its most important systems. Originally used to lift mounted implements only, hydraulic systems are now

Fig. 15.67. Tractor PTO shaft.

used for many control and power applications. In addition to controlling the action of both mounted and trailing implements, hydraulic systems are used to operate the steering system, assist in braking, adjust the operator's seat, change wheel spacing, change gear ratios in the transmission, engage the PTO drive clutch, actuate the traction-assist system, and power remote hydraulic motors. Fig. 15.68 shows the portions of the tractor involved with the hydraulic system. The machinery manager should have a general understanding of this complex system to use the tractor most effectively.

The heart of the hydraulic system is its pump. Tractors may have a single pump to provide for all hydraulic needs or multiple pumps, each responsible for its own part of the system. Gear pumps, plunger pumps, and vane pumps may be used, but piston pumps delivering against pressures as much as 17 MPa [2500 psi] are most popular. Systems with a common pump often have a variable-displacement pump design (Fig. 15.69) in which flow rate is reduced when system demand is low with a consequent saving in power. (Refer to Eq. 2.3.) Other systems use a smaller displacement pump that keeps a hydraulic pressure accumulator charged to meet sudden peak flow demands greater than the pump capacity. The accumulator in a hydraulic system is analogous to the battery in a tractor's electrical system.

A reservoir provides a reserve oil supply and expansion relief for the system. It may contain a filter to trap dirt. Some oil cooling is accomplished in a reservoir, but most cooling is done by a radiator located in front of the cooling water radiator. Many tractors use the transmission housing as a reservoir; then the hydraulic fluid must also serve as the transmission lubricant. In such cases the tractor manufacturer's specifications should be followed closely when hydraulic fluid is purchased.

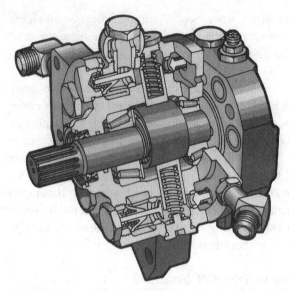

Fig. 15.69. Variable-displacement, multipiston hydraulic pump. With low system demand, pressure builds up in pump's crankcase, raises pistons off driving cam, and reduces piston stroke. Hydraulic output and power requirement are consequently reduced.

The utility of a hydraulic system is limited only by the ingenuity of its control valve arrangements. Valve action may be controlled by the tractor operator, by automatic pressure sensing devices, or by other valves. Each control valve is usually associated with a single actuator that is either a cylinder or a hydraulic motor. A system of valves and actuators may seem hopelessly complex when viewed as a whole. Complexity can be reduced by dividing the system into sections, each with its own actuator and control valve. (This procedure is followed in discussing Fig. 15.71 and 15.75.)

Fig. 15.70 illustrates an early design for remote cylinder operation. The control valve is manually operated, but the system does have some automatic sequencing features. The control lever is moved to position (1) and held there by the latch while the cylinders extend. When the cylinders complete their strokes, the control lever is automatically released to the normal or neutral position. When the control lever is pushed to (5), pressure is relieved and the cylinders retract. The particular design shown has a provision for delayed or sequential operation—the front cylinders can be made to act prior to the rear cylinder.

The cylinders in Fig. 15.70 are *single acting* in that pressure acts on only one side of the piston head. Single-acting cylinders are commonly used to actuate mounted implement lift arms, as the weight of

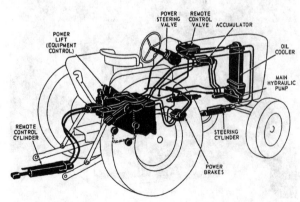

Fig. 15.68. Extent of hydraulic system in black.

the implement is available to retract the cylinder. *Double-acting* cylinders can apply pressure to either side of the piston head. These cylinders can apply force while retracting, but the potential force is less than that for extension because the effective area of the piston head is reduced by the area of the rod.

Some ASAE standards for remote cylinders are

1. A 203.2-mm [8-in.] or 406.4-mm [16-in.] working stroke depending on the drawbar pull capacity of the tractor
2. Mounting pins 25.4 mm [1 in.] in diameter for a 203.2-mm [8-in.] stroke cylinder and 31.75 mm [1 in.] for the 406.4-mm [16-in.] stroke cylinder
3. Operating time 1 1/2 to 2 s for the 203.2-mm [8-in.] stroke cylinder and 3 to 5 s for the 406.4-mm [16-in.] stroke cylinder

Other recommendations cover hose length, hose supports, and cylinder mounting clearances.

Self-sealing release couplings are used for remote cylinders. Provisions for limiting the effective stroke of the remote cylinder range from a simple mechanical stop to an electric solenoid control.

In contrast to the rather simple circuits in Fig. 15.70, Fig. 15.71 shows the intricacies of a portion of a modern tractor's hydraulic system. The details of operation of the components of this system are not covered in this text; however, a general description of the function of each component should be helpful to the machinery manager. The pump (P) draws oil from the reservoir (R) through the screen (S) and supplies fluid flow to the power steering system, the power brake system, the underdrive actuators, the differential lock, the seat cylinder, and the lines for lubricating the power transmission train.

A constant amount of oil is metered through the orifice plug (OP) and supplies the power steering system independent of the needs of the rest of the system. The flow divider (FD) diverts a variable amount of the flow past a bypass valve (BP) through

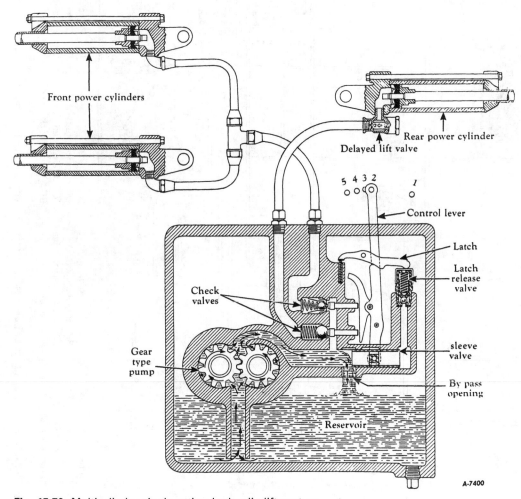

Fig. 15.70. Multicylinder single-action hydraulic lift system.

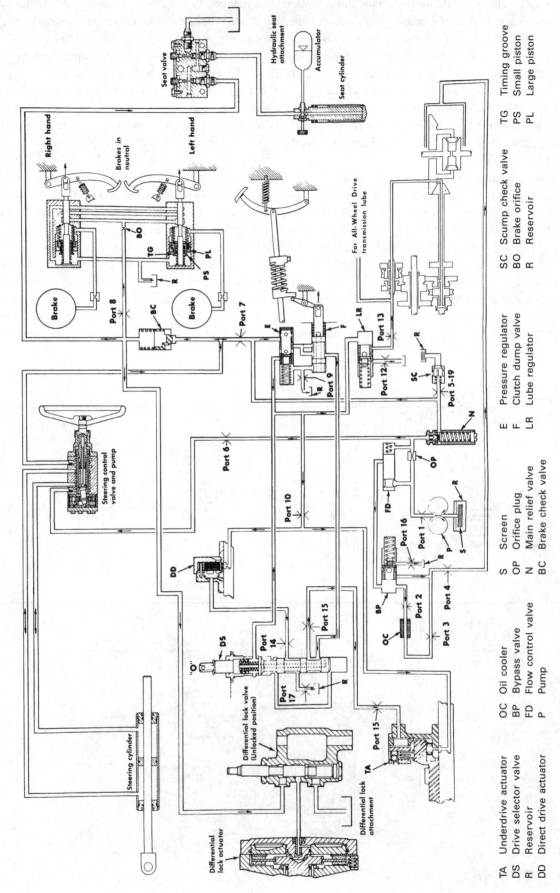

Fig. 15.71. Schematic diagram of portions of a tractor's hydraulic control and power train lubrication systems.

TA	Underdrive actuator	OC	Oil cooler	S	Screen	
DS	Drive selector valve	BP	Bypass valve	OP	Orifice plug	
R	Reservoir	FD	Flow control valve	N	Main relief valve	
DD	Direct drive actuator	P	Pump	BC	Brake check valve	
E	Pressure regulator	SC	Scump check valve	TG	Timing groove	
F	Clutch dump valve	BO	Brake orifice	PS	Small piston	
LR	Lube regulator	R	Reservoir	PL	Large piston	

the oil cooler (OC) to the differential and final drive gears for use as a lubricant. The main pressure relief valve (N) protects the system from excessive pressures.

The steering wheel of the tractor is connected to a rotary pump, which is capable of operating the steering cylinder alone if the main hydraulic pump should stop. With the pump running however, the steering wheel pump merely shifts the control valve to admit pump pressure for operating the steering cylinder.

The excess flow from the power steering components is led to the braking and differential lock mechanisms through the brake check valve (BC) to the drive control through a pressure regulator (E) and to a sump check valve (SC). The outlet of SC is pictured as being submerged in the reservoir (R). Fluid tends to seep out of the power steering components during long periods of idleness. A few turns of the steering wheel with a "dead" engine will refill the passages by sucking fluid from the reservoir past SC.

The brake control valves are fed through the brake orifice (BO). The brake foot pedals operate the valves, which have a small piston (PS) inside a large piston (PL). A timing groove (TG) gradually decreases as the two pistons move at different speeds when actuated by the pedal-operated control valve. The effect is to cause a gradual increase in the pressure applied to the brake cylinders. The operator experiences the same "feel" for the braking effort as is felt with a pure mechanical system. The check valve (BD) holds fluid trapped in the braking components to provide effective braking force even though the pump may have stopped.

Fluid for actuating the underdrive passes through the pressure regulator (E) to the clutch dump valve (F). Valve F is operated by a linkage from the main clutch pedal and releases all the flow to the reservoir anytime the clutch pedal is depressed; otherwise the pressure is conducted around the spool valve in F to the drive selector valve (DS). The drive selector is set manually in one of two positions: direct drive (DD) in which pressure is applied to clutch plates that lock the underdrive mechanism into a 1:1 gear ratio, or underdrive (TA) wherein the pressure actuates a different set of clutch plates to cause a gear reduction. In either situation the pressure regulator (E) holds a constant pressure on the drive actuator pistons.

The pressure maintained by E is high enough so that its piston uncovers the lubrication supply line. The bypass flow from E lubricates the underdrive parts, and after passing over the lubricating regulator valve (LR) it lubricates the gear transmission.

The hitch-related portion of the system is described in the discussion of traction-assist systems.

Traction

Traction, the ability of a tractor to develop a drawbar pull, depends on many factors. The type of traction device, the amount of ballast, the lug design, and the hitch mechanism all contribute to a tractor's ability to pull. Ballast and type of hitch are variables in Chapter 2, where the drawbar performance of two-wheel drive, rubber-tired tractors is predicted.

The design of the crawler tractor is generally acknowledged to provide superior traction on soils. But when compared to rubber-tired wheels, the steel track design is expensive to purchase and repair, is limited in forward speed, and has reduced performance on paved roadways.

An alternative rubber track design was introduced to agriculture in 1987 (Fig. 15.72). A rubber belt track is driven by a toothed rear sheave over a smooth front sheave. This design combines the performance of the steel track design with the ability to operate on paved roads at speeds up to 29 km/hr (18 MPH). The crawler tractor's utility in farming operations has increased markedly with this development.

Four-wheel drive tractors (Fig. 15.8) have superior traction compared to 2-wheel drive tractors, because they have greater powered wheel-to-soil contact area. But like crawler tractors, their economy of operation is less than for 2-wheel drive tractors except under heavy pull or soft soil conditions.

Fig. 15.72. Crawler tractor with rubber tracks.

Auxiliary drives can convert 2-wheel drive tractors and SP machines to 4-wheel drive. Most tractor manufacturers provide models that have mechanical front-wheel drive (MFWD), which can be engaged for heavy pull or poor traction situations and disengaged when not needed to permit less friction HP consumption (Fig. 15.73). Fig. 15.74 shows a hydrostatic drive added to an SP combine. Power is provided to the steering wheels by hydraulic motors. Traction tires replace the steering tires. Such additions are especially helpful in muddy conditions.

Tire size affects the tractive ability of wheel tractors. When soil shear strength is a limiting factor, large diameter, wide, drive-wheel tires have a greater soil contact area and are able to develop more pull than smaller tires. Large size has a favorable effect on steering tires too. The rolling resistance of front wheels subtracts from the thrust obtained from the drive tires; therefore large diameter tires having less rolling resistance are desirable. Also, large front tires can carry more weight than small tires (Table 2.1), an important matter for tractors equipped with front loaders.

Radial drive tires have a positive effect on the tractive ability of tractors. L. F. Bohnert and T. D. Kenady (B. F. Goodrich Tire Co.) report radials have improved tractive efficiency by 12% on tilled soil and 4.1% on sod. They also report reduced slip and fuel consumption, longer wear, and an improved ride.

Fig. 15.74. Increased traction from powered steering wheels.

Drive-wheel tire tread design has a highly variable effect on traction. Deep, self-cleaning lugs are advantageous on soft soils. Worn or even smooth tires are superior on firm surfaces.

Ballasting

Weighting or ballasting tractors is an acknowledged way to increase traction but at the expense of increased rolling resistance. Ballasting is limited to the weight-carrying capacity of tires (Table 2.1). Cast-iron weights and fluid in the tires are popular forms of ballast. A calcium chloride-water solution is a favorite ballast since it is inexpensive, heavy, will not freeze, and does not interfere with the tractor's clearance dimensions. Table 15.6 estimates the added fluid weight for several tire sizes.

Proper attention must be given to the placement of the ballast on the tractor frame to ensure steering control under heavy pull conditions. More front

Fig. 15.73. Two-wheel drive tractor equipped with mechanical front wheel drive (MFWD).

TABLE 15.6. Calcium Chloride Solution for Tractor Tires

Tire Size	Volume/Tire[a] L [gal]		Added Weight/Tire kg [lb]	
6.00-16	6	[1.6]	27	[60]
7.50-16	9	[2.4]	41	[91]
11.00-16	23	[6.1]	105	[232]
11.2-38	37	[9.8]	170	[374]
12.4-28	35	[9.3]	161	[354]
13.6-38	57	[15.1]	216	[475]
15.5-38	66	[17.5]	303	[666]
16.9-34	81	[21.4]	373	[820]
18.4-28	84	[22.2]	386	[850]
18.4-34	100	[26.4]	459	[1010]
24.5-32	170	[45.0]	782	[1720]

Note: 23% solution, 0.3 kg, 77% $CaCl_2$/litre H_2O [2.65 lb/gal]. Protects to -34°C [-30°F].
[a]For 75% (valve level) fill. For 90% fill add 20% more to these values.

weight is required for tractors pulling mounted implements since the traction-assist mechanisms transfer considerable weight from the front to the rear axle. Safe ratios of front-to-rear static weights are about 0.5 for mounted implements, 0.4 for semimounted implements, and 0.3 for trailing implements.

Traction-assist systems are built into most mounted implement hitches of tractors. The general operating principle is to shift weight from an attached implement and from the front of the tractor to the drive wheels as the drawbar pull increases. The traction-assist systems are used to augment the weight transfer that occurs naturally because of drawbar pull. Similar increases in traction could be obtained by adding deadweight to the tractor but power is required to move deadweight against rolling resistance. A more efficient practice is to keep the weight of the tractor as low as possible while causing a higher percentage of that weight to bear on the drive wheels when increased traction is needed. Some speak of this weight transferred by the traction-assist system as *live weight*.

A description of the action of the traction-assist system should permit the machinery manager to optimize its operation. Fig. 15.75 shows the action of both the traction-assist and the implement lift systems. The system illustrated is of the lower link sensing type.

The actuating signal for the traction-assist system in Fig. 15.75 originates as an implement pull on the lower link of the hitch. This pull is converted into a slight rotation at the center of the crank arm tube. The lower sensing arm transmits the movement through the draft control pickup arm to the control valve link pin. Heavy draft loads will cause this control pin to be moved left toward the rear of the tractor because of the action of the tension spring. The control pin is attached to the spool in the main control valve and movement to the left causes hydraulic pressure to be admitted to the piston head of the lift cylinder. The piston rod moves to the left, rotates the rockshaft, and starts to lift the lower links, all effectively causing more weight to bear on the tractor's rear drive wheels. But rotation of the rockshaft also causes the rockshaft cam to rotate against a rocker arm, which cancels out the initial displacement of the valve spool and limits the lower link lift action to a very small movement. A continually increasing draft, however, will cause a continual lifting of the arms.

The relaxation of draft causes a reverse mechanical action and the spool in the main control valve is moved to the right to unseat the ball in the drop check valve pilot. The trapped oil at the piston head and behind the drop check valve is released, the drop check valve moves to the right, and the trapped oil is released to the reservoir through ports in the action control valve. Part of this escaping fluid is used to lubricate the lift cylinder. The weight on the hitch causes the lower links to drop until the draft increases to a point where equilibrium is established once again.

The sensitivity of the traction-assist system is controlled by the tractor operator. By adjusting the draft control lever on the quadrant, the pivot point of the draft control pickup arm is shifted either up or down, and the relationship between the movement of the upper arm and the link point is changed. For maximum sensitivity (most hitch lift action per pound of draft force) the pivot for the draft control pickup arm must be in its lowest position.

The implement lift system is operated independently of the traction-assist system. Movement of the position control lever to the "up" position pulls the spool in the main control valve to the left, or raise, position. The rockshaft rotates and the position control linkage moves the center of the walking beam to the right. When the hitch has reached its fully raised position, the lower pin in the walking beam has traveled in its slot and pushed the spool in the main control valve back to its neutral position. The reverse mechanical movements will cause the hitch to drop.

The setting of the position control lever determines the speed of drop of the implement and the speed of action of the traction-assist system. When this lever is set in the lower or forward third of the quadrant, it permits the action control valve to work. The position of the tapered spool produces a related setting of the drop-retarding piston. The variable choking effect by this piston on the discharge flow from the lift cylinder gives a variable speed to the retraction of the lift cylinder.

Action control should be used only for mounted implements, since a restricted flow would also affect the action of any auxiliary valves placed between the unloading valve and the reservoir. One or more of these valves may be stacked together to operate one or more remote hydraulic cylinders or motors. It is usually desirable to have full flow available to such actuators.

The cushion relief valve located in the implement lift cylinder body is a protection device. It releases excessive pressures trapped above the piston head

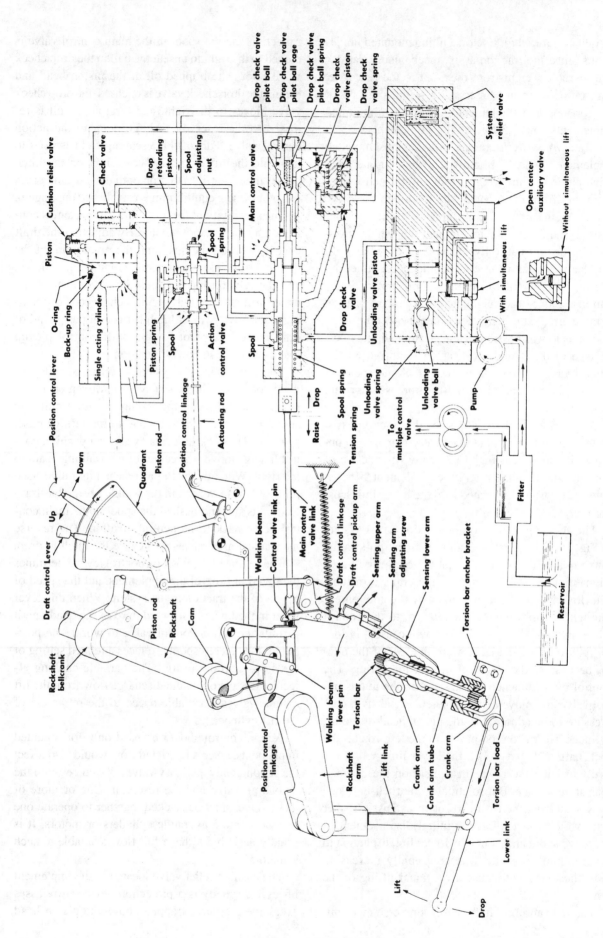

Fig. 15.75. Traction-assist and implement lift systems for the tractor in Fig. 15.71.

because of shock loads on the hitch and thermal expansion. The released fluid is dumped into the reservoir.

The second pump shown in Fig. 15.75 is delivering to a multiple-control valve. This pump is the same one shown in Fig. 15.71.

In some tractors, torque is used to control the traction-assist system. The torque in the transmission drive line is sensed and this signal actuates the hydraulic lift mechanism.

The amount of traction-assist system response has to be optimized by the tractor operator. Light drawbar pulls do not cause much control valve spool travel; thus the sensitivity control lever can be set in its most sensitive position for the traction-assist system to operate. Of course with a light drawbar load, traction is not usually a problem and the traction-assist system may not be needed. If the sensitivity control is left in its extremely sensitive position for high drawbar pull loads such as plowing, the system will hold the plow to a very shallow and erratic depth; for large drawbar pulls the sensitivity control should be set in a less sensitive position. The operator must find by trial in a specific field condition the sensitivity and speed of response settings that will give adequate traction-assist with a minimum amount of undesirable fluctuation in operating depth of the implement.

Operations

Several operating controls and adjustments are available to drivers of tractors and self-propelled machines. The machinery manager should be sure that the driver understands the response characteristics of each control and the purpose of each adjustment. Operator safety and economic benefit result.

Steering

The most obvious operator control is the steering wheel or steering levers. The vehicle seat should be adjusted to provide the operator with an easy control of the steering mechanism. Most large tractors and self-propelled implements have power-assisted steering mechanisms to reduce operator fatigue and thus increase operator alertness and work efficiency.

Each type of tractor and self-propelled machine has its own special steering response. Compared to smaller machines, the long wheelbase of large tractors and SP machines gives a slower response to steering inputs by the operator while making a turn on headlands or in transit. Many SP machines have rear-wheel steering. Such vehicles require greater outside turning clearance as opposed to greater inside turning clearance for front-steering vehicles. (See Machine Maneuverability in Chapter 1.) Skid-steer loaders, crawler tractors, and articulated-steer tractors (Fig. 13.15, 15.72, and 15.8) require an operator response quite different from axle-steered vehicles.

Effective steering is a skill acquired through both motivation and experience. An owner-operator is self-motivated because poor steering can mean substantial economic loss. Motivation for hired operators may be enhanced by flying over row crop fields and observing the preciseness of the row spacing and the number of missing row spots where the cultivator wandered. The machinery manager should provide a gradual learning experience and not place a new driver on harvesting or cultivation equipment that requires

Practice Problems
(Power Transmission)

15.20. If a single-plate disk clutch has an effective area of 775 cm² [120 in².] and an effective torque radius of 100 mm [4 in.], what torque can the clutch transmit without slipping if the total force is 2.67 kN [600 lb] and the coefficient of friction is 0.6?

15.21. What is the theoretical drawbar pull available from a tractor engine producing 37 kW [50 HP]

at 2400 rpm, a 10:1 transmission gear ratio, a 6:1 pinion-ring gear reduction, and a 5:1 final drive reduction? The tire size is 15 × 38.

15.22. If a tractor's hydraulic system can produce 10.3 MPa [1500 psi], what force can a remote cylinder 100 mm [4 in.] in diameter with a 25-mm [1-in.] piston rod provide while extending.

precise steering. Developments in the future may include complete automatic guidance.

Gear Ratio and Engine Speed

The modern tractor operator, by adjusting the transmission gear ratio to get any of a number of forward speeds, can shift into most of these gear ratios with little trouble. Within any gear ratio is further adjustment with the speed control or governor-setting lever.

Most tractor field speeds are determined by the implement and not by the power ability of the tractor; yet considerable fuel saving may be possible depending on how the operator arrives at the required speed. The fuel savings from proper governor settings at part loads are discussed in Chapter 2. To review, most tractors have the characteristic that fuel efficiency is improved with reduced governor settings at part load (refer to Fig. 2.14, 2.15, and 2.16). Significant increases in fuel efficiency may be expected, particularly at less than half load, if the speed control lever setting is reduced and a faster gear ratio is selected. However, the good operator will not let the tractor "lug down" below the point where damage can occur.

Fuel Setting

The adjustment with the most direct effect on fuel efficiency is the load mixture control on the carbureted engine. The load mixture adjustment is accomplished by turning the power needle. This mixture control needle valve is best set while the tractor is loaded by a dynamometer, as shown in Fig. 2.6. Most implement dealers have such dynamometers and can offer this service. Diesel engine fuel injection adjustments are intended to be a dealer service function.

Exhaust smoke is indicative of the operating conditions for engines. Blue smoke from either an SI or CI engine implies combustion of lubricating oil that enters the combustion chamber past the piston rings, past worn valve guides, or from worn turbocharger shaft seals. For CI engines, a white smoke consists of unburned fuel mist and occurs with misfiring during cranking and under low temperature operations. Black smoke from either engine arises from incompletely burned carbon because of a rich mixture for SI engines and because of worn injectors or over-fueling for CI engines. Caterpillar tests showed a slight gain in power from over-fueling (Fig. 15.76), but most CI fuel injection systems are set at a light gray smoke limit.

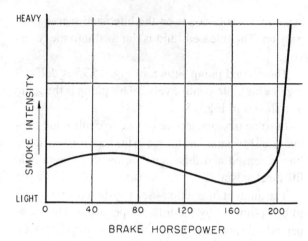

Fig. 15.76. Effect of over-fueling on smoke and power output.

Temperature Control

The operating temperature of the engine, as indicated by coolant temperature, can also affect power and efficiency. Table 15.5 shows 80°C [180°F] to be the optimum operating temperature for gasoline engines. The optimum temperature for a specific engine will vary from this value depending on the engine design and the fuel used. Heavy, poorly vaporizing fuels will require temperatures up to 90°C [194°F]. Highly volatile fuels (LP gas) need high temperatures not for vaporization but to reduce wear from poor lubrication. Temperatures higher than 93°C [200°F] for any engine will result in coolant loss and the burning of oil.

The tractor operator does not usually have direct control over the engine's temperature, as the thermostat maintains a constant temperature level automatically. The duty of the operator is to watch the temperature reading and take proper action if the indicated temperature varies from the manufacturer's recommendations.

Tire Wear

Tires deteriorate with use and exposure. Wheel slippage wears away the lugs on drive tires. Sidewalls can buckle and break from overloading and/or underinflation. Oil, oxygen, and sunlight cause rubber to craze, crack, and weaken.

Slippage may be reduced with smaller drawbar pulls. To retain the same field capacity with reduced pull, the forward speed must increase. This adjustment results in the type of operation most suited to rubber-tired tractors—light loads and high speeds.

Unfortunately, the tractor is not free to be operated at its optimum speed but must be regulated according to the needs of the implement that it is serving.

When heavy drawbar loads are necessary, slippage can be reduced with proper weighting. Added weight, either *deadweight* or *live weight* (obtained through traction-assist systems), is limited by the load-carrying ability of the tire, which in turn depends on the inflation pressure. Adherence to the Tire and Rim Association's recommendations for weight and inflation pressure should ensure long life for a tire's sidewall (Table 2.1).

Slippage is also reduced with good lug bar contact. Fig. 15.77 shows the limited contact obtained by the lug bars on hard surfaces when the tires are overinflated. The good power manager will vary the air pressure in traction tires as loads on the wheel are changed so that effective lug bar action can be secured.

Protection from petroleum products and sunlight is increased by providing and using proper storage space.

T. E. Tramel and J. T. Long (Mississippi State University) estimate tractor tire life for Mississippi's brown loam section:

| Tractor size | Tire life, yr | |
	front	rear
1-row	7.0	8.5
2-row	4.5	5.9
4-row	2.9	4.2

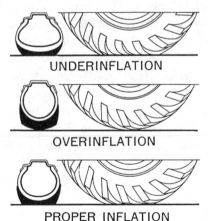

UNDERINFLATION

OVERINFLATION

PROPER INFLATION

Fig. 15.76. Effect of inflation on tire cross section.

Avoiding Power Waste

One way to improve power efficiency is to avoid waste. Removing the added weights from the tractor when they are not needed will reduce the force required to overcome rolling resistance. In plowed fields each additional 100 units of unnecessary weight requires about 30 additional units of tractive force. At 6 km/hr [3.75 MPH] this additional tractive force will amount to 1-kW [1-HP] waste for every 200 kg [333 lb] of surplus weight.

Maintenance

Tractors start deteriorating immediately after construction. Some deterioration is due to chemical combination with oxygen in the air (rust, etc.). Paint, grease, and sheltered storage will retard this type of deterioration. But most deterioration of the tractor is caused by use. The process of retarding or correcting deterioration is called *maintenance*. The reasons for deterioration in engine components are quite well known. If the tractor operator understands these reasons, maintenance may not be neglected.

Manufacturers design farm equipment for effective and easy maintenance of certain critical parts. They recommend maintenance intervals, which if followed should ensure long mechanism life. The machinery manager hopes not only to provide adequate maintenance but to avoid expensive overmaintenance.

Prevention of deterioration or *preventive maintenance* is the guiding principle behind many maintenance activities. Changing lubricating and hydraulic oils and replacing air, oil, and fuel filters at planned intervals are ways of preventing damage from acids and dirt. Cleanliness in and around the tractor is a very important maintenance activity.

The machinery manager is interested in performing those repair and maintenance activities which will insure reliability of the machine. The costs of downtime during critical field operations may appear to be so great that over-maintenance and over-repair is looked upon as prevention and worth the cost.

The tractor driver can undertake the maintenance activities when the machine is in use. Daily checks (sometimes even hourly) should be made for fuel, oil, and coolant levels, tension in belt and chain drives, loose bolted connections, needed grease applications, plugged radiator cores, and for excessive build-up of dirt, dust and crop materials around the engine and in the implement mechanisms. The machinery manager should emphasize to the operator that many of these checks have personal safety implications.

Repair of farm machines may be done on the farm and is done in the event of an emergency break-down. As farm machines and tractors become more complex, machinery managers may refer the machines to the equipment dealer for annual or biennial comprehensive service. Remoteness from the dealership may cause the machinery manager to maintain a well-equipped shop and experienced mechanics on the farm. Only the largest of farms should consider this investment. A mechanically skilled tractor driver or machinery manager may economically undertake on-farm repairs for the smallest of farms. If a substantial farm shop is to be maintained, thought should go toward the extent of a parts and maintenance supply items that may be economically worthwhile to avoid costly machine break-down time.

Scheduled changing of crankcase oil is a necessary maintenance activity. As the oil circulates through the engine it becomes contaminated with water, fuel, acids, dirt, soot, and metallic particles. Slow accumulation of contaminants is normal and is corrected by periodic changing of the oil.

Examination of the drained oil by sight, smell, and feel can alert the machinery manager to abnormal conditions in the engine. A greasy feel of oil between the thumb and fingers indicates viscosity. Diesel engine crankcase oil turns black quite quickly from fuel soot, while the oil in SI engines in good condition will lose its original color more slowly. The LP gas engine will retain the original oil color and clearness over the whole change interval. Oil from a heavily loaded engine may have a burnt odor, particularly if the oil level is low. In such a case the viscosity of the drained oil could be higher than the original oil. Oil that feels thin may have the odor of fuel and indicates fuel contamination.

Quick heating of a representative sample of the drainage will produce a crackling sound for even small amounts of entrapped water. Water contamination can come from low engine operating temperatures as well as from coolant leakage.

Commercial analysis of oil samples taken from an engine crankcase can be a useful tool for engine maintenance. Contaminants and degradation of the oil can cause costly and rapid wear of the engine parts. A viscosity decrease indicates fuel dilution and oil breakdown due to shearing of the viscosity index improvers.

A gradual increase in viscosity with service time and a deviation of less than 10% from the original viscosity should not be of concern. Metal and silicon contaminant analysis and measures of acidity can only be provided by a commercial testing laboratory.

Rapid oil analysis is a service provided by Texaco Inc. and others. Fleet owners make regular use of such analyses to detect impending failures that can be avoided by prompt corrective action. The farm machinery manager should consider such analyses if a sensual examination indicates an abnormality.

The *air cleaner* can act as a large throttling valve on the air intake of an engine; if not properly cleaned and serviced, the air cleaner will restrict airflow to the engine and full power will not be attained. This throttling builds up so gradually that the tractor operator is unaware of any decrease in engine performance. Clean air is particularly important for turbocharged engines. The compressor blades are susceptible to erosion, bending, and breakage from ingested dirt and other particles. Only a planned inspection schedule of at least once a week will avoid the power loss from a restricted air cleaner.

The delivery of *fuel injectors* varies with wear and dirt buildup. Checking the performance of injectors and pumps is a job for the specialist, but the machinery manager should see that the job is done periodically to avoid a drop in tractor performance.

Tappet clearance can be adjusted quite easily. As the hours of use accumulate, an engine's valve linkage will wear, causing a gradual increase in tappet clearance. The valve range is reduced with increased tappet clearance; thus there will be a power loss. Tappet clearance should be checked once each season. Hydraulic valve lifters automatically correct for wear.

Governor wear is another important contributor to loss of power of a tractor engine. Wear in the governor itself and in the linkage joints will cause the tractor engine to drop below its rated speed. In addition, response to overloading is sluggish and hunting may also occur. Power is lost and fuel efficiency drops.

An occasional check of the engine's no-load speed will give an indication of governor linkage wear. Special attention should be given to the oil supply to the governor, as inadequate lubrication encourages rapid wear.

Engine *overheating* because of scale in the cooling system will reduce power and increase deterioration. Engine temperatures can be maintained at the proper level (Table 15.5) only with rapid heat transfer from the cylinders into the coolant. As little as 1.5 mm [0.06 in.] of deposits on the cylinder walls is equivalent to a heat flow resistance of an additional 75 mm [3 in.] of cast iron. The use of soft water and scale inhibitor additives is preventive maintenance.

Tractor Modifications

Machinery managers are constantly tempted to modify their tractors or to add newly developed devices to improve the performance. It is true that many of the past modifications have been incorporated into new tractors by manufacturers. High compression, additional transmission gear ratios, power steering, dual drive tires, and turbochargers are examples of mechanisms that were once field modifications for tractors. But many other modifications such as high-lift cams, overbored cylinders, high-altitude pistons, and increased fuel flow have not been so successful. Some problems associated with field modifications have been ring sticking; overheating; oil burning; and failures in bearings, drive lines, valves, valve seats, pistons, and gasketed joints.

The machinery manager should be reluctant to modify current tractors that have a balanced design with little excess strength or capacity. Any modification tending to increase the output of one segment of the system is very likely to overload another.

EQUIPMENT SELECTION

Farm equipment selection is often a blending of what the farmer thinks is needed and what the manufacturers think the farmer should need from the line of equipment that they, the manufacturers, can produce at a profit. To a great extent the farmer's wants are created by what the manufacturers offer—and the manufacturer builds only what the farmer will buy.

Such an ingrown cycle cannot be efficient if equipment selection is based on impulse buying and promotions appealing to pride. The costs of agricultural production can spiral and the economic health of farmers will suffer. The equipment manufacturers' economic health ultimately depends on the economic health of their customers, the farmers; therefore, an efficient method of matching equipment to the true needs of agriculture will benefit both farmers and manufacturers.

The economic selection of field equipment is a complex problem that has some unique characteristics compared with other industries. First, farms are comparatively small-scale operations, have diversified enterprises, and are subject to many special local conditions; thus each farm must be treated as a special problem. Second, since agricultural production is seasonal, equipment will necessarily stand idle much of the time. Also, most field implements are operated by a shared power unit, the tractor; and a change in one tractor-implement operation will affect the whole system. Consequently, the complete system of implements must be considered. Third, the supply and ability of farm labor, which usually includes management personnel, is quite varied. Finally, a characteristic that is widely recognized but difficult to analyze is the need for timely operations because of the seasonal requirements of crops.

The problem of selecting field machinery efficiently is therefore one of adjusting the factors of implement performance, power availability, labor, timeliness, and costs until an optimum economic return results.

The ability of the farm power and machinery manager to select proper equipment is a most valuable function. Many activities relate to this function. Data must be collected and assessed on such widely varying subjects as the life cycle of nematodes to economic machine design for mass production. Significant trends in the technology of agricultural production must be anticipated and recognized. The manager must have the objectivity and courage to be committed to a considered economic line of action and the drive to follow through.

Part V provides the machinery manager information to keep with analyzing, organizing, and selecting appropriate machinery. Selection is properly the final subject, as it must rely on concepts developed in earlier parts of the book—performance characteristics, costs, operations, and power. One additional topic, the cost of timeliness, is introduced at the point at which it would normally enter into consideration.

Only capacity selection techniques are discussed in this section. Functional selection is best made after field trials with specific machines in actual field situations.

Chapter 16

MACHINERY SELECTION

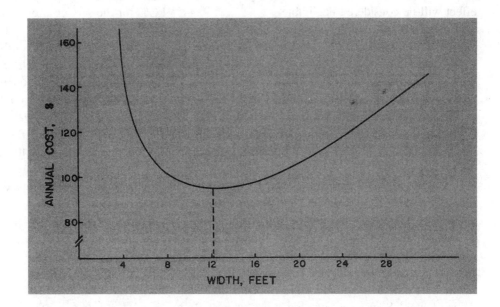

Size selection of machinery must necessarily be based on anticipated performance and anticipated costs. Since these future values can never be known exactly, selection must proceed with a liberal or flexible view toward some of the relationships among the pertinent variables. Some of the rigid relationships may have to be relaxed in the interest of arriving at a general, workable method for selection. Based on the above philosophy, the approximate cost analysis in Chapter 4 is considered an adequate description of the economics of machinery operation. While one should be cautious about making arbitrary judgments, one should not be reluctant to accept approximations since the future can never be more than an estimation.

In field machinery selection, the most pertinent variable is size or capacity of the machine. Although forward speed and power availability affect field capacity, initially it is assumed that power is not lacking and the forward speed is the maximum value that does not reduce the effectiveness of operation.

The symbol, w, will be used to represent the effective width of action of all field implements. It would appear to be difficult to use w to represent the capacity of a baler, but a moment's reflection about windrow swath widths and yields of material on an area basis will show that such capacity can be represented in terms of w.

As a first trial, the basis for selecting the proper capacity of a field machine is to find the minimum-cost machine by using Eq. 4.7 and the minimization method of Appendix B.

Before such a procedure can be applied, all variables in Eq. 4.7 that depend on the size of the machine must be expressed in terms of w. The major variable so dependent is the purchase price, P. It is now necessary to use a new statement of purchase price for capacity selection considerations, as the purchase price cannot be known until after the size of the machine is known. Let p be understood to be the purchase price per additional unit of width; then the P in Eq. 4.7 can be written as pw (or P = pw).

The forward speed will be constant with different sizes of machines as long as power is not limiting.

As discussed in Chapter 1, the size of a machine may have an effect on the field efficiency, e, but the effect will be considered negligible.

The repair and maintenance costs must also be expressed in terms of w. Let it be understood that the yearly cost of repair and maintenance, (R&M)P, now will be replaced by rmpw, where rm is the value of repair and maintenance per hour expressed as a decimal of the purchase price, pw.

Fuel and oil costs per hour are known to be definitely proportional to the size of equipment. For simplicity it is assumed that they are directly proportional to size; therefore the variables O and F can be expressed as ow and fw where o and f refer to the oil and fuel costs per hour per unit of implement width.

The cost of labor, L, is readily recognized as being essentially independent of the size of the machine. The cost of tractor rent, T, is assumed to be a function of time only and independent of the size of the implement. Such an assumption is subject to challenge but will be honored for this analysis.

Eq. 4.7 thus is transformed into Eq. 16.1, a statement of the annual cost of a machine where the appropriate variables are expressed on a basis of unit of machine width.

$$AC = \frac{(FC\%)pw}{100} + \frac{cA}{Swe}(rmpw + L + ow + fw + T)$$

$$(16.1)$$

The diagram at the beginning of this chapter represents a plot of Eq. 16.1 for implements priced at \$p per unit width and used 120 ha [300 a] annually. Note there is a definite size that produces a minimum cost.

Eq. 16.2 defines the lowest point on the cost curve represented by Eq. 16.1,

$$w = \sqrt{\frac{100\,cA}{(FC\%)\,pSe}(L+T)}$$

$$(16.2)$$

Typical values of p may be estimated from Appendix D. The proper value of p to use would be one calculated from a price quotation for a specific implement that would appear to be a good candidate for selection.

The selection of the proper FC% is vital to the validity of Eq. 16.2. Because of the effect of depreciation on costs, some anticipated service life will have to be estimated. Table 16.1 lists the values for FC% for the service life of the machine in question. To review, the total service life of a machine is the length of time in years until the machine has only salvage value. If the machinery manager anticipates replacing the machine before it reaches salvage value, a value for service life in Table 16.1 should be selected that is about midway between the actual service life and the age of the machine when traded. Such a hedge will cover the difference between the straight-line depreciation remaining values and the actual trade-in value allowed by the dealer of used machinery.

Note that repair and maintenance, fuel, and oil costs drop out of consideration in Eq. 16.2 because they are proportional only to area covered and would have the same dollar per year cost regardless of machine size.

Timeliness

At this point timeliness of a field operation must be considered to have an economic value. Timeliness costs arise because of the inability to complete a field operation in a reasonably short time. These are not out-of-pocket costs but reductions in potential return, as when the yield and quality of a crop are reduced

TABLE 16.1. Values for Fixed Cost Percentage

Service Life, yr	Value of FC%
1	100
2	53
3	37
4	29
5	25
6	22
7	20
8	18
9	17
10	16
11	15
12	14
15	13
20	12

because of delays in harvesting. Delays due to bad weather cannot be charged to the machine, but delay in harvesting the last part of a field because the machine has low capacity *is* a cost that should be borne by the machine. Timeliness costs are so important that in the machinery selection process they must be evaluated quantitatively and considered as a valid cost of field machinery operation.

Several measures of timeliness costs are available from field experiments. Wheat has been reported to suffer a 46% reduction in yield for each week of delay in planting. In the northern Corn Belt it is estimated that each day of delay in planting corn after May 15 can decrease yields 63 kg/ha [1 bu/a]. Timely rotary hoeing in Iowa while the weeds were in the "white" stage increased yields 300 kg/ha [5 bu/a] above that for hoeing when the weeds were in the 1–3 leaf stage. Harvest losses for corn combining in the Midwest can be 125 kg [2 bu]/week of delay past the optimum harvest moisture (26%). In Ohio it was reported that shatter losses increase 67 kg/ha [1 bu/a] for every 5 days after wheat reaches maturity (30% moisture), and test weights decrease about 1.6% every 4 days (Fig. 10.14). Fig. 16.1 shows the feeding value of hay harvested at different dates in Minnesota. About 1% loss/day occurs on either side of the optimum time of June 17. The loss increases to about 3% per day after the initial 10-day period. Michigan State University dairy specialists report that each day of delay in cutting alfalfa after June 1 reduces its value 2.5%. T. H. Garner and J. F. Dickson (Agricultural Engineering Department, Clemson University) could not find statistically significant effects for timeliness in cotton defoliation; but they did note that planting date markedly affected yield, lint qual-

ity, and harvesting efficiency. T. E. Corley (Auburn University) found a 5% loss with each 4-week delay in cotton harvesting.

Crop varieties can make a difference in timeliness losses. In the Illinois evaluation of timeliness for combining soybeans (Fig. 10.15 and discussion), late-planted Harosoys had a measured linear economic loss of 0.0125 of the gross value of the crop per day of delay. This decimal of the gross value is called a *timeliness factor*, K, and has units of 1/day. For the other varieties a more typical loss was K = 0.0045. The penalty for early harvest (before 13% moisture) is greater, K = 0.017. The losses in this combining evaluation included marketing penalties for moisture, damage, and foreign matter as well as the losses from shatter and the combine. The K factor for planting Chippewa soybeans at Urbana was about 0.007, while for Clark was 0.002. Fig. 16.2 shows some varietal effects on the timely planting of soybeans in Missouri.

Enough evidence is available to estimate timeliness factors for most machine operations. Table 16.2 lists some typical factors. These values should be considered conservative. The timeliness factor for specific

Table 16.2. Timeliness Loss Factors

Operation	K
Tillage	0.0001-0.002
Seeding	0.002-0.006
Cultivation, spraying	0.010
Harvesting	
corn	0.003
cotton	0.008
green forage	0.001
hay (alfalfa)	0.010
small grain	0.004
soybeans	0.005

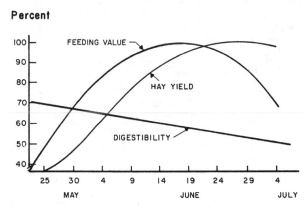

Fig. 16.1. Timeliness costs in hay harvesting.

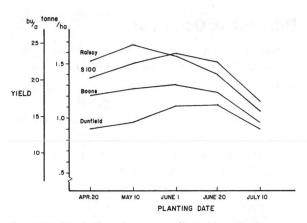

Fig. 16.2. Yield of soybeans as affected by planting date. Four varieties tested.

situations may be much greater. The K factors in Table 16.2 are based on crop area. If a crop requires multiple operations (2 cultivations, 3 cuttings of alfalfa, etc.), the total machine area should be divided by 2, 3, etc., when used in selection analyses.

These K factors apply directly to a small operation or to any large operation handled as an entity, but they should be modified for most large areas. For maximum returns, timeliness considerations require that a 1000-ha corn growing operation would need to be planted all in one day, cultivated all in one day, and harvested all in one day. The machinery system that would even approach this capacity would be prohibitively expensive. Instead, three different varieties of corn could be scheduled for planting at their various optimum times. The cultivation and harvesting timeliness requirements would now be presumed to be one-third as critical, and the annual area should be divided by three. Good farm machinery management includes the dispersion of optimum operating times to reduce the capacity requirement of the machine system.

Total timeliness costs for an operation depend on the scheduling of operations with respect to the optimum time and on the duration of the operation. Three types of scheduling are recognized. One might be typical for harvesting perishables where the operation commences as soon as the crop is mature and then proceeds as quickly as possible before deterioration becomes excessive. Such timing can be called *delayed scheduling*. *Premature scheduling* describes the situation where the machine operation is planned to be completed by the optimum time. *Balanced scheduling* describes the situation where the machine operation time extends equally on each side of the optimum time.

Duration of Operation

The total time required for a field machine operation depends on the capacity of the machine and the number of available working days. Each region of the country has a unique climate, and different machine operations will have different criteria as to what constitutes a working day. Wet soil and wet crops are the usual deterrent to field machine operations.

Duration of field operations can be shortened by extending the working hours per day if multiple operators are available. Night-time operations have always been an option (Fig. 16.3) when weather, soil, and crop conditions were favorable. With the advent of GPS, wide tillage and seeding machines can now

Fig. 16.3. GPS can lengthen the working day.

be accurately steered without the sighting problems associated with darkness.

B. Bolton, J. B. Penn, F. T. Cooke, Jr., and A. M. Heagler (Department of Agricultural Economics, Louisiana State University) have estimated the days suitable for working soil in the Mississippi River Delta cotton area (Table 16.3). This table is based on a soil moisture accounting for the top 15 cm [6 in] of soil considering rainfall, drainage, runoff, and evapotranspiration. Harvesting days available are based on a

Table 16.3. Soil Working Days Available (85% probability, Mississippi River Delta)

Months	Clay	U	Sandy Soil	U
March	0	0	3.5	0.11
April	4.0	0.13	9.5	0.32
May	10.5	0.34	14.0	0.45
June	13.5	0.44	16.0	0.52
July	10.0	0.31	16.0	0.42
August	15.5	0.50	17.0	0.55
September	18.5	0.62	17.5	0.58
October	15.0	0.48	14.0	0.45
November	2.0	0.07	3.0	0.10

Harvesting Days Available (85% probability, Mississippi River Delta)					
Harvest	Date	Clay	U	Sandy Soil	U
Corn	Aug.16-Sep.14	18.5	0.62	19.0	0.63
Early soybeans	Sep.10-Oct.9	19.0	0.63	20.5	0.68
Medium soybeans	Oct.1-Oct.30	18.5	0.62	18.0	0.60
First cotton	Sep.21-Oct.20	18.0	0.60	19.0	0.63
Second cotton, late soybeans	Oct.21-Nov.19	11.0	0.37	12.5	0.42

slightly wetter soil moisture criterion. The working days available can only be expressed on a probability basis because of the randomness of weather. The 85% probability can be interpreted as meaning that the listed days or more could be expected in 85 years out of 100. The *utilization*, U, is the ratio of expected working days to total days. The values in Table 16.3 are multiplied by 0.86, 0.85, 0.84, and 0.83 for non-working Sundays and for 1, 2, and 3 holiday months, respectively.

Similar evaluations of working days have been made in other areas of the country. G. E. Ayres and C. V. Fulton (Agricultural Engineering Department, Iowa State University) determined the probabilities for climatic weeks in Ames (Appendix H). A. M. Feyerheim, L. D. Bark, and W. C. Burrows have published a series of wet and dry day probabilities for the Midwest states as a part of the activities of the NC-26 Technical Committee, a cooperative activity of thirteen state agricultural experiment stations in the North Central region and the USDA. Table 16.4 is derived from seventeen years of data in which local observers over the state reported the number of field working days every week. The time utilization factor, U, is somewhat greater for Illinois than Mississippi because of lower rainfall.

The timeliness loss factor, machine capacity, and expected duration time are brought together in Eq. 16.3 to predict the timeliness cost (TC) in \$/machine operation.

$$TC = \frac{K\,Y\,V\,A^2}{(sc)\,(nt)\,UZ} \qquad (16.3)$$

where K = timeliness loss factor, $\dfrac{1}{day}$

Y = potential crop yield, $\dfrac{kg}{ha}\left[\dfrac{\$}{bu}\right]$, etc.

V = value of the crop, $\dfrac{\$}{kg}\left[\dfrac{\$}{bu}\right]$, etc.

A = crop-area involved ha [a]

U = fractional utilization of total time, decimal

Z = effective machine capacity, $\dfrac{area}{day}$

sc = 2 for premature or delayed schedules, 4 for balanced

nt = number of times A should be divided because of dispersed optimum times

Table 16.4. Estimated Working Days in Illinois (90% probability)

Month	Average Probability, 5 mm [0.2 in.] Rain or Less	Working Days per Month	U
April, May	0.75	14.6	0.48
June	0.83	16.6	0.55
July, August, September	0.86	18.5	0.59
October, November	0.84	17.4	0.58

Optimum Width

Eq. 16.2 can now be modified to include a charge for timeliness.

$$w = \sqrt{\frac{100\,cA}{(FC\%)\,pSe}\left(L + T + \frac{K\,Y\,V\,A}{(sc)\,(nt)\,Uh}\right)} \qquad (16.4)$$

where h = hours worked per day. Designated as the *optimum width equation*, Eq. 16.4 is a valuable aid in selecting the most economical implement. It is plotted as curve (b) in Fig. 16.4 and illustrates the difference when Eq. 16.2 is modified by a timeliness factor. Also implied is the fact that the optimum width of implement is always greater than the least-cost width. One should not be concerned about relative vertical position of the curves. The adjusted curve (b) includes cost of poor timing of field operations; curve (a) does not.

Solving Eq. 16.4 will produce a very precise mathematical answer. The practicality of this precision will depend on the degree of sharpness of the annual cost curve at its minimum point. The range in allowable w for a preselected difference in annual cost is given by Eq. 16.5 and illustrated in Fig. 16.4.

If the annual cost of operation of an implement were allowed to vary as much as \$10 above the minimum cost (d = 10), the resulting range in w might be quite large for spike-tooth harrows and quite small for combines, as the purchase price of the implement is quite important. Eq. 16.5 is valuable because it allows more flexibility in implement size selection while maintaining a close rein on the annual costs.

For very large acreages the optimum width formula yields an implement size so large as to be impractical and unavailable as a single machine. In such circumstances multiple units are required. Dividing the largest practical and available implement width into the optimum width answer will produce the required number

$$w_{1,2} = w + \frac{100\,d}{2(FC\%)\,p} \pm \sqrt{\frac{100\,d}{(FC\%)\,p}\left(w + \frac{100\,d}{4(FC\%)\,p}\right)}$$

(16.5)

where $w_{1,2}$ = double answer obtained that defines a range in implement widths wherein annual costs of operation are approximately minimum

d = arbitrary number of dollars above the minimum annual cost that the machinery manager is willing to accept to get a range in implement size

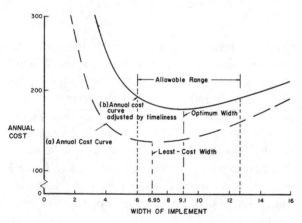

Fig. 16.4. Comparison of least-cost and optimum widths.

of units. The values of L and T must be recalculated if more than one tractor and operator are involved.

Inserting new values of L and T into Eq. 16.4 may give an even greater optimum width that could cause the addition of another unit. An exact solution results from trial-and-error methods. It is unwise to insist on exactness, since the approximation after two trials is probably adequate.

An example problem should be helpful. Notice that Eq. 16.4 can be used even though the machine is used on two or more crops.

Select the optimum width of sprayer to use on 40 ha [100 a] of wheat and 30 ha [75 a] of barley. Field speeds are 12 km/hr [7.5 MPH]. The field efficiencies are 0.7. The expected life of the sprayer is 10 years. Tractor fixed costs and operator labor are $6/hr and $7/hr, respectively. The values of the crops (YV) are $500/ha [$200/a] for wheat and $400/ha

[$160/a] for barley. The expected time utilization factor is 0.5. Ten-hour days are permissible. A balanced timeliness schedule is planned: (sc) = 4. The price of a tractor mounted sprayer is $3000 + $35/m [$10/ft], Appendix D. Assume an interest rate of 10%. The FC% is 16 (Table 4.7) + 2.5 or 18.5%. The value of p, price per additional unit, is $35 [$10]. The timeliness value, K (from Table 16.2) is 0.010 for both crops. Ideally, all the crop must be sprayed at once: (nt) = 1.0.

Eq. 16.4 is rearranged to accommodate the use of the machine on two separate crops.

$$w^2 = \frac{100c}{(FC\%)\,p}\left\{\frac{Aw}{S_w e_w}\left(L_w + T_w + \frac{K_w Y_w V_w A_w}{(sc)_w (nt)_w U_w h_w}\right)\right.$$

$$\left. + \frac{A_b}{S_b e_b}\left(L_b + T_b \frac{K_b Y_b V_b A_b}{(sc)_b (nt)_b U_b h_b}\right)\right\}$$

Solving numerically in SI units:

$$w^2 = \frac{100 \times 10}{18.5 \times 35}\left\{\frac{40}{12 \times 0.7}\left(7 + 6 + \frac{0.010 \times 500 \times 40}{4 \times 1 \times 0.5 \times 10}\right)\right.$$

$$\left. + \frac{30}{12 \times 0.7}\left(7 + 6 + \frac{0.010 \times 400 \times 400 \times 30}{4 \times 1 \times 0.5 \times 10}\right)\right\}$$

$$w = 16.6\,m\,[54\,ft]$$

The size of the optimum width may be surprising to some for such a small annual use. But the price for an additional metre of boom is small and the timeliness costs are large. One must conclude that a wide boom is economic.

Eq. 16.5 gives a range of sizes for which the costs, d, are less than $5 annually from the minimum.

$$\frac{100\,d}{(FC\%)\,p} = \frac{100 \times 5}{18.5 \times 35} = 0.772$$

$$w_{1,2} = 16.6 + \frac{0.772}{2} \pm \sqrt{0.772\left(16.6 + \frac{0.772}{4}\right)}$$

$$= 17.37 \pm 3.6 \quad or \quad 13.8\,to\,21\,m$$

The main use of Eq. 16.4 is to select a single implement to fit into an already existing system. In such an instance the value for T is already known or can be approximated. If self-propelled equipment is being selected, Eq. 16.4 can be used independently from the rest of the machinery system because T = 0.

Eq. 16.4 has only limited value when used for heavy-draft tillage implements such as plows, disk harrows, chisel plows, and subsoilers. These implements are relatively inexpensive for the amount of power they require. The use of Eq. 16.4 will result in very large sizes for these implements as it is assumed that T is not affected by the size of the implement. In actuality, it is the cost of power, T, more than the cost of the implement, p, that determines the optimum size for heavy tillage implements. (Selecting optimum power is the subject of Chapter 17.) Eq. 5.2 can be used for determining a least-cost width for tillage machines if a required capacity is known and a true value for T is available.

The cost minimization procedures used for field machines apply to crop dryer selection also. One difference is that the dryer is assumed to operate 24 hr/day while the harvesters operate fewer hours. The timeliness loss factor to use in drying is that of the harvester but with values of U = 1 and h = 24 in the timeliness cost equations.

An equation of annual cost is written, using dryer capacity terms:

$$AC = \frac{(FC\%)\,p\,c}{100} + \frac{B}{c}\left(L + F + T + \frac{KVB}{(sc)\,(nt)\,Uh}\right) \quad (16.6)$$

where the new symbols are
  B = total quantity to be dried
  c = actual quantity per hour capacity needed

For any specific drying system, the moisture removal efficiency is about constant; therefore the fuel consumption of the burner and the tractor need not be considered in the selection problem. The optimum capacity of the dryer is given by

$$c = \sqrt{\frac{100B}{(FC\%)\,p}\left(L + T + \frac{KVB}{24\,(sc)\,(nt)}\right)} \quad (16.7)$$

As an example, consider the selection of a continuous dryer for removing an average of 5 points of moisture from 1000 t [1100 T] of $100/t [$91/T] shelled corn. A 75 kW [100 HP] tractor having a fixed cost charge of $8.00/hr will be used to power the dryer. Supervisory labor is estimated at $0.50/hr. The annual dryer fixed costs are assumed to be 16% of the purchase price. The price of the dryer is based on the capacity of removing 10 points of moisture or of drying 25% corn. If 20% corn is to be used, the price per capacity, p, must be changed; a comparison of the heat requirements for continuous dryers (Fig. 12.17) shows that 5/4.375 or 1.15 times as much heat is needed to dry a unit of corn from 20% to 15% W.B. A moisture calculation table shows the moisture contents based on a unit of wet grain:

%	units H_2O +	units d.m. =	units total
25	0.25	0.75	1.00
20	0.19	0.75	0.94
15	0.13	0.75	0.88

There is 0.06/0.12 or half as much water removed in drying from 20% to 15% as from 25% to 15%. The capacity of a dryer operating in this situation is therefore $1/0.5 \times 1/1.15$ or 1.74 times as great as that operating under standard ASAE conditions (10 points removed).

The price for continuous dryers is based on standard 10-point removal; thus the price must be converted to a new base.

$$\frac{\$3060}{t/hr} = \frac{\$2500}{m^3/hr} \times \frac{1\,m^3\,corn}{719\,kg\,at\,15\%}$$

$$\times \frac{0.88\,kg\,at\,15\%}{1\,kg\,at\,25\%} \times \frac{1000\,kg}{t}$$

$$corrected\ price = \frac{\$3060}{t/hr} \times \frac{1}{1.74} = \frac{\$1759}{t/hr}$$

Assuming sc = 4 and nt = 1:

$$c = \sqrt{\frac{100 \times 1000}{16 \times 1759}\left(0.5 + 8 + \frac{0.003 \times 100 \times 1000}{24 \times 4 \times 1}\right)}$$

$$c = 6.43\ t/hr$$

$$\left[c = \sqrt{\frac{100 \times 1100}{16 \times 1589}\left(0.5 + 8 + \frac{0.003 \times 91 \times 1100}{24 \times 4 \times 1}\right)}\right.$$

$$\left. c = 7.1\ T/hr \right]$$

Practice Problems

16.1. What is the least-cost width for a spike-tooth harrow used on 120 ha [300 a] annually? Labor cost = $8/hr, tractor fixed costs = $6/hr, speed of operation = 8 km/hr [5 MPH], field efficiency = 0.70, and the price of the harrow is $165/m [$50/ft]. Use a 10-yr life and 10% interest.

16.2. Approximate a timeliness factor, K, for a combine from the following data:

Oct. 15   113 kg/ha [1.8 bu/a] total loss
Nov. 15   396 kg/ha [6.3 bu/a] total loss
Dec. 15   679 kg/ha [10.8 bu/a] total loss

The gross yield of the field is 5.66 t/ha [90 bu/a].

16.3. Select for a farm in Iowa the optimum width of PTO mower to cut 10 ha [25 a] of alfalfa during each of the climatic weeks 15, 20, and 28. Use the 0.88 working day probability in Appendix H. Use FC% = 18; S = 9.6 km/hr[6MPH]; e = 0.80; L and T = $6/hr; and YV = 250 [100], 200 [80], and 200 [80] for the three cuttings each year. Labor is limited to 10 hr/day. Assume sc = 4, nt = 1, and obtain the rear-mounted cutter bar mower price from Appendix D.

16.4. If the machinery manager does not care within $10/yr what the cost of the mower in 16.3 is, what is the permissible range in size for the mower?

16.5. What is the optimum use for a $12,000, 8-row rear-mounted cultivator working in 1-m [39-in.] spaced rows? Average speed is 5.5 km/hr [3.44 MPH], L = $6, T = $9, and FC% = 20. Half the area is expected to be planted to soybeans worth $550/ha [$220/a] and half to corn worth $600/ha [$240/a]. Use sc = 2, nt = 1, U = 0.6, and h = 10.

Chapter 17

POWER SELECTION

Selecting the proper power level for a farm is a most involved problem; yet, because the cost of power is a significant item in many operations some logical procedure must be found.

If all implements were self-propelled the analysis would be much easier, but in reality most implements are powered by tractors. Economic production is obtained by using an interchangeable power unit—the tractor. As these tractors are used with many implements, one is forced to examine the problem from a cost-of-the-whole-system viewpoint. As in the selection of individual field implements, the criterion for tractor selection is an economic one. The capacity or size of the power unit must be matched with the amount of work to be done as dictated by timeliness and cost.

As in the implement selection problem, the amount of use is most important. Area covered is an adequate statement of use for implements, but *total energy requirement* is thought to be the proper expression for tractor use. Unfortunately, the amount of power and the amount of energy required for performing various farm tasks are neither precise nor constant.

Several inherent "slack" factors will permit the use of average estimates of energy. The load factor used for designating usable power, the fact that the tractor can operate at various speeds, and the fact that the problem requires total, not instantaneous, energy all contribute to the use of average values.

The average annual energy requirements for the proposed work of a tractor must be known before an optimum power level can be selected. Field operation energy can be estimated for any given farm from the performance of an existing implement. For example, a 75 kW [100 HP] PTOP tractor seems about half-loaded when traveling in tilled soil at 8 km/hr [5 MPH] and while pulling a 3-m [9.84-ft] implement. Half load is recognized by noting on the tachometer that the engine speed is halfway between the no-load speed and rated speed, all at wide open governor control. Engine power would be 37.5 kW [50 HP] PTOP of which only a portion (96%, 36 kW [48 HP]) is available at the axle and even less at the drawbar. Entering Fig. 2.11 with an estimation of drive-wheel slip at 5% or less defines a ratio of drawbar power to axle power as about 0.5, which means that the actual drawbar power exerted is 18 kW [24 HP]. The implement energy required, E, can then be calculated from the drawbar power as

$$7.5 \frac{kW \cdot hr}{ha} = 18 \text{ kW} \times \frac{1}{3 \text{ m}} \times \frac{1 \text{ hr}}{8 \text{ km}} \times \frac{1 \text{ km}}{1000 \text{ m}} \times \frac{10{,}000 \text{ m}^2}{1 \text{ ha}}$$

$$\left[4 \frac{HP \cdot hr}{a} = 24 \text{ HP} \times \frac{1}{9.84 \text{ ft}} \times \frac{1 \text{ hr}}{5 \text{ mi}} \times \frac{1 \text{ mi}}{5280 \text{ ft}} \times \frac{43{,}560 \text{ ft}^2}{a} \right]$$

Typical draft and energy requirements per acre of many implements are given in Table 2.5. Any rolling resistance for PTO-operated machines or attached transport wagons must be added to the values in Table 2.5 to get actual tractor work loads. The effects of speed on draft are discussed in the tillage sections. For selection purposes the energy at typical speeds is assumed for each operation. The most important point about energy requirements per area is that they are essentially constant regardless of the size of the tractor or implement when typical field speeds are used.

Table 17.1 lists the energy requirements, G, for several processing operations on farms. If the tractor is used for this type of work too, such values cannot be left out when looking at the total energy required on a farm.

A third area requiring tractor energy is the transport of materials. The amounts required are computed by assuming that 1 unit of mass (includes tractor) is needed to transport 2 units of mass one way. Assum-

Table 17.1. Farm Processing Energy Requirements

Crop Handling and	G Factor	
	kW · hr/t	HP · hr/T
Processing Operations		
Loading manure	0.16	0.2
Shelling corn	1.0	1.2
Grinding		
ear corn	4.5	5.5
shelled corn	6.6	8
oats	12	15
Blowing silage	1.2	1.5
Crop drying	2.3	2.8

ing an average rolling resistance of 5% produces a total transport energy of

0.27 kW·hr to transport 1 t for 1 km
[0.53 HP·hr to transport 1 T for 1 mi]

The energy required for making the empty return trip is included.

The technique for determining the optimum amount of horsepower is similar to that for determining the optimum size of implement—minimize the annual cost with respect to a pertinent variable. In this case engine power is the pertinent variable where pwr is that unknown amount of power that will produce minimum annual power costs.

In general the annual cost, AC, for a farm power unit can be expected to consist of

$$AC = \frac{(FC\%) P}{100} + hr \text{ used } (RMP + L + O + F + \text{timeliness})$$

The costs of repair and maintenance, fuel, and oil were considered in Chapter 16 to be a direct function of area; therefore they are also a direct function of energy expended to cover the area. As the energy requirement is expected to be constant for any specific farm operation, repair and maintenance, oil, and fuel will have no influence on the optimum size of the power unit.

The only important variables are the labor cost and the timeliness cost. After all, a 5 kW garden tractor could provide all the energy needed on many farms by working continuously day and night. The costs for labor and timeliness would be quite high, however. Powerful tractors are required only because work has to be done in a limited time.

More specifically, the annual cost for tractor power is expressed as the sum of the fixed costs and the total labor and timeliness costs for each of the three types of farm jobs—field work, transport work, and processing work:

$$AC = (FC\%)\, P/100 + \text{hr field work} (L + \text{timeliness})$$
$$+ \text{hr transport work} (L) + \text{hr processing work} (L)$$

Probably there should be a timeliness charge for both the transport work and the processing work, but for simplicity it will be assumed that these do not limit field operations.

The annual cost equation must be written in terms of the tractor size variable, pwr, and then minimized by the techniques in Appendix B. The purchase price, P, is replaced by t pwr, where t is the price of the tractor per maximum PTOP.

The annual hours of work required for each class of tractor operation are given by Eq. 17.1, 17.2, and 17.3

$$\frac{\text{field hr}}{\text{yr}} = \frac{A\ \text{ha}}{\text{yr}} \times \frac{E\ \text{kW}\cdot\text{hr}}{\text{ha}} \times \frac{1}{r_1\,(\text{pwr})\,\text{kW}}$$

$$\left[\frac{\text{field hr}}{\text{yr}} = \frac{A\ a}{\text{yr}} \times \frac{E\ \text{HP}\cdot\text{hr}}{a} \times \frac{1}{r_1\,(\text{pwr})\,\text{HP}} \right] \tag{17.1}$$

$$\frac{\text{transport hr}}{\text{yr}} = \frac{0.27\,\text{kW}\cdot\text{hr}}{t\cdot\text{km}} \times \frac{W\ t}{\text{yr}} \times \frac{1}{r_2\,(\text{pwr})\,\text{kW}} \times \frac{D\ \text{km}}{1}$$

$$\left[\frac{\text{transport hr}}{\text{yr}} = \frac{0.53\,\text{HP}\cdot\text{hr}}{T\cdot\text{mi}} \times \frac{W\ T}{\text{yr}} \times \frac{1}{r_2\,(\text{pwr})\,\text{HP}} \times \frac{D\ \text{mi}}{1} \right]$$
$$\tag{17.2}$$

$$\frac{\text{processing hr}}{\text{yr}} = \frac{G\ \text{kW}\cdot\text{hr}}{t} \times \frac{1}{r_3\,(\text{pwr})\,\text{kW}}$$

$$\left[\frac{\text{processing hr}}{\text{yr}} = \frac{G\ \text{HP}\cdot\text{hr}}{T} \times \frac{WT}{\text{yr}} \times \frac{1}{r_3\,(\text{pwr})\,\text{HP}} \right] \tag{17.3}$$

where r_i is a typical ratio of tractor output power to PTOP. The r_1 and r_2 values for drawbar loads can be estimated from Fig. 2.11. For strictly PTO loads in processing jobs, $r_3 = 1.0$.

The annual cost for tractor power in terms of the power level, pwr, is given by Eq. 17.4, which sums the hours of work required for each tractor operation and multiplies by the cost per hour:

$$AC = \frac{(FC\%)t}{100}(\text{pwr}) + \sum \left\{ \frac{A_i E_i}{r_i\,(\text{pwr})} \left(L_i \frac{K_i Y_i V_i A_i}{(\text{sc})_i\,(\text{nt})_i\, U_i h_i} \right) \right.$$

$$\left. \times \frac{L_i}{(\text{pwr})} \left(\frac{c D_i W_i}{r_i} + \frac{G_i W_i}{r_i} \right) \right\} \tag{17.4}$$

where previously undefined terms are

   i = subscript identifying specific operation, areas, energy values, labor cost, etc.

   Σ = sum evaluated for all i operations

   c = constant, 0.27 [0.53]

The optimum power level is determined by minimizing the costs:

$$\text{pwr} = \sqrt{ \sum \left\{ \frac{100 A_i E_i}{r_i (FC\%) t} \left(L_i \frac{K_i Y_i V_i A_i}{(\text{sc})_i\,(\text{nt})_i\, U_i h_i} \right) \right\} + \sum \left\{ \frac{100 L_i}{(FC\%) t} \left(\frac{c D_i W_i}{r_2} + \frac{G_i W_i}{r_3} \right) \right\} } \tag{17.5}$$

The machinery manager's duty in equipment selection is to minimize the cost of the complete system. Seldom does it happen that the tractor size necessary to pull the economically optimum plow or other large tillage implement is also the optimum size for the system as a whole. The value of Eq. 17.5 is that it includes the effects of all the operations in arriving at an optimum tractor size. The following example problem demonstrates the use of this equation.

Example

A farm manager wants to determine the optimum machinery system that will include a large tractor for heavy field work, a smaller tractor for transport operations and utility work around the farmstead, and a self-propelled combine. The data sheets in Laboratory Exercise 19 serve to outline the solution procedure.

Cropping practice

Crop	Annual area	Value
Corn grain to be combined	100 ha [247 a]	$800/ha [$325/a]
Corn for ensilage	10 ha [24.7 a]	$800/ha [$325/a]
Soybeans to be combined	100 ha [247 a]	$500/ha [$200/a]
Hay, baled, 3 cuttings	10 ha [24.7 a]	$250/ha [$100/a]

Machines required

All corn area is to be processed, after combining, with a stalk chopper and disk harrow pulled in tandem (treated as one machine). Moldboard plowing for the soybean seedbed and chisel plowing for the corn seedbed is required. A field cultivator precedes the row crop planting. All row crops are rotary hoed once and cultivated once. The growing corn is sprayed once. The hay is to be made with a mower-conditioner, rake, and baler. The silage is cut with a forage harvester. Two varieties each of corn and soybeans are planted at different times to spread the timeliness demands on the machinery system.

The large tractor is used annually for

Manure loading	200 t	[220 T]
Crop drying	820 t	[900 T]

The small tractor is used annually to transport the following materials an average of 0.8 km [0.5 mi].

Fertilizer, Chemicals	82 t	[90 T]
Hay	111 t	[122 T]
Silage	300 t	[330 T]
Manure	200 t	[220 T]
Shelled corn	820 t	[900 T]
Soybeans	269 t	[296 T]

Load factors of 0.8 will be used for the large tractor and 0.9 for the small. The maximum size tractors available for purchase that meet the farm needs will be assumed to be 135 kW [180 HP] for the large tractor and 65 kW [87 HP] for the small. Diesel tractors will be purchased.

Large tractor selection (only Parts I and III apply)

Part I. For field operations; L = $8.00/hr, t = $465/kW [$347/HP], FC% = 16

Machine	Crop	A		E		r	K	YV		sc	nt	U	h	$\dfrac{100\,AE}{r\,(FC\%)t}\left(L + \dfrac{K\,(YV)\,A}{(sc)\,(nt)\,Uh}\right)$	
Plow	soybeans	100	[247]	36	[19.5]	0.68	0.002	500	[200]	2	1	0.7	10	1078	[1921]
Chppr & disk	soybeans	100	[247]	16	[8.8]	0.77	0.002	500	[200]	2	1	0.6	10	456	[825]
Chisel	corn	110	[272]	31	[17]	0.68	0.002	800	[325]	2	1	0.6	10	1528	[2784]
Field cltvtr	soybeans	100	[247]	4.4	[2.4]	0.54	0.002	500	[200]	4	2	0.5	8	122	[219]
Field cltvtr	corn	110	[272]	4.4	[2.4]	0.54	0.002	800	[325]	4	2	0.5	8	163	[294]
Forage	corn silage	10	[25]	46	[25]	0.8	0.001	800	[325]	4	1	0.6	10	65	[117]
Baler	hay	30	[74]	9.2	[5]	0.3	0.010	250	[100]	2	3	0.55	12	122	[219]
Rotary hoe	soybeans	100	[247]	2	[1.1]	0.38	0.010	500	[200]	4	2	0.55	12	124	[224]
Rotary hoe	corn	110	[272]	2	[1.1]	0.38	0.010	800	[325]	4	2	0.55	12	192	[351]
													Total	3850	[6954]

Part III. For farmstead operations

Operation	G		W		r	L	$\dfrac{100\,GWL}{(FC\%)tr}$	
Dry corn	2.3	[2.8]	820	[900]	1	0.50	13	[23]
Load manure	0.16	[0.20]	200	[220]	0.5	8.00	7	[13]
						Part III Total	20	[46]

Optimum power $= \sqrt{I + III} = \sqrt{3869} = 62$ kW [84 HP]

Purchased PTO power (at 0.8 load factor $= 62\,[84]/0.8 = 78$ kW [105 HP]

Number of tractors required $= 1$

Small tractor selection (only Parts I and II apply)

Part I. For field operations; L = $8.00/hr, t = $465/kW [$347/HP], FC% = 16

Machine	Crop	A		E		r	K	YV		sc	nt	U	h	$\dfrac{100\,AE}{r\,(FC\%)t}\left(L + \dfrac{K\,(YV)\,A}{(sc)\,(nt)\,Uh}\right)$	
Row planter	soybeans	100	[247]	5.7	[3.1]	0.53	0.003	500	[200]	4	2	0.5	8	183	[329]
Row planter	corn	110	[272]	5.7	[3.1]	0.53	0.003	800	[325]	4	2	0.5	8	258	[467]
Mower & cndtnr	hay	30	[74]	14	[7.6]	0.82	0.010	250	[100]	4	3	0.6	12	61	[109]
Rake	hay	30	[74]	0.9	[0.5]	0.20	0.010	250	[100]	4	3	0.6	12	16	[30]
Row cltvtr	soybeans	100	[247]	2.4	[1.3]	0.58	0.010	500	[200]	4	2	0.6	10	102	[182]
Row cltvtr	corn	110	[272]	2.4	[1.3]	0.58	0.010	800	[325]	4	2	0.6	10	161	[290]
Spray	corn	110	[272]	0.5	[0.3]	0.44	0.010	800	[325]	2	1	0.55	10	148	[295]
													Total	929	[1702]

Part II. For transport operations; L = $8.00/hr, c = 0.27 [0.53]

Hauling	D		W		r	$100cDWL/(FC\%)tr$	
Fertilizer	0.8	[0.5]	82	[90]	0.8	2.38	[4.30]
Hay	0.8	[0.5]	111	[122]	0.8	3.22	[5.82]
Silage	0.8	[0.5]	300	[330]	0.8	8.71	[15.75]
Manure	0.8	[0.5]	200	[220]	0.8	5.81	[10.50]
Shelled corn	0.8	[0.5]	820	[900]	0.8	23.81	[42.96]
Soybeans	0.8	[0.5]	269	[296]	0.8	7.81	[14.13]
					Total	51.74	[93.46]

Optimum power $= \sqrt{I + III} = \sqrt{980.74} = 31.3$ kW [42.0]

Purchased PTO power (at 0.9 load factor) $=$

$$31.3\,[42.0]/0.9 = 34.8 \text{ kW [46.7 HP]}$$

Number of tractors required $= 1$

Results

	Large Tractor	Small Tractor	
Purchase price	$36,270	$16,182	
Estimated annual use, hr	300	250	
Estimated value of T, $/hr	$19.34	$10.35	
Acutal T(found after selecting machinery)	$18.54	$10.44	Close enough! No need to recalculate.

Implement selection

1. Implements for large tractor—field speed, field efficiency, price, R&M

	Moldboard Plow	Stlk Chppr & Disk	Chisel Plow	Field Cltvtr	Forage Hrvstr	Baler	Rotary Hoe
S, km [mi]/hr	8 [5]	9.6 [6]	5.4 [3.4]	6.4 [4.0]	6.4 [4.0]	6.4 [4.0]	19 [12]
e	0.8	0.85	0.8	0.9	0.6	0.75	0.75
p, $/m [ft]	1722 [525]	3625 [1090]	600 [180]	1000 [305]	12 530 [3820]	3500 [1063]	600 [180]
R&M	0.000 29	0.000 25	0.000 06	0.000 01	0.0003	0.0004	0.000 94

2. Implements for small tractor—field speed, field efficiency, price, R&M

	Row Planter	Mower & Cdtnr	Rake	Row Cltvtr	Sprayer
S, km [mi]/hr	8 [5]	8 [5]	8 [5]	6.4 [4]	8 [5]
e	0.65	0.8	0.75	0.75	0.60
p, $/m [ft]	2624 [800]	3000 [914]	937 [285]	1443 [440]	35 [10]
R&M	0.0014	0.0007	0.0007	0.000 01	0.0003

3. Values of $\frac{100c}{(FC\%)p} \times \frac{A}{(Se)}\left[(L + T + KYVA/X)\right]$ (where $X = (sc)(nt)Uh$)

a. Big tractor; L + T = $27.34, c = 10

Implement	$\frac{100c}{(FC\%)p}$	$\frac{A}{Se}\left(L+T+\frac{KYVA}{X}\right)=$ Total Crop CORN SILAGE	$\frac{A}{Se}\left(L+T+\frac{KYVA}{X}\right)=$ Total Crop CORN GRAIN	$\frac{A}{Se}\left(L+T+\frac{KYVA}{X}\right)=$ Total Crop SOYBEANS	Sum, w^2	w, m [ft]
Moldboard plow	0.0363			$\frac{100}{(8)(0.8)}(27.34+7.14)=19.6$	19.6	4.42 [14.5]
Stlk chppr & disk	0.0172			$\frac{100}{(9.6)(0.85)}(27.34+8.33)=7.5$	7.5	2.75 [9.02]
Chisel plow	0.1042	$\frac{10}{(5.4)(0.8)}(27.34+1.33)=6.92$	$\frac{100}{(5.4)(0.8)}(27.34+13.33)=98.1$		105.2	10.2 [34]
Field cltvtr	0.1736	$\frac{10}{(6.4)(0.9)}(27.34+0.50)=8.39$	$\frac{100}{(6.4)(0.9)}(27.34+5.00)=97.5$	$\frac{100}{(6.4)(0.9)}(27.34+3.13)=91.8$	197.7	14.1 [46.1]
Forage hrvstr	0.0050	$\frac{10}{(6.4)(0.6)}(27.34+0.33)=0.36$			0.36	0.60 [2.0]
Rotary hoe	0.1042	$\frac{10}{(19)(0.75)}(27.34+1.52)=2.11$	$\frac{100}{(19)(0.75)}(27.34+15.15)=31.1$	$\frac{100}{(19)(0.75)}(27.34+9.47)=26.9$	60.1	7.8 [25.4]
Crop HAY						
Baler	0.0179	$\frac{30}{(6.4)(0.75)}(27.34+1.89)=3.26$			3.26	1.81 [5.9]

b. Small tractor; L + T = $18.35, c = 10

Implement	$\frac{100c}{(FC\%)p}$	$\frac{A}{Se}\left(L+T+\frac{KYVA}{X}\right)=$ Total Crop CORN SILAGE	$\frac{A}{Se}\left(L+T+\frac{KYVA}{X}\right)=$ Total Crop CORN GRAIN	$\frac{A}{Se}\left(L+T+\frac{KYVA}{X}\right)=$ Total Crop SOYBEANS	Sum, w^2	w, m [ft]
Row planter	0.0238		$\frac{100}{(8)(0.65)}(18.35+7.50)=11.8$	$\frac{100}{(8)(0.65)}(18.35+4.69)=10.5$	23.2	4.81 [15.8]
Row cltvtr	0.0433		$\frac{100}{(6.4)(0.75)}(18.35+16.67)=31.6$	$\frac{100}{(6.4)(0.75)}(18.35+10.42)=26.0$	59.4	7.7 [25.3]
Sprayer	1.7857	$\frac{10}{(8)(0.6)}(18.35+7.27)=95.3$	$\frac{100}{(8)(0.6)}(18.35+72.72)=3388$		3483	59 [194]
Crop HAY						
Mower & cndtnr	0.0208	$\frac{30}{(8)(0.8)}(18.35+0.87)=1.9$			1.9	1.38 [4.5]
Rake	0.0667	$\frac{30}{(8)(0.75)}(18.35+0.87)=6.4$			6.4	2.53 [8.3]

Results

Implement	$W_{1,2,}$[a] metre	Size Selected	Hours of Use	Power[e] Req'd, kW	Costs,[f] $/area	Custom Rate, $/area
Moldboard plow	4.8-4.0	1.5[b]	104	62.0	41.26	37.00
Chopper and disk	3.0-2.5	2.8	44	56	34.85	33.00
Chisel plow	11.3-9.2	2.6[b]	98	62.0	35.29	28.00
Field cultivator	17.9-10.4	11.9[b]	15	62.0	14.22	22.00
Forage harvester	0.7-0.5	0.6	29	40.5	285.17[g]	110.00
Rotary hoe	8.8-7.0	6.2	24	62.0	7.21	5.00
Baler	2.0-.6	1.8	35	35.3	79.20[g]	40.00
Row planter	5.2-4.5	3.6[b]	112	31.3	27.00	25.00
Row cultivator	8.3-7.1	3.6[c]	121	9.5	17.28	18.00
Sprayer	70-50	15[d]	15	13.6	8.18	8.00
Mower and conditioner	1.6-1.2	2.1[d]	22	28.7	53.25[g]	38.00
Rake	3.0-2.2	2.1[d]	24	7.6	27.89[g]	13.00

a. Range found using d = $10.
b. Limited by tractor power.
c. Limited to match planter.
d. Limited to available width.
e. 0.36 E w S/(3.6 r).
f. Fuel consumption from Table 2.3, 31¢/L; 5¢ and 8¢/hr for oil; R&M from Table 4.5 (average/ hr over accumulated use limit).
g. Machine not selected, custom cost much cheaper.

Recalculation (excluding operations done by custom operator)

1. Large tractor Sum of field hours
   Sum of farmstead hours 285
   (r_3 = 1 for crop drying, 0.1 for manure loading)

   a. Drying corn:

   $$\frac{2.3 \text{ kW} \cdot \text{hr}}{t} \times \frac{820 \text{ t}}{\text{yr}} \times \frac{1}{1 \times 78 \text{ kW}} = \quad 24$$

   b. Loading manure:

   $$\frac{0.16 \text{ kW} \cdot \text{hr}}{t} \times \frac{200 \text{ t}}{\text{yr}} \times \frac{1}{0.1 \times 78 \text{ kw}} = \quad 4$$

   Total 313 hr

2. Small tractor
   Sum of field hours 248
   Sum of transport hours

   $$\frac{0.27 \text{ kW} \cdot \text{hr}}{t \cdot \text{km}} \times \frac{1782 \text{ t}}{\text{yr}} \times \frac{1}{0.8 \times 35 \text{ kW}} \times \frac{0.8 \text{ km}}{1} = \quad 14$$

   Total 262 hr

The hours of use required, 313 and 262, match quite closely the estimated hours on which T was based, 300 and 250. Recalculation of tractor power required, omitting forage machinery, lowers the power levels by 1 kW.

This tractor-implement system is both economic and timely. The forage operations should be done by a custom operator to save over $3800/yr. Several operations have costs somewhat above custom but should be retained because of convenience and timeliness (crucial). With the exception of chisel and moldboard plowing, all other operations can be completed within a week (given average weather) since each row crop is divided into two noncompeting enterprises by staggered plantings. The hay operation times are divided into three cuttings during the year.

Self-propelled combine selection

The selection of the self-propelled combine with its base unit and two interchangeable heads involves a rather complex application of Eq. 16.4. The base unit size or capacity is expressed by the power (pwr) of its engine. The size or capacity of the heads is expressed by their effective width of action. Using subscripts c and s to represent the corn head (or corn operations) and the soybean head (or soybean operations), respectively, the annual costs of the combine in terms of its engine power are

$$AC = \frac{(FC\%)_b\,P_b\,(pwr)}{100} + \frac{(FC\%)_c\,P_c\,w_c}{100} + \frac{(FC\%)_s\,P_s\,w_s}{100} + \frac{E_c A_c}{rc(pwr)}\left(L + \frac{K_c\,(YV)_c\,A_c}{(sc)_c\,(nt)_c\,U_c h_c}\right) + \frac{E_s A_s}{r_s\,(pwr)}\left(L + \frac{K_s\,(YV)_s\,A_s}{(sc)_s\,(nt)_s\,U_s h_s}\right)$$

The relationship among the three expressions for capacity, pwr, w_c, and w_s must be known before the optimum width can be determined. The width of the head mounted per power of the base machine varies with the crop, yield, and engine power. Typical ratios at 75-cm [30-in.] row spacing are 22.5 kW/m for corn heads and 20 kW/m for cutter bar heads. From such ratios all head widths can be determined in terms of power required.

$$W_s = (pwr)/20.0 \quad w_c = (pwr)/22.5$$

Substituting into the annual cost equation and using the methods in Appendix B produces an expression for the optimum width in terms of (pwr).

$$(pwr) = \left\{ \frac{E_c A_c}{r_c Z}\left(L_c + \frac{K_c\,(YV)_c\,A_c}{(sc)_c\,(nt)_c\,U_c h_c}\right) + \frac{E_s A_s}{r_s Z}\left(L_s + \frac{K_s\,(YV)_s\,A_s}{(sc)_s\,(nt)_s\,U_s h_s}\right) \right\}^{1/2}$$

where Z is the fixed cost factor,

$$Z = \frac{(FC\%)_b\,P_b}{100} + \frac{(FC\%)_c\,P_c}{100\,(22.5)} + \frac{(FC\%)_s\,P_s}{100\,(20)}$$

The evaluation of the equation is made using the following data:

$(FC\%)_{b,c,s} = 16$	$(YV)_c = 800,\ (YV)_s = 600$
$A_{c,s} = 200$	$(sc)_c = 4,\ (sc)_s = 2$
$R_{c,s} = 0.60$	$(nt)_{c,s} = 2$
$E_c = 22.0,\ E_s = 20.0$	$U_{c,s} = 0.58$
$L_{c,s} = 12.0$	$h_c = 10,\ h_s = 8$
$K_c = 0.003,\ K_s = 0.005$	$S_c = 4.0,\ S_s = 4.8$

Prices obtained from Appendix D are

Base, p_b	39000 + 600/kW
Soybean head, p_s	4500 + 980/m
Corn head, p_c	1500 + 3600/row

Only the second part of the price function, the price per *added* unit of width, is pertinent to the selection equation.

Numerically, $Z = 138$ and (pwr) $= 49.72$ kW. The optimum-size heads would be

$49.72/22.5 = 2.21$ m or 2.95 rows or 3 row corn head
$49.72/20 = 2.49$ m of soybean head

The purchase prices would be

Base	$39000 + 670 \times 49.72 =$	\$68,832
Cutter bar head	$4500 + 980 \times 2.49 =$	\$ 6,940
Corn head	$1500 + 3600 \times 3 =$	\$12,300
	Total $=$	\$88,072

The operating costs of the combine can be estimated using an R&M value of 0.00001 (from Table 4.5 at the accumulated use limit), 5¢/hr for engine oil, 2.18 kW·hr/L fuel efficiency at 60% load (Table 2.2) and at a price of 31.0¢/L.

For corn harvesting,

$$0.16 \times \left(\frac{68832}{2} + 12{,}300\right) + \frac{22.0 \times 200}{0.60 \times 49.72}$$

$$\times\,(0.00013 \times 81132 + 12 + 0.05 + 7.07) = \$11{,}850$$

For soybean harvesting,

$$0.16 \times \left(\frac{68832}{2} + 6940\right) + \frac{20 \times 200}{(0.60 \times 49.72}$$

$$\times\,(0.00013 \times 75772 + 12 + 0.05 + 7.07) = \$10{,}501$$

	Cost per area $/ha	Custom rate $/ha
Corn harvesting	59.25	67
Soybean harvesting	52.51	53

The annual costs are less than the custom costs and this machine should be purchased. The size of this machine is small by most standards. The total work hours required are 222 for the corn and 178 for the soybean operations. But the price of the combine is so high compared to the value of the crops that timeliness losses have little influence on the optimum.

The machinery manager is cautioned not to overestimate the monetary value of timeliness.

Selection by Computer

The computational work in solving Eq. 16.4, 17.5, and other related ones for a large system is both complex and tedious. Accuracy, speed, and ease are improved with the use of computers. Programmable calculators can be used for segments of the solutions. Large digital computers can be programmed to give a solution with just one entry of the data.

Computer solutions permit much more precise system selection determinations. The data contained in Fig. 2.11 on tractor performance can be stored and used as an upper constraint on the economic optimum size of machinery. Optimum field speeds can be determined. The output can include such additional data as tractor ballast additions, average power, expected drive-wheel slippage, and fuel used. Custom operations can be considered and the system modified to include custom operations where they are more economic. Expected costs and investments can also be determined.

Fig. 17.1 depicts an output sheet for a computer solution of the example farm operation. There are some differences between the computer and hand solutions because of the differing solution approaches. The computer selects slightly smaller tractors, 70 and 30 kW, than does the hand solution, which picks 78 and 34.8 kW. The computer model selects the least expensive of arbitrarily selected trial power levels, while the hand solution is the first approximation of the optimum power level. The power limit constraint in the computer selection is determined by the relationships in Fig. 2.11. Eq. 17.5 uses r values that are averages of the relationships in Fig. 2.11. These differing power constraints produce a significant difference in the field cultivator size selected.

There are differences in implement selection too. The computer model generally picks slightly smaller implement sizes and consequently requires slightly longer hours of annual use. The annual costs per area for the computer selection are consistently less than those for the hand solution. Both indicate that the forage operations should be leased to custom operators. The computer program allows a manager to select the moldboard plow, chisel plow, and rotary hoe even though their costs are slightly above custom costs. The computer solution does not constrain the row cultivator to match the row planter width—a circumstance that is not always practical.

The computer selection program allows for the comparison of the costs at various power levels, as indicated in Table 17.2. As with the relationship between annual cost and effective width of machines, the costs near the optimum do not change much with changes in machine size.

The large-tractor results are rather erratic because of the custom alternative. At both high and low power levels many of the operations prove to be more economical by custom operators with consequent low use and high cost for the tractor. The power level selected, 70, is $7 more costly than the 80 kW level; but it was selected because the moldboard plow, chisel plow, and rotary hoe were retained as tractor-owned operations. The fact that the area costs for these rather sizable operations were matched by custom costs may mean that the published custom rates lag real-world inflation.

Of special interest to the machinery manager is the effect of size or scale on the optimum power level for a farm. Fig. 17.2 shows a computer plot of such an analysis. The total system cost is divided by the crop area to give values to the ordinate. The abscissa is the power level for the farm as indicated by maximum PTOP of its tractors.

Table 17.2. Cost Results from Different Power Levels

Power Level kW [HP]	Annual Cost $	System Hours	Investment $
Large tractor			
10 [13]	23,471	438	10,565
20 [27]	23,295	393	19,056
30 [40]	22,820	347	26,755
40 [54]	22,069	675	36,011
50 [67]	20,867	569	41,791
60 [80]	20,363	501	47,162
70 [94]	19,971	395	48,874
80 [107]	19,964	138	46,185
90 [121]	20,041	67	42,761
Small tractor			
15 [20]	15,735	461	20,384
20 [27]	14,323	362	25,055
25 [34]	13,817	304	29,626
30 [40]	13,651	216	31,267
35 [47]	13,711	198	34,430
40 [54]	13,846	184	37,593
45 [60]	14,070	183	39,824
50 [67]	14,281	183	41,899
55 [74]	14,503	183	43,974

LARGE TRACTOR

	Tractor Power	No. of Tractors	Investment
System Annual Cost (Includes Timeliness) $19971.61	70 [93.8]	1	$48873.70

System Hours 394.7

Implement	Optimum Size	Range Permitted	No. Required	Purchase Price
Plow	1.3 [4.3]	0.8 [2.6] - 1.3 [4.3]	1	1337.75
Chop-disk	1.7 [5.6]	1.4 [4.6] - 2.1 [6.9]	1	7574.54
Chisel	2.3 [7.5]	1.6 [5.2] - 2.3 [7.5]	1	2324.09
Field, cultivator	5.1 [16.7]	3.4 [11.2]- 5.1 [16.7]	1	5095.53
Field, harvester	0.0 [0.0]	0.0 [0.0] - 0.0 [0.0]	0	0.0
Baler	0.0 [0.0]	0.0 [0.0] - 0.0 [0.0]	0	0.0
Rotary, hoe	5.8 [19.0]	4.3 [14.1]- 7.9 [25.9]	1	3491.80

Operation	Annual Cost $	Speed km/HR [MPH]	Power kW [HP]	Slip %	Pull kN [lb]	Fuel L [gal]	Custom Cost $	Annual hr	$/Area
Plow beans	4697.77	7.2 [4.5]	55.7 [74.7]	9.5	18.9 [4249]	2660 [704]	0	116.8	37.36
Chop-disk beans	3290.59	9.3 [5.8]	40.1 [53.8]	3.5	3.8 [854]	1744 [461]	0	71.4	26.74
Chisel corn	5132.37	4.9 [3.0]	52.5 [70.4]	9.5	28.0 [6295]	2639 [698]	0	109.6	30.52
Field cultivator, beans	1355.81	5.5 [3.4]	34.4 [46.1]	14.5	10.8 [2428]	602 [159]	0	34.1	12.31
Field cultivator, corn	1597.14	5.5 [3.4]	34.4 [46.1]	14.5	10.8 [2428]	635 [168]	0	37.5	12.33
Chop silage	1100.87	0.0 [0.0]	0.0 [0.0]	0.0	0.0 [0]	0 [0]	1100.00	0.0	110.00
Bale hay	1239.46	0.0 [0.0]	0.0 [0.0]	0.0	0.0 [0]	0 [0]	1200.00	0.0	40.00
Rotary hoe, beans	693.31	17.8 [11.1]	57.0 [76.4]	6.1	4.1 [1371]	237 [72]	0	12.1	5.72
Rotary hoe, corn	864.30	17.8 [11.1]	57.0 [76.4]	6.1	4.1 [1371]	301 [80]	0	13.3	5.72

SMALL TRACTOR

	Tractor Power	No. of Tractors	Investment
System Annual Cost (Includes Timeliness) $13651.15	30 [40.2]	1	$31267.48

System Hours 215.9

Implement	Optimum Size	Range Permitted	No. Required	Purchase Price
Planter	3.2 [10.5]	2.5 [8.2] - 4.3 [10.5]	1	6583.30
Mower-conditioner	0.0 [0.0]	0.0 [0.0] - 0.0 [0.0]	0	0.00
Rake	0.0 [0.0]	0.0 [0.0] - 0.0 [0.0]	0	0.00
Cultivator	6.0 [19.7]	5.0 [16.4]-6.1 [20.0]	1	8709.17
Sprayer	15.0 [49.2]	7.0 [23.0]-24.6 [80.7]	1	3525.00

Operation	Annual Cost $	Speed km/HR [MPH]	Power kW [HP]	Slip %	Pull kN [lb]	Fuel L [gal]	Custom Cost $	Annual hr	$/Area
Planter beans	2284.89	6.8 [4.3]	19.2 [25.8]	14.5	5.1 [1147]	593 [157]	0	60.8	19.51
Plant corn	2797.76	6.8 [4.3]	19.2 [25.8]	14.5	5.1 [1147]	652 [172]	0	66.9	19.56
Mow & cond. hay	1143.39	0.0 [0.0]	0.0 [0.0]	0.0	0.0 [0]	0 [0]	1140.00	0.0	38.00
Rake hay	410.67	0.0 [0.0]	0.0 [0.0]	0.0	0.0 [0]	0 [0]	390.00	0.0	13.00
Cultivator, beans	1700.16	5.7 [3.5]	15.2 [20.4]	11.5	5.4 [1214]	256 [68]	0	34.7	12.92
Cultivator, corn	1870.18	5.7 [3.5]	15.2 [20.4]	11.5	5.4 [1214]	282 [75]	0	38.2	12.92
Spray corn	3444.11	7.3 [4.6]	13.3 [17.8]	8.3	3.0 [674]	107 [28]	0	15.3	7.08

Fig 17.1 Computer output for example tractor selection problem.

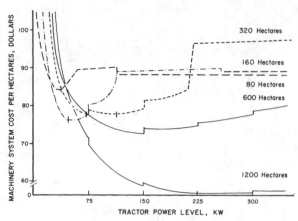

Fig. 17.2 Optimum tractor power level and size of farm.

The general tendency of the curves is to be U-shaped. The least-cost values are quite precise for the smaller farms but have an ever increasing range as the farms grow larger. Occasionally there are severe jumps in the curves. Some of the discontinuities represent the sudden rise or drop in costs as the number of tractors, and tractor operators, increases by one. The jumps within a tractor number interval illustrate a "snowball" effect. As the power level rises, the costs of the system also rise to a point where one operation can no longer compete with custom costs. The program drops this implement from consideration and recalculates costs. Since the tractor use declines, its costs per hour rise, which causes another implement operation to be dropped, etc. In the example, the last operation to be dropped proved to be the spraying operation.

Note that the larger farms have the lowest costs—at least as far as equipment costs are concerned. Note also that some farm sizes are just not as economic as those either slightly larger or slightly smaller. Some combinations of available equipment and size of operation apparently produce an uneconomic match.

The only differences in the input data for these farms are the total crop areas, increased tonnages of yield, and changes in timeliness considerations. The 1200-ha farm, for example, could not expect to plant all its corn in one week and have an economical equipment system. Therefore, staggered plantings of different hybrids having various lengths of growing seasons were assumed to reduce the cost of timeliness.

This analysis used an upper limit of 75 kW as the largest available row crop tractor. If the optimum power should come out to be 130 kW, two tractors would be required—two 65-kW tractors or one 100-kW and one 30-kW, or any other combination of two tractors that adds up to 130. Three or more tractors are *never* assumed to be less costly if two are adequate. If the system solution requires multiple tractors, the machinery manager is free to adjust the size of individual tractors and their associated equipment to optimize other factors that cannot be given a dollar value.

The optimum power level is indicated by a hash mark at the lowest point of each curve. While some variation exists because of unique combinations of area and equipment widths, one can conclude that about 100 kW is a least-cost and optimum power level for each 300–400 ha of cropland for these farms.

The foregoing computer analyses are presented to show concepts and possibilities. The data used apply only to a specific farm situation. As farms differ widely in enterprise and practice, these results cannot be used as generalities. This conclusion really defines the challenge of machinery management: Each farm is a special case and requires its own unique system of machines and its own unique management.

USED EQUIPMENT

Equipment selection also involves some decision about the economy of purchasing used machinery. Anytime the price of used equipment drops below its actual worth as determined by its performance capability, an economic opportunity exists for the employment of a used machine.

Cost estimations based on averages produce a strong case for the purchase of used equipment—particularly for tractors and implements subject to a very rapid and perhaps artificial depreciation during their early life. G. H. Larson, in a doctoral thesis (Michigan State University 1955), developed the analysis presented in Fig. 18.1 depicting the annual costs and replacement times for tractors purchased new and at ages 1 through 7. Larson used "as-is" values from dealers' price guides for depreciation rates and purchase prices. He obtained other fixed costs and the repair rates from previous tractor cost records and assumed that all tractors, new or used, would follow the same depreciation and repair schedules.

Fig. 18.1 suggests three interesting conclusions: (1) a new tractor should be replaced at the end of 9 years, (2) used tractors purchased as late as 6 years of age can have lower operating costs than a new tractor, and (3) a tractor purchased at age 3 and sold at age 6 has the lowest cost.

Used farm implements offer even greater opportunities for reduced costs. An implement that has had proper maintenance and not much use should prove to be very economical to a second owner. Unfortunately the implements available at the estimated value depreciation price have usually experienced considerable use. Furthermore, implements

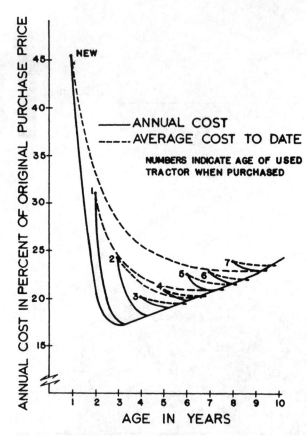

Fig 18.1. Cost comparisons for new and used equipment.

Source: Larson 1955.

that have operated satisfactorily are not generally offered for sale after only two or three years of age.

The ability of machinery managers to recognize discrepancies between price and value is all-important. They must be able to evaluate wear, recognize the adequacy of prior maintenance, judge the deterioration due to weathering and age, and know the typical performance of the various models of machinery. Probably they must be both skilled and experienced in machinery repair. If a manager lacks many of these skills, the profitability of used equipment may in fact be only a matter of chance.

Many buyers consider factors other than economic ones when purchasing new rather than used machinery:

1. Probable greater reliability of new equipment
2. Improved design of new equipment resulting in greater capacity, more efficiency, and less labor required
3. Personal preference because of ease of operation, increased bodily comfort, and prestige

While there are some genuine economic opportunities in the used equipment market, their occurrence may be such a matter of chance that the farm machinery manager of a large commercial farm should not purchase used equipment.

An individual farmer, however, should never overlook the potential cost reductions possible with used equipment.

Part VI

LABORATORY EXERCISES

Learning undertaken in the laboratory and the field where actual machines are used in realistic situations contributes greatly to understanding and retention of the material studied in the classroom.

Following are some recommended laboratory exercises that present the student with definite challenges. Answering these challenges should amplify and fix the concepts of farm machinery management in the student's mind, and more important, produce a confidence in the student that will lead to positive management actions.

The order in which the laboratory exercises are listed is meant to indicate a lesson plan that follows a straight-through study of the text material. If the laboratory exercises are followed in order, no activity will involve material that is not common knowledge, has not been presented in the laboratory, or has not been discussed previously in the classroom.

Obviously more laboratory exercises are included than can be completed on a one-per-week schedule on either a quarter or semester system. It may be possible to combine some of the calibration and harvesting laboratories. On the other hand, the machinery selection and the fuel comparison studies should extend over more than one period each.

1. Problem Solving
2. Tractor Study
3. Machine System Performance
4. Tractor Power Performance
5. Tillage
6. Volume Seeders
7. Single-Seed Planters
8. Dry Chemical Distributors
9. Liquid Chemical Distributors
10. Small-Grain Harvesting
11. Corn Harvesting

Variation in laboratory results is normal and to be expected. Variation occurs because of small errors in the measuring instruments, because of small errors in reading those instruments, and because seldom can all the variables in an experiment be controlled. Sometimes the variation is insignificant for machinery management purposes. Other times it is not; to ignore the effects of natural variation shows negligence on the part of the machinery manager.

Variation should be acknowledged and laboratory exercises should be designed to reveal the amount of variability in the results. The only way to determine the amount of variability is to repeat the test or measurement several times and compare the results. Repeated tests for the purpose of statistical analysis are called replications.

Once significant variability is discovered, it should be reported. The average value is an important statistic that is widely recognized and easily calculated, but it does not convey any information about the variability present. In fact, the average is designed to remove the effects of variability from data.

A combination of the average and the amount of variation in the data conveys the most meaning. The simplest measure of variability is to report the complete range in the sample. But chance conditions may produce a point that is atypical of the data. A widely recognized measure of variability that is not greatly affected by a single improbable point is the *standard deviation, S.D.*, defined as the square root of the sum of the squares of the deviations of the points about the average. For easy calculation the formula below is often used:

$$S.D. = \sqrt{\frac{\sum x_i^2 - (\sum x_i)^2 / N}{N-1}}$$

where N = number of points in the sample population

x_i = value of the ith point

$\sum$ is summation sign

Some hand calculators are programmed to find the S.D. of entered data with but one key stroke.

The range established by (average + S.D.) and (average - S.D.) will include about 68% of all the expected data points in a normal distribution. This range is an adequate report of the variability discovered.

The *coefficient of variation, C.V.*, is a unitless indication of variation sometimes used to report relative variability. It is computed by dividing the standard deviation by the average. A C.V. of one reports a situation where the variability is so great that about 1/3 of the data values lie outside the interval from zero to twice the average value. Such a report also indicates that the distribution of the data is not a normal distribution.

PROBLEM SOLVING

Unit Factor System

The problems arising in the machinery management area are not difficult mathematically, but often the student becomes confused with the analysis of the problem. The following material is presented with the intention of making the analysis of these problems simple.

A few general rules apply to the solution of any problem. In order, they are:

1. Read *carefully* the information given.
2. Ascertain what answer is required.
3. Jot down any formulas that apply.
4. Determine which data are pertinent.
5. Make a unit analysis.
6. Make necessary mathematical calculations.

The first three rules are rather obvious; but the fourth requires a general knowledge of the subject matter of the problem. This knowledge may be gained from the lecture, readings, and practical experience in the laboratory and the field.

The fifth rule is most important to the solution of problems found in agriculture. Many problems simply involve units. For example, the information may be gained that a tractor will travel 100 m in 40 s. The *speed* of the tractor *is* 100 m in 40 s, but such units will not convey meaning as well as the more familiar term, kilometres per hour. The problem then is one of converting units, which requires the use of *unit factors*. Unit factors are defined as ratios whose actual value is unity, or one; for example: 60 s/1 min. The numbers and labels above the line are actually equal to the numbers and labels below the line and the actual value of the factor is one. However, the *numerical* figures are not equal to one, but are introduced only as conversion factors for the units.

To use the data given before, find the speed of the tractor going 100 m in 40 s in terms of kilometres per hour.

After reading the problem and ascertaining the units required for the answer, make a unit analysis. First, put down a blank for the answer with the correct units as shown.

$$\underline{\hspace{2cm}}\ \frac{km}{hr}$$

Then place an equality sign after the answer and insert the pertinent information required for solving the problem.

$$\underline{\hspace{2cm}}\ \frac{km}{hr} = \frac{100\,m}{40\,s}$$

The rest of the example problem is merely one of converting the basic data into the correct units using unit factors.

$$\underline{\hspace{2cm}}\ \frac{km}{hr} = \frac{100\,m}{40\,s} \times \frac{1\,km}{1000\,m} \times \frac{60\,s}{1\,min} \times \frac{60\,min}{1\,hr}$$

The three unit factors, 1 km/1000 m, 60 s/1 min and 60 min/1 hr, serve to cancel out the unwanted units—that is,

$$\underline{\hspace{2cm}}\ \frac{km}{hr} = \frac{100\,\cancel{m}}{40\,s} \times \frac{1\,km}{1000\,\cancel{m}} \times \frac{60\,s}{1\,\cancel{min}} \times \frac{60\,\cancel{min}}{1\,hr}$$

Now the units on the right side of the equation are equal to the units on the left side. The student should now be confident that the proper mathematical operations are indicated.

The sixth rule, the mathematical work, is also made easier because of the possibilities for numerical cancellation. For example, in the problem considered:

$$\underline{\hspace{2cm}}\ \frac{km}{hr} = \frac{\cancel{100}\,\cancel{m}}{\underset{2}{\cancel{40}}\,s} \times \frac{1\,km}{\cancel{1000}\,\cancel{m}} \times \frac{\overset{3}{\cancel{60}}\,s}{1\,\cancel{min}} \times \frac{\cancel{60}\,\cancel{min}}{1\,hr}$$

The reduced equation can be solved without a calculator.

The answer to the problem is 9.0 km/hr.

Some problems are not solved completely by the foregoing method. Problems involving addition and subtraction must be organized differently, but the principle of carefully labeling each number with its appropriate unit still applies.

Some problems are solved by a formula. A formula problem is already organized, has the mathematical operations indicated, and is correct in its unit relations. Formulas are developed to speed up solutions. Proper use of the unit factor system will allow students to develop their own formulas for problems that recur.

The student should work out the following Unit Factor Practice Problems.

Unit Factor Practice Problems:

Data:
1. Two-row harvester working in 1-m rows
2. Travels 25 m in 30 s
3. Harvests 200 kg in 50 s
4. Density of yield is 800 kg/m³

5. Time loss, 30%
6. Works 7 hr/day
7. Tractor uses 12 L/hr of fuel
8. Fuel costs 20¢/L

1. What is the capacity of the harvester in kilograms per minute?

$$\underline{\hspace{2cm}} \frac{kg}{min} = \underline{\hspace{2cm}} \times \underline{\hspace{2cm}}$$

2. What is the speed of the harvester in kilometres per hour?

$$\underline{\hspace{2cm}} \frac{km}{hr} = \underline{\hspace{1.5cm}} \times \underline{\hspace{1.5cm}} \times \underline{\hspace{1.5cm}} \times \underline{\hspace{1.5cm}}$$

3. What is the theoretical field capacity of the harvester in hectares per hour?

$$\underline{\hspace{2cm}} \frac{ha}{hr} = \underline{\hspace{1.5cm}} \times \underline{\hspace{1.5cm}} \times \underline{\hspace{1.5cm}} \times \underline{\hspace{1.5cm}} \times \underline{\hspace{1.5cm}}$$

4. How many hectares will this harvester process in one day?

$$\underline{\hspace{2cm}} \frac{ha}{day} = \underline{\hspace{1.5cm}} \times \underline{\hspace{1.5cm}} \times \underline{\hspace{1.5cm}}$$ (% is a unitless ratio times 100)

5. How many tonnes are harvested in 1 km of harvester travel?

$$\underline{\hspace{2cm}} \frac{t}{km} = \underline{\hspace{1.5cm}} \times \underline{\hspace{1.5cm}} \times \underline{\hspace{1.5cm}} \times \underline{\hspace{1.5cm}}$$

6. What is the fuel consumption of the tractor in kilometres/litre of no-stop harvesting?

$$\underline{\hspace{2cm}} \frac{km}{L} = \underline{\hspace{1.5cm}} \times \underline{\hspace{1.5cm}} \times \underline{\hspace{1.5cm}} \times \underline{\hspace{1.5cm}} \times \underline{\hspace{1.5cm}}$$

7. What is the yield of the field in tonnes per hectare?

$$\underline{\hspace{2cm}} \frac{t}{ha} = \underline{\hspace{1.5cm}} \times \underline{\hspace{1.5cm}} \times \underline{\hspace{1.5cm}} \times \underline{\hspace{1.5cm}} \times \underline{\hspace{1.5cm}} \times \underline{\hspace{1.5cm}}$$

8. What is the fuel cost per tonne harvested, assuming no time loss?

$$\underline{\hspace{2cm}} \frac{¢}{t} = \underline{\hspace{1.5cm}} \times \underline{\hspace{1.5cm}} \times \underline{\hspace{1.5cm}} \times \underline{\hspace{1.5cm}} \times \underline{\hspace{1.5cm}}$$

9. What is the fuel cost per hectare, assuming no time loss?

$$\underline{\hspace{2cm}} \frac{¢}{ha} = \underline{\hspace{1.5cm}} \times \underline{\hspace{1.5cm}} \times \underline{\hspace{1.5cm}} \times \underline{\hspace{1.5cm}} \times \underline{\hspace{1.5cm}} \times$$

$$\underline{\hspace{2cm}} \times \underline{\hspace{1.5cm}}$$

Computer Solutions

Machinery management involves many computations and analyses that lend themselves to computer solution. Some sample uses of computer programs using the widely known BASIC language are presented in the hope that the student will be stimulated to develop others.

The following program illustrates the use of a program to determine the costs of operation of a single field machine. This program can sum the costs for multiple operations per year. If the machine is tractor operated, the fixed and R&M costs of the tractor must be known and entered as "Tractor costs." A sample solution is presented using METRIC units.

```
' This program calculates the cost of operation of a field machine.
' It can sum the costs for several different operations.If tractor operated,
' the tractor fixed and R & M costs per hour must be known and totaled for
' input.
'
10      CLS
20      PRINT
30      PRINT "FARM MACHINE OPERATION COSTS "
40      PRINT
50      PRINT
100     AT = 0
105     X = 0
110     INPUT "Type METRIC if using SI units, otherwise press ENTER        ", m$
125     PRINT
130     INPUT "Name of machine                                             ", n$
150     IF m$ = "METRIC" THEN GOTO 180
160     C = 8.25
170     GOTO 190
180     C = 10
190     INPUT "Enter purchase price, $                                      ", P
200     INPUT "Enter selling price, $                                       ", S
210     INPUT "Enter annual interest rate, decimal                          ", I
220     INPUT "Enter ownership life, years                                  ", L
230     INPUT "Effective width, (ft or m)                                   ", W
240     INPUT "Enter R&M, decimal                                           ", RM
300     CRF = I * (1 + I) ^ L / ((1 + I) ^ L − 1)
310     CC = (P − S) * CRF + S * I
315     PRINT
320     PRINT "ENTER SINGLE OPERATION COSTS"
330     GOSUB 700
335     PRINT
340     INPUT "If additional operations, Type y or n ", X$
360     IF X$ = "n" THEN GOTO 390
370     GOSUB 700
380     GOTO 340
390     TC = CC + X
400     TCA = TC / AT
410     PRINT
420     PRINT
430     PRINT USING "Total Annual Costs, $#######.##"; TC
440     PRINT
```

```
450     PRINT USING "Average cost/area, $  ####.##"; TC / AT
460     END
' Hourly Cost Subroutine
700     PRINT
710     PRINT
800     INPUT "Area of operation (acres or hectares)          ", A
820     AT = AT + A
830     INPUT "Field speed (km/h or MPH)                       ", S
840     INPUT "Field efficiency (decimal)                      ", E
870     INPUT "Labor charge ($/h)                              ", LA
890     INPUT "Fuel costs ($/h)                                ", F
910     INPUT "Oil costs ($/h)                                 ", O
930     INPUT "Tractor costs ($/h)                             ", T
950     X = X + (RM * P + LA + O + F + T) * C * A / (S * W * E)
960     RETURN
```

FARM MACHINE OPERATION COSTS EXAMPLE SOLUTION

Type METRIC if using SI units, otherwise press ENTER

Name of machine	SP Combine
Enter purchase price, $	122000
Enter selling price, $	8000
Enter annual interest rate, decimal	.09
Enter ownership life, years	15
Effective width, (ft or m)	4.8
Enter R&M, decimal	.0002

ENTER SINGLE OPERATION DATA

Area of operation (acres or hectares)	200
Field speed (km/h or MPH)	4.6
Field efficiency (decimal)	.85
Labor charge ($/h)	7
Fuel costs ($/h)	1.2
Oil costs ($/h)	.2
Tractor costs ($/h)	0
If additional operations, Type y or n	y
Area of operation (acres or hectares)	20
Field speed (Kmh or Mph)	4.3
Field efficiency (decimal)	.9
Labor charge ($/h)	7
Fuel costs ($/h)	1.1
Oil costs ($/h)	.18
Tractor costs ($/h)	0
If additional operations, Type y or n	n
Total Annual Costs, $	18709.88
Average cost/area, $	85.04

Computer solutions are most helpful for problems requiring repeated computations. An optimum machinery replacement time illustrates this advantage. An example produces the accumulated machine costs per unit of area annually. In the example below, the accumulated costs are still decreasing after 5 years and this machine should not be replaced unless a replacement can give lower costs either immediately or over a period of years.

'This program calculates an optimum replacement time for farm machinery.
'To get the summary, type y after the last year's data entry, otherwise ENTER.
'A table of accumulated cost per year of life will be displayed.
'The replacement year is the one having the least cost per unit of use.
'

```
10      PRINT
20      PRINT
30      PRINT                              "MACHINERY REPLACEMENT TIME"
40      PRINT
50      DIM X(35)
60      I% = 0: V = 0: W = 0: Z = 0
70      CLS
100     I% = I% + 1
110     PRINT "Year = "; I%
120     INPUT "Enter Depreciation                              ", D
130     V = V + D
140     INPUT "Enter Year's Repair Costs                       ", R
150     Z = Z + R
160     INPUT "Enter Year's Use (hours or area)                ", U
170     W = W + U
180     X(I%) = (V + Z) / W
190     S = (D + R) / U
200     PRINT USING "Yearly cost of use, $/area       ######.##"; S
220     PRINT
230     INPUT "Last data ? (y)"; y$
240     IF y$ = "y" THEN GOTO 270
250     PRINT
260     GOTO 100
270     PRINT
280     PRINT
290     PRINT TAB(15); "Year of Life"; TAB(33); "Accumulated $/unit"
300     PRINT
310     FOR A% = 1 TO I%
320     PRINT TAB(20); A%;
330     PRINT USING      "#################.##"; X(A%)
340     NEXT
350     END
```

MACHINERY REPLACEMENT TIME

YEAR = 1

Enter Depreciation	700
Enter Year's Repair Costs	70
Enter Year's Use (hours or area)	500
Yearly cost of use, S/area	1.54

YEAR = 2

Enter Depreciation	600
Enter Year's Repair Costs	120
Enter Year's Use (hours or area)	500
Yearly cost of use, $/area	1.44

YEAR = 3

Enter Depreciation	500
Enter Year's Repair Costs	150
Enter Year's Use (hours or area)	500
Yearly cost of use, $/area	1.30

YEAR = 4

Enter Depreciation	400
Enter Year's Repair Costs	200
Enter Year's Use (hours or area)	400
Yearly cost of use, $/area	1.50

YEAR = 5

Enter Depreciation	350
Enter Year's Repair Costs	227
Enter Year's Use (hours or area)	447
Yearly cost of use, $/area	1.29

Year of Life	Accumulated $/unit
1	1.54
2	1.49
3	1.43
4	1.44
5	1.41

A field capacity sensitivity program illustrates some advanced programming statements. To find the relative importance of the variables in the field capacity equation, a change is made in one of the variables while holding all others constant. Then the solution is repeated with a different variable changed and the resulting field capacity is compared with the first.

Total field time can be defined as the sum of the operating time, the turning time at the end of the rows, and the unproductive time needed to service the machine, fill or empty containers, etc. Using a rectangular field with headlands as in Eq. 1.3, a program was written to determine the sensitivity of the machine's capacity to a change in each variable.

The inverse of Eq. 1.3 has units of time per area.

The three times are summed and inverted to indicate the capacity of the machine in area per hour. In the example, making a 10% improvement in the factors indicated that either a 10% increase in speed or width had an equal effect on productivity. The student should try other data to produce different conclusions.

Data are entered horizontally across the chart as 5 characters per entry (including decimal point) as indicated by the headings. The data field remains blank until all 5 characters are entered. Add trailing zeros when necessary. Type y to confirm each data input or n to edit the entry. Type y after entering a line of data to be able to enter another line. Type n to quit entry and get an output and end the program. Note that this program uses COMMON units as requested at the beginning. This input must be in capital letters.

```
0       CLS
25      OPT = .0001: LOCATE , , 1: WIDTH 80
30      PRINT "Type METRIC if using SI units; otherwise type COMMON"
35      PRINT : m$ = INPUT$(6): PRINT : PRINT
40      PRINT "FIELD CAPACITY SENSITIVITY TO OPERATING VARIABLES"
45      PRINT ""; m$; "Units": PRINT : PRINT :
50      PRINT "Implement Implement Time Turn Field Coverage Field"
55      PRINT "Speed   Width   Delay   Time    Length  Width  Capacity"
60      PRINT "mph(km/h)      ft(m)      h/ac(h/ha)      secs    ft(m)   decimal ac/hr(ha/hr)"
65      PRINT
70      PRINT " All data entries must have 5 characters including decimal point."
72      PRINT " Type y(yes) or n(no) after each entry to validate or correct."
74      PRINT " Type y after entering a line for more data, n if no."
75      PRINT "_____          "
80      PRINT
85      IF m$ = "METRIC" THEN GOTO 95
90      X1 = 8.25: X2 = 12.1: GOTO 100
95      X1 = 10: X2 = 2.77778
100     s$ = INPUT$(5):
105     PRINT TAB(2); : PRINT USING "###.#"; VAL(s$);
106     y$ = INPUT$(1)
107     IF y$ = "n" THEN P1 = CSRLIN: P2 = POS(0) − 5: LOCATE P1, P2: GOTO 100
110     w$ = INPUT$(5)
115     PRINT TAB(12); : PRINT USING "###.#"; VAL(w$);
116     y$ = INPUT$(1)
117     IF y$ = "n" THEN P1 = CSRLIN: P2 = POS(0) − 5: LOCATE P1, P2: GOTO 110
120     d$ = INPUT$(5)
125     PRINT TAB(23); : PRINT USING "#.###"; VAL(d$);
126     y$ = INPUT$(1)
127     IF y$ = "n" THEN P1 = CSRLIN: P2 = POS(0) − 5: LOCATE P1, P2: GOTO 120
130     t$ = INPUT$(5)
135     PRINT TAB(33); : PRINT USING "###.#"; VAL(t$);
136     y$ = INPUT$(1)
137     IF y$ = "n" THEN P1 = CSRLIN: P2 = POS(0) − 5: LOCATE P1, P2: GOTO 130
140     1$ = INPUT$(5)
145     PRINT TAB(43); : PRINT USING "####."; VAL(1$);
146     y$ = INPUT$(1)
147     IF y$ = "n" THEN P1 = CSRLIN: P2 = POS(0) − 5: LOCATE P1, P2: GOTO 140
150     e$ = INPUT$(5)
155     PRINT TAB(53); : PRINT USING "#.###"; VAL(e$);
156     y$ = INPUT$(1)
157     IF y$ = "n" THEN P1 = CSRLIN: P2 = POS(0) − 5: LOCATE P1, P2: GOTO 150
160     a = 10 / (VAL(s$) * VAL(w$) * VAL(e$))
170     b = VAL(d$)
180     c = 2.7778 * VAL(t$) / (VAL(1$) * VAL(w$) * VAL(e$))
190     CAP = 1 / (a + b + c)
195     PRINT TAB(61); : PRINT USING "##.##"; CAP
200     IF CAP > = OPT THEN OPT = CAP: s = VAL(s$): w = VAL(w$): d = VAL(d$):
        t = VAL(t$): 1 =   VAL(1$): e = VAL(e$)
```

```
300    x$ = INPUT$(1)
310    IF x$ = "y" THEN GOTO 100: PRINT
400    PRINT : PRINT : PRINT "OPTIMUM VARIABLES": PRINT
410    PRINT "S = "; s; " W = "; w; " D = "; d;" T = "; t; " L = "; 1; " E = "; e
415    PRINT : PRINT "Optimum capacity = "; OPT
420    END
```

Implement Speed mph(km/h)	Implement Width ft(m)	Time Delay h/ac(h/ha)	Turn Time secs	Field Length ft(m)	Coverage Width decimal	Field Capacity ac/hr(ha/hr)

All data entries must have 5 characters including decimal point.
Type y(yes) or n(no) after each entry to validate or correct.
Type y after entering a line for more data, n if no.

5.0	7.0	0.100	10.0	1300.	.95	2.48
5.5	7.0	0.100	10.0	1300.	.95	2.66
5.0	7.7	0.100	10.0	1300.	.95	2.66
5.0	7.0	0.090	10.0	1300.	.95	2.54
5.0	7.0	0.100	9.0	1300.	.95	2.48
5.0	7.0	0.100	10.0	1430.	.95	2.48
5.0	7.0	0.100	10.0	1300.	.96	2.49

OPTIMUM VARIABLES

S = 5	W =	7.7	D = .1	T = 10	L = 1300	E = .95

Optimum capacity = 2.657229

Programming Practice

Develop a program to calculate the fuel consumption in L/hr for a 10,000 kg, 100 kW diesel tractor having a rolling resistance coefficient of .10. Assume a 7% slip and no power loss for tire deflection or soil displacement. The tractor pulls a 12,000 kg wagon having a rolling resistance coefficient of .15. Calculate for speeds of 2, 4, 6, 8, 10, and 12 km/hr. Assume level ground. Use the average values given in Table 2.2. Hint: Find the power required as a % of maximum PTOP and use IF statements to relate fuel consumption to the 5 ranges of loading.

TRACTOR STUDY

Objectives:

1. Become acquainted with sources of tractor information.
2. Become familiar with the construction and terminology of tractors and their accessories.

Procedure:

1. Complete the tractor terminology exercise.
2. Individually or in small groups, select a tractor for detailed study.
3. Collect information about the tractor selected. In general, the information on tractors can be divided into four categories:

   a. Specifications for chassis, engine, and accessories
   b. Operating and maintenance instructions
   c. Performance data
   d. Purchase price

   Usually no single source can supply a potential buyer with all this information. Information about tractors may be found:

   a. In company catalogs and advertising literature
   b. In the operator's and service manuals for the tractor
   c. In technical society publications:
      (1) Agricultural Engineers Yearbook, ASAE, 2950 Niles Rd., St. Joseph, MI 49085
      (2) SAE Handbook, SAE, 400 Commonwealth Dr., Warrendale, PA 15096
   d. In farm journals with machinery sections:
      Farm Journal, 230 W. Washington Square, Philadelphia, PA 19102-2181
      Prairie Farmer, 191 S. Gary Ave., Carol Stream, IL 60188-2089

Successful Farming, 1716 Locust St., Des Moines, IA 50309-3023
   e. In farm equipment trade journals:
      Implement and Tractor, P.O. Box 1420, Clarksdale, MS 38614
   f. In Nebraska Tractor Test publications.
   g. From the local dealer. While not present, the dealer should not be overlooked as an information source. The local dealer is able to make recommendations concerning performance and adaptation of the equipment in the local area, which may have special problems not covered by other information sources. Also note that a lower price is not always indicative of a good buy if essential dealer services—guarantees, repair parts, and skilled overhaul facilities—are lacking.

4. Record the information gained in response to the questions asked in the General Information section.
5. Examine the tractor if possible.
6. Share with other teams the unique specifications of the selected tractor by a 3-min oral presentation.
7. Individually rate the operator comfort and convenience features of the tractors available.
8. When practical, each student should operate each tractor around a closed course and shift into the various field speed gears, try the underdrive shift, use the brakes, engage the PTO, accelerate, etc. Upon finishing driving, the student should try to back the tractor to a line on the ground so that the lower hitch points are within but not beyond 5 cm [2 in.] of the ground line, and then stop the engine and instruct the next driver as to the location of controls and the starting procedure. *Each student must read and understand the OSHA regulations concerning tractor operations before operating a tractor.* Refer to Chapter 3 of this book.

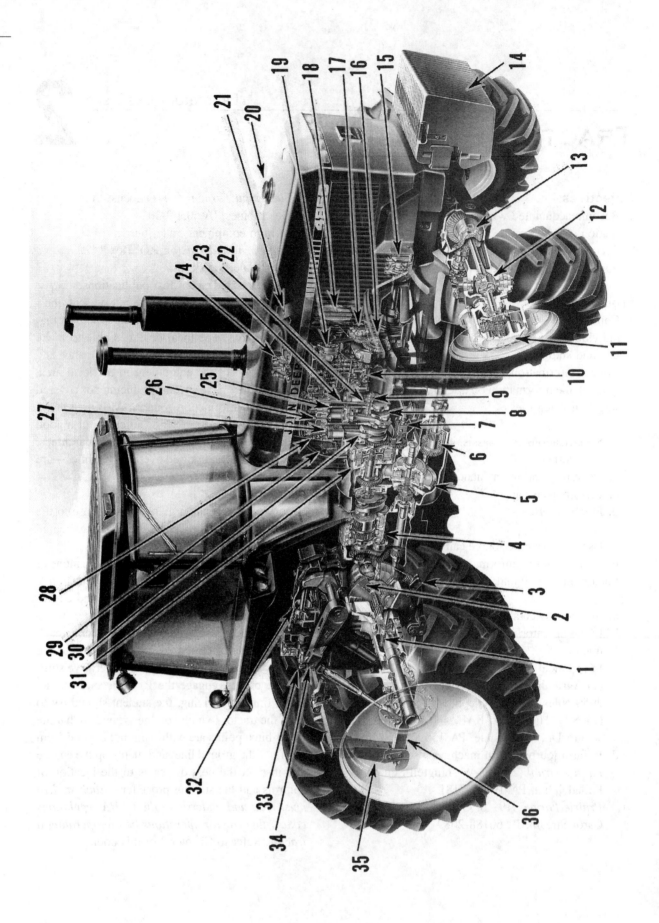

Tractor Terminology

Match the term listed with the key number on the opposite page.

_____ Alternator

_____ Ballast, front

_____ Brake disk

_____ Cam follower

_____ Camshaft

_____ Camshaft drive gear

_____ Connecting rod

_____ Coolant pump

_____ Crankcase oil pump

_____ Crankshaft

_____ Cylinder liner

_____ Differential, front axle

_____ Drive clutch

_____ Fan

_____ Fast hitch

_____ Front wheel drive clutch

_____ Fuel filter

_____ Fuel tank cap

_____ Hitch lift cylinder

_____ Hydraulic system pump

_____ Injection pump

_____ Injector

_____ Lower hitch link

_____ Piston

_____ Planetary gear reduction, front axle

_____ PTO clutch

_____ PTO drive shaft

_____ Pushrod

_____ Radiator

_____ Remote hydraulic outlet

_____ Rockshaft

_____ Traction-assist sensing link

_____ Transmission, hydraulic shift

_____ Turbocharger

_____ Universal joint

_____ Valve, exhaust

General Information:

1. Manufacturer of tractor _____ 2. Model _____ 3. Serial number _____

4. Standard wheel tread: front _____ rear _____ 5. Turning radius _____

6. Tire size: front _____ rear _____ 7. Shipping weight _____

Nebraska Publications:

1. Number of test _____

2. Maximum power

   a. PTO _____ b. drawbar _____

3. Drawbar performance data—maximum power in selected gears.

Gear	Power	Drawbar Pull	Speed	% Slip

4. Weight of tractor when tested _____

Engine:

1. Model number _____ 2. Number of cylinders _____

3. Bore _____ 4. Stroke _____ 5. Rated rpm _____

6. Removable liners? _____ wet or dry? _____

7. Number of main bearings _____ 8. Compression ratio _____

9. Turbocharged? _____ 10. Have intercooler? _____

Fuel System:

1. Fuel recommended _____ 2. Carburetor or injector? _____

3. Fuel pump? _____ how driven? _____

4. Number and location of fuel filters _____

_____

5. Engine have air precleaner? _____

6. Capacity of fuel tank _____

7. Fuel consumption—average from Varying Power Test, Nebraska Tractor Tests _____

8. Governor mounted where? _____

9. How does governor control speed? _____

Electrical Systems:

1. Ignition type (compression, spark) _____

2. Firing order _____ 3. Battery voltage _____

4. Location of charging regulator _____

5. Does the tractor have GPS capability? List and describe all electronic features. _____

_____

Cooling System:

1. Liquid capacity _____ 2. System pressurized? _____

3. List of drain ports _____

4. Locate thermostat. By-pass type? _____

Lubrication System:

1. Viscosity of oil recommended for engine? _____

2. Crankcase oil capacity _____

3. Locate engine oil filter, manufacturer? _____

4. Recommended oil change interval? _____

  filter change interval? _____

5. Does tractor have crankcase breather or ventilator? _____

6. Viscosity of oil recommended for transmission? _____

  differential case? _____

7. List of grease fitting points _____

  _____

Power Transmission System:

1. Diameter of main clutch? _____

2. What type of transmission? _____

3. How many gears forward? _____ reverse? _____

4. Can gears be changed without stopping forward motion of tractor? _____

Hydraulic System:

1. Check mechanisms powered hydraulically:

  steering _____ brakes _____ clutch _____ transmission shifting _____

  overdrive or underdrive actuation _____ rockshaft operation _____

  remote cylinder operation _____ traction-assist _____ PTO clutch _____

  list other _____

2. Fluid capacity of system _____ 3. Fluid specifications _____

4. Locate system filter. What is filter change interval? _____

5. Number and location of hydraulic pumps _____

6. What is system relief valve opening pressure? _____

7. Pump(s) flow capacity? _____

8. Number of control levers _____ 9. Variable traction-assist sensitivity? _____

10. Diameter and stroke of remote cylinders _____

Power Take-Off Drive:

1. Two-speed PTO standard or optional? _____

2. Type of drive (transmission, continuous, independent)? _____

3. PTO rpm at rated engine rpm _____

4. Engine rpm at ASAE PTO rpm _____

Hitch:

1. Does hitch include drawbar? _____ 3-point hitch? _____ semimounted hitch? _____
2. ASAE Hitch Category Number _____
3. Locate, describe draft sensing mechanism, if any. _____
4. Describe quick-hitch features, if any. _____
5. How are lower links blocked or allowed to sway? _____
6. List adjustments and controls on the hitch links. _____

Evaluation of Controls, Safety, and Convenience:

Assign a value from 1 (minimum) to 10 (maximum) to the following points:

1. Is this tractor easy to mount? Are the handholds in convenient positions? Does the cab door unlatch and open easily? _____

2. Is the seat comfortable and easily adjustable to your leg and arm dimensions? Is the seat belt easy to adjust and fasten? _____

3. Are the steering wheel adjustments easy to use? Do they allow adequate adjustment to fit your steering motion needs? _____

4. Is the clutch pedal located conveniently? Is it easy to operate? _____

5. Are the brake pedals easy to apply and easily locked together? Is the parking brake easy to apply? Are brakes subject to inadvertent application? _____

6. Is the governor control located conveniently? Is its range of movement comfortable? Can it be located easily by feel? Is it subject to inadvertent movement? _____

7. Are the most important instruments visible and easy to read? Are they grouped for quick scanning with minimum eye movement from steering needs? _____

8. Are the hydraulic controls convenient to use? Are they grouped for operation by touch alone? Are they located or separated so that identity by touch is not a problem? _____

9. Evaluate visibility from tractor seat to the front, the rear, hitch points, and PTO shaft. _____

10. Does the cab provide adequate protection from wind, cold, sun, rain, heat, noise, and overturning? _____

Total _____

MACHINE SYSTEM PERFORMANCE

Objective:
Develop skill in analyzing the performance of a field machine system.

Procedure:
1. Observe the events of an ongoing field operation. A harvest operation, particularly forage harvesting, is a desirable system for study.

2. Time and record the events in each machine's cycle over several cycles.

3. Average the times and construct a cycle diagram.

4. Determine the average system capacity in tonnes [tons] per hour.

5. Find the average Labor efficiency in tonnes [tons] per work-hour.

6. Make recommendations for improvements.

Specification Sheet

Machine name						
Make						
Model						
Size						
Power						
No. used						

Data Sheet

EXAMPLE

Machine ___wagon___   Machine _____   Machine _____

Cycle Activity	Elapsed Time (min:s)	Av. (min)	Cycle Activity	Elapsed Time (min:s)	Av. (min)	Cycle Activity	Elapsed Time (min:s)	Av. (min)
Road travel	3:12 3:17l, 3:24	3.29						
Open, close gate	1:20 1:00, 1:00	1.11						
Field travel	1:15, 1:10	1.2						
Wait for harvester	0, 0:30	0.25						
Hitch to harvester	0:45, 0:35	0.67						
Fill	26:50, 24:10	25.5						
Unhitch	0, 0	0						
Wait for tractor	0, 0	0						
Hitch to tractor	1:00, 0:40	0.83						
Field travel	1:25, 1:20	1.38						
Open, close gate	1:10, 1:20	1.25						
Road travel	3:30, 3:50	3.67						
Wait to unload	0, 0	0						
Unload	17:35, 18:10	17.88						

TRACTOR POWER PERFORMANCE

Objectives:

1. Measure the power and fuel efficiency of a farm tractor.
2. Become familiar with tractor testing procedures using a dynamometer.

Procedure:

1. Connect a tractor to a PTO dynamometer of known brake arm length and scale tare weight, if any.
2. Allow the engine to reach the operating temperature before recording data.
3. Measure the density of the fuel with a hydrometer. Make temperature corrections if necessary. (A standard API hydrometer is recommended.)

   a. The specific gravity, sg, of a petroleum fuel in terms of degrees API is

   $$(sg)_f = \frac{141.5}{131.5 + API°}$$

   b. The actual density of the fuel is its specific gravity, $(sg)_f$, times that for water, 1 kg/L [8.33 lb/gal].

   c. The indicated energy content, E, of a petroleum fuel is given by the formula:
   $$E \text{ MJ/kg} = 51.8 - 8.77 \ (sg)_f^2$$
   $$[E \text{ Btu/lb} = 22{,}320 - 3780 \ (sg)_f^2]$$

   d. The price of the fuel should be obtained from a local wholesaler.

4. With full-open governor control, apply increments of load from zero to an amount that causes the engine to produce peak torque. At least ten tests should be made. Record the individual test values on the data sheet. If time permits, each test should be repeated to get an indication of the variability involved.
5. Calculate average values of fuel efficiency and plot them as a function of power.
6. Plot engine performance curves of torque and power versus engine speed.
7. Note on the graphs in (5) and (6) the governed and overload ranges of operation.
8. Find the percent thermal efficiency of the most efficient fuel test.

313

Tillage

Objectives:

1. Become familiar with several types of tillage implements.
2. Make preliminary shop adjustments on typical implements.
3. Make appropriate field adjustments.

Procedure:

1. Determine the specifications of several tillage implements by measurement and by use of the operator's manuals.
2. Position the tractor and implement in a working attitude on the shop floor. Adjust the hitch and the implement control links to produce a level implement trailing parallel to tractor travel.
3. Go to a practice field if possible. Vary the hitch and the implement adjustments and note their effect on performance of the implement.

Specification Sheet

Item	Implement #1	Implement #2
Type		
Make		
Model		
Size		
Describe: hydraulic lift control		
leveling adjustments		
depth adjustments		
hitch adjustments		
degree of pulverization adjustments		

VOLUME SEEDERS

Objectives:

1. Become familiar with the types of volume-seeding machines and volume-metering mechanisms.
2. Estimate field performance from shop calibration.
3. Measure the variability in seeding machine performance.

Procedure:

1. Examine several types of volume seeders and record their specifications.
2. Make a calibration check of metering rates.

    a. For ground-driven machines:

    (1) Measure the squash-radius of the drive wheel (center of axle to the ground surface). The measurement of the rolling circumference of the wheel on the actual surface to be seeded is preferred but is seldom possible when making a preseason calibration check.

    (2) Block up one drive wheel so that it may be motorized or turned by hand. From one or more openings, catch the delivery for a known number of drive-wheel revolutions.

    (3) Test at least four settings from the minimum to the maximum rate setting. Replicate each test at least three times. One series of tests should check the effects of variable speeds.

    (4) Transform the test data into area rates, kg/ha [lb/a], for each rate setting.

    (5) Develop a calibration curve of actual rate versus rate setting showing the curve of the averages and the curves indicating one S.D. on each side of the average.

    b. For PTO-driven machines:

    (1) Set up the machine to operate without forward movement.

    (2) Collect the output for a given interval of time instead of a representative distance.

    (3) Continue with (3) under (a).

    c. For broadcasters:

    (1) Provide enough floor space for the complete delivery pattern to be displayed. Operate without forward motion for a known time interval.

    (2) Sweep the material into a row perpendicular to machine travel by carefully sweeping only parallel to machine travel.

    (3) Measure the quantities obtained per increment of pattern width. Determine a proper overlap.

    (4) Continue with (3) under (a).

SINGLE-SEED PLANTERS

Objectives:

1. Become familiar with seeding machines having single-seed metering devices.
2. Estimate field performance from a shop calibration.
3. Determine the variability in machine performance.

Procedure:

1. Examine several types of single-seed metering machines and record their specifications.
2. Match the seed plate or drum to the seed.

   a. Measure the squash-radius of the drive wheel.
   b. Block up the drive wheel so that it may be motorized or turned by hand.
   c. Slowly turn the drive wheel far enough to pass 10 cells in the seed plate or drum. Repeat at least three times. A perfect seed plate-seed match will give exactly 10 seeds for each trial.

3. Make a calibration check.

   a. With the seeder blocked up, arbitrarily select a test duration time or definite number of drive-wheel revolutions per test.
   b. Turn the drive wheel for that distance or time and collect the seed from one or more opener outlets. Be sure the seed drops onto a soft surface.
   c. Determine the number of seeds caught and calculate the actual seeding rates and the S.D. for each run.

4. Prepare a calibration chart of the manufacturer's rate setting versus the actual seeding rates found. Show a curve of the averages and the curves 1 S.D. on each side of the averages.

5. Questions:

   a. What is the percent deviation of run # _____ based on the manufacturer's rate setting?
   b. Examine several samples of seed metered. What is the percent damage?

DRY CHEMICAL DISTRIBUTORS

Objectives:

1. Become familiar with the types of dry chemical distributors (including fertilizer and granular pesticide attachments for seeders).
2. Estimate field performance from a shop calibration.
3. Measure the variability in performance of machines operating on slopes and with varying tire pressures.

Procedure:

1. Examine several types of fertilizer distributors and their metering mechanisms. Record their specifications.
2. a. Refer to Procedure 2 in Exercise 6, Volume Seeders, for methods of calibrating ground-driven versus PTO-driven and multimetering versus broadcast machines.

   b. If possible, down-the-row variation of a single-metering mechanism should be checked. Set the metering rate at its maximum and operate the machine over the shop floor. Measure carefully the quantities delivered at several points along the row of chemical deposited.

3. Vary the air pressure in the tire of a ground drive wheel. Apply different levels of mass to the distributor (riding students) and measure the change in machine advance per wheel revolution by marking beginning and end positions on the shop floor. If possible, operate in the field to include the amount of wheel slippage that occurs in soft soils. Develop a chart of percent wheel advance versus added mass for correct, high, and low air pressures.

4. Mount a distributor (granular applicator) so that it may be operated in three inclined positions—forward, back, and to the side. Check performances both for inclination and for depth of material in the box.

Specification Sheet

Item	Machine #1	Machine #2	Machine #3
Type			
Make			
Model			
Size—width covered			
Type of metering device			
No. possible metering rates			

LIQUID CHEMICAL DISTRIBUTORS

Objectives:
1. Become familiar with liquid chemical distributing machines and mechanisms.
2. Estimate field performance with a shop calibration.
3. Determine the effects of some operating variables of sprayers.

Procedure:
1. Examine several types of distributors. Record their specifications.

2. Make a calibration using the procedures in 2(b), Exercise 6, Volume Seeders, for sprayers and gravity-flow distributors; and in 2(a) for machines with ground-driven pumps.
3. Determine the delivery pattern of adjoining nozzles on a sprayer's boom. Use a calibration stand to collect and indicate the spray pattern, Fig. 9.24. Conduct tests at different nozzle heights.

Specification Sheet

Item	Distributor #1	Distributor #2	Distributor #3
Type			
Make			
Size			
Type of pump if any positive or nonpositive?			
Mechanical agitation, bypass agitation or both?			
Type of pressure regulator, if any			
Metering method			
Types of nozzles, if any			
Ground clearance			
Outlet spacing			
Power source			
No., positions of filters or screens			

10

SMALL-GRAIN HARVESTING

Objectives:

1. Become familiar with combine mechanisms.
2. Conduct a material efficiency test.

Procedures:

1. Examine a combine equipped for harvesting small grain. Write specifications for its functional parts.
2. Establish and lay out field test areas.
3. Glean a representative area for shatter (preharvest) loss. If possible, use this gleaned area for the cutter bar test also.
4. Operate the combine at a steady speed over the cutter bar test area.
5. Stop forward motion without any combine discharge contaminating the cutter bar test area, raise the header, and allow the combine to clear itself.

Back off the test area and glean the cutter bar loss (or cutter bar plus shatter).

6. While operating typically and at a steady speed, catch a yield sample over a marked distance.
7. During the yield test unroll a canvas, attach a bag, or place a panel to catch the combine discharge over a known distance.
8. Gather and weigh all losses.
9. Separate a harvested grain sample into threshed, unthreshed, and damaged grain portions. Thresh the unthreshed grain by hand and weigh the chaff and hulls along with any other foreign matter.
10. Record the data and determine the material efficiency. Tests can be run at different forward speeds and different machine settings. If time permits, the tests should be replicated and tests under varying crop moistures should be made.

CORN HARVESTING

Objectives:

1. Become familiar with row crop harvesters and mechanisms.
2. Conduct a material efficiency test.

Procedure:

1. Examine a harvester equipped with a corn head. Write specifications for its functional parts.
2. Establish and lay out the field test areas.
3. Glean from the test area all loose ears (ears detached from stalks).
4. a. Operate the harvester at a steady speed over a cleared subtest area.

   b. Stop forward motion without any harvester discharge contaminating the test area, raise the header, and allow the combine to clear itself.

   Back off the test area and glean the header shelled corn loss.

   c. Continue harvesting, catching a yield sample over a known distance.

   d. During the yield test unroll a canvas, attach a bag, or place a panel to catch the harvester discharge over a known distance.

5. Gather and weigh all losses and the net yield.
6. Separate the harvest sample into its components, weigh, and find the proportions of each.
7. Record the data and determine the material efficiency. Tests can be run at different forward speeds and for different machine settings. If time permits, the tests should be replicated and tests at different crop moistures should be made.

Specification Sheet

Item	Harvester #1	Harvester #2	Harvester #3
Make and model			
Type and size			
Gathering device:			
floating points?			
to adjust snapping rolls			
has stripper plates?			
If picker:			
type of husking bed			
husking rolls, material, number			
shelled corn saver?			
to increase husking effectiveness			
what other attachments?			
If combine:			
to improve shelling effectiveness			
to adjust cleaning sieves			
to adjust fan blast			
what other attachments?			

HARVESTING RATE

Objective:

Determine the effects of harvesting rate on material efficiency, power requirements, and operator control.

Preparation:

The engine must have been previously calibrated for power versus either manifold vacuum (SI) or fuel flow rate (CI). A corn combine is suggested since the picker ear loss is easily determined and skill is needed to keep the machine on the row at high speeds.

Procedure:

1. Divide into test groups.
2. Assign an operating forward speed (or transmission gear setting) to each test group.
3. While each group operates the harvester in turn, the other groups check losses and power requirements.
4. Record the data. Replicate the tests if time.
5. Prepare a graph of average loss versus forward speed.
6. Prepare a graph of accumulated power versus forward speed. Show the contributions of each component. See Fig. 10.19.

Data Sheet

Harvester make, model, size _____ Crop and soil conditions _____

Preliminary readings: The governor control lever should remain in full open position with no crop in the harvester.

Specification Sheet

Test	Vacuum or Fuel Flow Value	Power
Engine alone with clutch depressed		
Engine with cutch released, neutral		
Operating harvester		
harvester plus header		
Operating unloader only		
empty		
full		
operating with forward motion harvester disengaged (rolling resistance)		
harvester engaged		

FORAGE MOWING

Objectives:

1. Gain a working knowledge of mowing machines and their mechanisms.
2. Conduct a field performance test.

Procedure:

1. Examine several types of mowers and forage conditioners. Record the specifications.
2. Operate the machine in the field.
3. Immediately fork the cut material off the swath and:

   a. Determine the average and the standard deviation of the stubble height.
   b. Gather missed stems and clippings too small to be removed by the fork from a representative area.

4. Evaluate the effects of various test treatments on performance. Different forward speeds and guard angularity should be checked. Replicate the treatments if possible.

Specification Sheet

Item	Mower #1	Mower #2	Mower #3
Make and model			
Type and size			
Cutter operating speed, cycles/min			
How driven?			
Slip clutch?			
Range in heights of cut			
Breakaway release mechanism description			
For cutter bar mowers only:			
guard spacing			
stroke dimension			
to change mower's register			
describe balanced drive mechanism			
range of guard angularity			

RAKES

Objectives:

1. Gain a working knowledge of rakes and their mechanisms.
2. Conduct a field performance test.
3. Determine the effects of raking variables.

Procedure:

1. Examine several rakes and write specifications.
2. Place numbered tissue papers in the mown hay and locate their initial positions in the swath relative to a fixed point.
3. Operate the rake at the manufacturer's recommended speed and settings.
4. Record the distance between the initial and final tissue paper positions.
5. Glean from a representative area any desired materials missed by the rake. Determine the amount on an area basis.
6. Examine the windrow. Sort out any undesirable materials picked up by the rake and determine their percentage of the total mass.
7. Check the effects of forward speed, various reel speed/forward speed ratios, various teeth inclinations, and various teeth heights on performance. If time permits, replicate and try a range of forage moisture contents.

Specification Sheet

Item	Rake #1	Rake #2	Rake #3
Make, model, type			
Width of action			
How driven?			
Differential action on wheels (if ground driven)?			
Number of reel bars			
Range in height adjustment			
Teeth spacing To adjust angel of the teeth			
Tedding attachement?			
Bar positioning device description			

BALERS

Objectives:

1. Gain a working knowledge of balers and baling mechanisms.
2. Check rectangular baler timing.
3. Determine the effect of adjustments and operations on bale dimensions.
4. Determine the baler capacity at various forward speeds.

Procedure:

1. Examine several balers and write specifications.
2. Refer to the manufacturer's service bulletin and check the timing for:

   a. Feeder fingers related to the plunger
   b. Needle entrance to the bale chamber related to the plunger
   c. Twine disk holder or wire clamp related to the needle

3. Operate the baler on uniform windrows. Vary forward speed from very slow to very fast and observe the effect if any on bale shape and bale density.

4. If a rectangular baler:

   a. Adjust the feeder mechanism's positioning of the material in the bale chamber and observe the effect on bale shape.
   b. Adjust the density control over its range and observe its effect on the mass of individual bales.

5. If a round baler, check the effects on bale shape of operating with:

   a. Small windrows fed into the exact center, the left side, and the right side of the pickup
   b. Full windrows that fill the pickup width

6. Operate the baler over several full, field-length windrows and find both area-related and mass-related capacities. Time the operation and determine its field efficiency. Replicate tests if time permits.

FORAGE HARVESTERS

Objectives:

1. Gain a working knowledge of forage harvesters and their mechanisms.
2. Evaluate variability in the harvester's length of cut.
3. Determine field capacity.

Procedure:

1. Examine several harvesters and write their specifications.
2. Collect a sample of the harvested material.

   a. Measure the length of 100 particles.

   b. Find the average length and the standard deviation.
   c. Divide the total range of the particle measurements into equal increments of particle size and plot a frequency distribution.
   d. Compare with the theoretical length of cut, if any.

3. Operate the harvester in typical field conditions. Harvest several loads and find both area-related and mass-related capacities. Time the operation and find its field efficiency. Replicate tests if time permits.

Specification Sheet

Item	Harvester #1	Harvester #2	Harvester #3
Make and model			
Power source			
Type of cutter			
Number of knives			
Throat area			
Shear bar: adjustable?			
replaceable?			
To sharpen knives			
Knives adjustable?			
Has overrunning clutch?			
Has quick stop and reversal of feed?			
Number of lengths of cut available			

17

OPTIMUM GOVERNOR SETTINGS

Objective:

Determine the effect on fuel efficiency of reduced governor setting at part loads.

Procedure:

1. Determine the fuel efficiency of a farm tractor at various engine loadings for each of several governor control settings.

   a. Place the tractor, any fuel-burning type, on a PTO dynamometer and measure the fuel consumption and power for each test.

   b. Initially set the governor control in full-open position and gradually increase loads from no load to a load near the torque peak of the engine. The increments should be approximately one-tenth the total range. Record load, speed, fuel consumption, and governor setting.

   c. Repeat for governor settings of ¾, ½, and ¼ open. The number of tests may be decreased with each governor control setting to about five at ¼ open.

2. Report

   a. Compute the power output and fuel efficiency in kW·hr/L [HP·hr/gal] for each test.

   b. Plot the values for each governor setting on a graph of power versus engine speed. Note the fuel efficiency value in pencil at each point.

   c. Plot lines of constant fuel efficiency (10, 9, 8, etc.). Some interpolation will be necessary. The lines should appear as in Fig. 2.14.

   d. Under what loading condition was the tractor tested most efficient? What reservations should one have about operating a tractor at this loading?

OPERATION ENERGY REQUIREMENTS

Objective:
Determine energy requirements for selected tractor-powered implements.

Equipment:
Implements, a high-powered tractor equipped with a recording vacuum meter or fuel flowmeter, and a PTO dynamometer.

Procedure:
1. Establish the vacuum or fuel flow versus power calibration curves for the tractor with the meter and the dynamometer. (This data may be obtained concurrently with the data obtained during the Tractor Power Performance test, Laboratory Exercise 4.)

   a. Attach the meter to the tractor for use during both the dynamometer and the field tests.
   b. Conduct a wide-open governor control, variable-load dynamometer test. Record the meter readings for each load. Ten levels of load should be used between no load and the load at rated speed.
   c. Plot a curve of meter reading versus power.
   d. Remember to find a meter reading for the engine under no-load conditions.

2. Obtain the tractor engine's meter readings while operating the various implements under field conditions (PTO-powered implements and heavy-draft tillage implements make good test implements).

   a. Operate the tractor in each forward gear over the soil surface to establish the tractor's rolling

resistance energy requirement. Wide-open governor setting must be maintained.
   b. Attach the implement and operate in several gears and/or at various depths. Measure actual forward speed, width of implement action, and depth of tillage if appropriate. Record the implement parameters. The engine must not be pulled down to rated speed.

3. Subtract tractor rolling resistance from the total indicated draft force and find the drawbar draft per unit of implement width for each test. Replicate.
4. Determine the implement energy, kW·hr/ha [HP·hr/a], required for each test.
5. Submit a written report. If three or four students work as a team for testing, a joint report is appropriate.

   a. Prepare a title page and a statement of the objective of the test.
   b. Describe the equipment and the test procedures.
   c. Include the original data sheet and notes.
   d. Present results either in tabular or graphical form.
   e. Discuss results and form conclusions concerning the effects of speed and/or draft on the energy requirements of the implement tested.
   f. Include sample calculations in an Appendix.

Question
Why must one subtract the power loss due to traction-wheel slippage from the power indicated by the meter to arrive at implement draft and energy requirements?

MACHINERY SELECTION

Objective:
Determine the optimum machinery system for a given farm.

Procedure:
1. Establish cropping practices and the expected crop yields and values.
2. Establish the required machine operations and secure real or representative data for energy requirements (Table 2.5).
3. Evaluate Eq. 17.5 for tractor selection (Parts I, II, and III of this exercise).
4. Use Eq. 16.4 to select individual implements.
5. Find the permissible range in implement sizes, make a feasible size selection, and find the time and power required for each implement. Use Eq. 4.7 to find the expected annual costs and compare with custom costs.
6. If the power required is too great or the custom cost is cheaper, limit implement size to the maximum permissible for the power level and drop expensive implements from the system in favor of custom work if such an alternative is truly reasonable. After making these changes, the whole system must be recalculated since a new optimum must be found with these new considerations.
7. Present a detailed report as if from a consultant to a farm manager.

Data Sheet

1. Cropping Practice:

Crop	Annual Area	Expected Yield	$ Value/Unit of Yield

2. Establish the relationship between crop culture and the *necessary* machine operations. For example: the number of diskings before planting corn, the required tillage before seeding alfalfa, whether the oats will be top-dressed with fertilizer, the number of cuttings for hay, etc.

Part I. For field operations:

Machine	Operating on Crop	A	E	r	K	YV	sc	nt	U	h	$\frac{100\,AE}{r_i\,(FC\%)t}\left(L+\frac{K\,(YV)\,A}{(sc)_i\,(nt)_i\,U_i\,h_i}\right)$

Part I total = _____

Part II. For transport operations (c = 0.27 [0.53]):

Operation	D	W	r	$\frac{100\,cDWL}{(FC\%)tr}$

Part II total = _____

Part III. Farmstead operations:

Operation	G	W	r	L	$\dfrac{100\ \text{GWL}}{(\text{FC\%})\text{tr}}$

Part III total = _____

Part III total = _____

Optimum power = $\sqrt{\text{I} + \text{II} + \text{III}}$ = _____

Maximum PTO power _____

Number of tractors required _____

4. Results:

	Tractor _____	Tractor _____	Tractor _____	Tractor _____
Purchase price				
Estimated annual use, hr				
Estimated value of T, $/hr				
Actual T (found after selecting machinery)				

5. Implement Selection. Use typical field speeds, S, and field efficiencies, e. Let $X = (sc)(nt)Uh$. The price per added width, p, can be estimated from Appendix D.

Implement	_____	_____	_____	_____	_____	_____	_____
S							
e							
p							
R & M							

Implement	$\dfrac{100\,c*}{(FC\%)p}$	Crop _____ $\dfrac{A}{Se}\left(L+T+\dfrac{KYVA}{X}\right)$ = Total	Crop _____ $\dfrac{A}{Se}\left(L+T+\dfrac{KYVA}{X}\right)$ = Total	Crop _____ $\dfrac{A}{Se}\left(L+T+\dfrac{KYVA}{X}\right)$ = Total	w^2, Sum of Totals	w

*c= 10 [8.25]

5. Results. Use Eq. 16.5 to find range in sizes with d=$ _____. Custom rates may be found in Appendix C.

Implement	$w_{1,2}$	Size selected	Hours of Use	Power Required	Cost/Area	Custom Rate, $/Area

7. Recalculations: Compare the total field hours required for tractors with the estimated annual hours in the Implement Results table (Step 6). If the sum of the actual field hours and the estimated transport and farmstead hours is significantly different from the estimated annual hours of tractor use, the value of T must be changed in the Implement Selection table and that table checked to see if any significant changes are required.

The final check for the machinery manager is the scheduling check. The hours required by the machine can be converted to days of duration of operation by dividing by Uh. These calendar days must closely relate to the optimum operation times and must not overlap or compete with other operations. If they do, the machinery manager must resolve the conflict by reassessing the various K values and reworking the selection problem for new economic answers. Or the manager may arrange for temporary additional power and labor, hire custom operators, work longer hours per day, gamble on the U confidence level for the predicted weather, or buy larger than economic machinery to provide capacity "insurance" for the farm operations. Before such purchasing the true cost of such an alternative should be carefully evaluated.

APPENDICES

Appendix A

Areas and Values of U.S. Crops (USDA 1997)

Crop	Area, ha [a]	(millions)	$(billions)
Corn	29.8	[73.7]	24.4
Soybeans	28.3	[69.9]	17.7
All Wheat	25.7	[63.5]	8.6
Hay Crops	24.6	[60.8]	13.4
Cotton	5.4	[13.3]	6.1
Sorghums	3.8	[9.4]	1.5
Barley	2.6	[6.4]	0.9
Oats	1.2	[2.9]	0.3
Rice	1.2	[3.0]	1.7
Dry, Edible Beans	0.7	[1.8]	0.6
Peanuts	0.6	[1.4]	0.9
Potatoes	0.6	[1.4]	2.6
Sugar Beets	0.6	[1.4]	1.2
Sugar Cane	0.4	[0.9]	0.8
Tobacco	0.3	[0.8]	3.0

Appendix B

Maximum or Minimum by Differential Calculus

Differential calculus is one method of finding a maximum or minimum value of an algebraic equation with respect to one of its variables. A derivative of y with respect to x is expressed as dy/dx and is defined as

$$\frac{dy}{dx} = \frac{\Delta y}{\Delta x} \text{ as the limit of } \Delta x \text{ approaches zero}$$

where $\Delta y/\Delta x$ = the small change in y as x changes a small amount

As an example consider the equation

$$y = \frac{x^2}{x} - x + 1$$

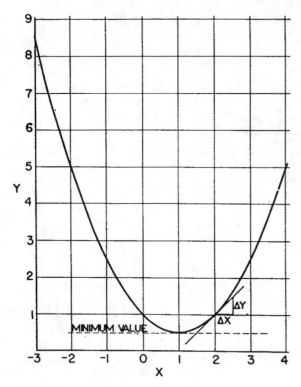

Fig. B.1

Figure B.1 is a plot of the equation showing clearly that the minimum value occurs at x = 1 where the slope of the tangent to the curve is exactly horizontal (the slope is zero) and the value of $\Delta y/\Delta x$ or dy/dx is equal to zero. Note that at y = 1 and x = 2, the tangent is inclined to the horizontal and $\Delta y/\Delta x$ has a value other than zero. The minimum point of an equation may be determined without plotting by differentiating the equation.

The rule for differentiating here and in all the examples in the text is

$$\text{if } y = x^n \text{ then } \frac{dy}{dx} = nx^{n-1}$$

The derivative of the sample equation is

$$\frac{dy}{dx} = \frac{2x^{2-1}}{2} - 1x^{1-1} + 0 = x - 1 \text{ (since } x^0 = 1)$$

The derivative is equal to 0 at the minimum point :

$$0 = x - 1 \text{ or } x = 1$$

The student should be cautioned that not all algebraic equations have a maximum or minimum and that some have more than one.

Appendix C

Machinery Cost Estimates, Illinois, 1997

Includes labor at $10.00/hr, fuel at $0.26/L [$1.00/gal], and equipment replacement costs for ranges of annual use. Excludes management overhead, risk, and profit. Custom rates may be either higher or lower depending on the annual use of the custom machine.

(From W.R.Harryman, J.C.Siemens, and B.Kirwan. Machinery Cost Estimates. Farm Business Management Handbook, University of Illinois)

Tractors
Rent w/o labor or fuel
2WD: $0.24/PTO kW·hr [$0.18/PTO HP·hr]
MFWD: $0.25/PTO kW·hr [$0.19/PTO HP·hr]
4WD: $0.19/PTO kW·hr [$0.14/PTO HP·hr]

Implements
Includes tractor, fuel and labor. Data arranged in order of low use, medium use (**high-lighted**), and high use.

	$/ha	[$/a]
Moldboard plowing	99 **45** 37	43 **18** 15
Chisel plowing	61 **27** 21	25 **11** 9
Tandem disk harrowing	37 **16** 13	15 **6** 5
Field cultivating	26 **12** 10	11 **5** 4
Row-crop planting	40 **17** 14	16 **7** 6
No-till planting	49 **20** 15	20 **8** 6
Grain drilling	40 **17** 14	16 **7** 6
No-till drilling	67 **27** 20	27 **12** 8
Rotary hoeing	11 **5** 4	5 **2** 1
Row-crop cultivating	37 **16** 12	15 **7** 5
Conventional knife mowing	29 **17** 15	12 **7** 6
Mower-conditioning	45 **25** 20	18 **10** 8
Disc mowing	27 **16** 13	11 **7** 5
Raking	28 **17** 14	12 **7** 6
Small rect. baling with ties	126 **72** 60	51 29 24
Large rect. baling with ties	116 **64** 52	47 **26** 21
Round baling with ties	128 **72** 59	52 **29** 24
Forage harvesting, row crop	224 **117** 80	91 **48** 33
Forage harvesting, windrowed	200 **99** 67	80 **40** 27
Field spraying, 9m boom	12 **6** 5	5 **2** 2
Field spraying, 24m boom, SP	5 **3** 2	2 **1** 1
Combining corn, SP	74 **62** 50	30 **25** 20
Combining beans, small grain,SP	62 **53** 50	25 **22** 20
Rotary weed cutter	56 **27** 23	23 **11** 9
Flail shredder	36 **19** 14	15 **8** 6
Applying Anhydrous Ammonia	28 **14** 12	12 **6** 5

Appendix D

Representative Farm Machinery Prices, 1999

Prices quoted are list prices for commonly equipped machines. Where applicable, prices are given per capacity of the machine (width, power, volume, etc). Prices do not include freight charges or taxes.

n = number of rows, bottoms, openers, etc., otherwise a machine's width is expressed in m [ft]

Machine	Dollars
Balers	
Rectangular, PTO	
competitive, twine	10,800
regular, twine	12,500
add for wire tying	1,000
commercial, wire	16,500
Round, PTO	
1.5 m [5 ft] wide by 1.5 m [5 ft] diam.	15,000
1.5 m [5 ft] wide by 1.8 m [6 ft] diam	21,000
add for surface wrap unit	3,000
Chisel plows (standards spaced 0.3 m [1 ft])	
Mounted	
rigid standards	520 + 710/m [210/ft]
spring reset standards	520 + 930/m [280/ft]
Pull	
rigid standards	-2700 + 2100/m [640/ft]
spring reset standards	-2700 + 2300/m [710/ft]
High capacity (>13 m [40 ft])	
spring reset	16,000 + 1100/m [360/ft]
add chisel points at $8 each or sweeps	
at 5.50 + 0.1/cm [0.27/inch] width	
tine harrow attachment	-230 + 230/m [68/ft]
Deep rippers	
shear bolt standards	-860 + 700n
trip standards	- 860 + 1300n
Spring reset standards	1500 + 1650n
Combines	
Base machine	
pull. PTO	65,000
self-propelled	39000 + 610/kW [450/HP]
add for hillside models	78,000
Heads	
grain (with reel)	4500 + 980/m [300]
pickup	9100
corn	1500 + 3600n
Corn pickers, PTO	12,500 +2850n
Cotton pickers, self-propelled	
2 – row	130,000

Machine	Dollars
4 – row	220,000
less 4500 for low drum	
Cotton strippers, self-propelled	
4- row	105,000
Chemical applicators	
Self-propelled, high clearance	
Sprayer	39,000 + 560/m [132/ft]
Pull-type sprayer, PTO	
per L [gal] tank capacity	9100 + 0.9/L [3.5/gal]
Field distributers, pull	100/m [360/ft]
Broadcasters, PTO, pull	
(per t [T] tank capacity)	3800 + 1900/t [160T]
Anhydrous applicators,	1900L [500 gal] tank
pull	4800 + 960n
attachment for toolbar	960 + 1100n
Self-propelled, flotation tires,	91,000 + 5520/m [160/ft]
Field cultivators	
15 cm [6 in.] spacing, 18 cm [7 in.] sweeps	
Rear mounted	
rigid frame	1300 + 310/m [95/ft]
folding frame	2100 + 4470/m [145/ft]
Pull-type	
rigid frame	1200 + 600/m [162/ft]
folding frame	3600 + 430/m [117/ft]
flexible frame	3000 + 830/m [250/ft]
High capacity, pull, > 15m [40ft]	1500/m [470/ft]
Forage blowers	5000
Forage harvesters	
Flail	4500/m [470/ft]
Shearbar (per max. PTO power req'd)	
pull, base	160/kW [120/HP]
self-propelled, base	700/kW [520/HP]
Heads	
row crop	-4900 + 5000n
cutterbar	4000/m [1200/ft]
windrow pickup	
competition	3000
commercial (for self-propelled)	6100
Harrows	
Disk, tandem (0.23 m [9 in.] spacing, 45 cm [18 in.] diameter	
rear, mounted	810 + 450/m [140/ft]
pull, hydraulic lift	
competitive	1150 + 1140/m [350/ft]
regular	4000 + 900/m [270/ft]
folding	850 + 2300/m [800/ft]
high capacity > 7.5m [24 ft]	4400 + 2600/m [800/ft]
Roller, pull with tines	
rigid frame	2400 + 1600/m [490/ft]
folding frame	6900 + 1600/m [490/ft]
Spike-tooth	570 + 260/m [80/ft]

Manure spreaders,	
PTO, 2-wheel	
(per m³ [ft³] level capacity)	-2900 + 2400/m³ [68/ft³]
Middlebreakers, bedders	870n
with planting attach. (lister)	1480n
Moldboard plows	
Mounted	1000n
Semimounted	
economy	3700 + 920n
regular	2300n
two-way	1200 + 2400n
spring reset standards, add	400n
Mowers	
Cutterbar	
mounted	2800 + 3440/m [104/ft]
pull	3400 + 340m [104/ft]
Rotary	1000 + 420/m [130/ft]
Mower conditioners	
Cutterbar	-3600 + 6000/m [1850/ft]
Rotary cutters	-2450 + 5600/m [1700/ft]
Rakes	
Side-delivery, 2.6m [8.5 ft]	
mounted	-450 + 1800/m [570/ft]
pull	-1000 + 1800/m [570/ft]
Wheel (per wheel)	570n
Rotary Hoes	
Mounted	1390 + 780/m [240/ft]
Folding (>9 m [30 ft]	2400 + 780/m [240/ft]
Row cultivators	
(0.75 m [30 in] rows, 20 cm [8 in] sweeps	
Front-mounted	1240 + 750n
Rear-mounted	
rigid frame	42 + 640n
folding, n > 12	4300 + 600n
Seeders (per opener)	
Grain drills, plain, 20 cm [8 in] spacing	
mounted	2400 + 260n
pull	700 + 260n

folding, n>40	4300 + 600n
press wheels, add	130n
fertilizer attach. add	93n
grass, legume seed attach., add	230/m [70/ft]
Airflow seeder	3500 + 285n
Row planters,	
0.75 m [30 in] row spacing	
unit planters	680n
mounted	-2600 + 2300n
pull	3700 + 1950n
fertilizer attachs	
dry, add	645n
liquid, add	590n
insecticide, herbicide attachs	230n
Stack mover	2300
Stack wagon	27,000
Tractors, per kW [HP]	
Cab, diesel, mechanical shift transmission, PTO, lights, hydraulics, hitch rockshaft	
single axle drive	7000 + 390/kW [500/HP]
MFWD, add	5800 + 34/kW [45/HP]
tracks, add	19,000
two-axle drive	30,900 + 285/kW [390/HP]
Wagons	
Running gear (per t [T] load capacity)	240/t [220/T]
Flat bed	2300
Box bodies, per m³ [ft³] capacity	
forage unloading	2300 + 380/m³ [10.7 ft³]
gravity unloading	320/m³ [9 ft³]
Windrowers	
Self-propelled, base, diesel	600 + 1780/m [525/ft]
heads	
auger platform	6400 +820/m [250/ft]
belt platform	110 + 1320/m [400/ft]
conditioner, add	2300
Pull, PTO, belt platform	2700 + 1200/m [360/ft]

Appendix E

Changes in Machinery Use Data

Corn Belt Farms: 45% corn, 45% soybeans, 10% small grain acreages

	W & NW Ohio 1984-1990[a]	Central Illinois 1958-1974[b]
No. of farms in survey	375	45
Ave. farm size, ha [a]	354 [875]	202 [500]
(std. dev.)	(249)(615)	...
Tractors		
No. in survey	1892	151
Ave. age, hr of use	3165	2105
(std. dev.)	(2388)	...
Ave. annual use	290	388
(std. dev.)	(155)	(151)
S. P. Combines		
No. in survey	636	94
Ave. age, hr of use	1478	510
(std. dev.)	(824)	...
Ave. annual use, hr	246	121
(std. dev.)	(112)	...
Machinery repair and maintenance, $/ha [a]	29.33 [11.87]	4.50 [1.82]
(std. dev.)	(28.02)(11.34)	...

[a] J. A. Gliem, T. G. Carpenter, R. G. Holmes and G. S. Miller, 1991 "Seven Years of Agricultural Machinery Variable Cost Data," ASAE Paper No. 91-1091.

[b] D. R. Hunt, 1975 "Eight Years of Farm Machinery Cost Monitoring," ASAE Paper No. 75-1544.

Appendix F

Principal Machines on Farms

1990 estimates by I&T from U.S. Census data

Wheel tractors	4,622,155
Trucks	3,428,200
Moldboard plows	2,527,835
Mowers	1,335,605
Row planters	1,128,840
Side-delivery rakes	970,265
Manure spreaders	919,120
Balers	824,285
Grain drills	788,805
Self-propelled combines	671,205
Mower-conditioners (cutterbar)	657,465
Forage harvesters (shear bar)	290,355

Appendix G

Geographic Distributions

1990 estimates by I&T from U.S. Census data

Tractors		*Self-propelled Combines*	
Iowa	301,755	Illinois	61,010
Texas	287,105	Iowa	60,510
Minnesota	265,775	Minnesota	51,490
Illinois	262,845	Kansas	44,120
Wisconsin	250,105	Indiana	36,450
Ohio	197,320	Ohio	36,150
Missouri	195,955	Nebraska	35,125
Indiana	171,700	Missouri	34,015
Nebraska	171,195	N. Dakota	30,700
Kentucky	169,565	Wisconsin	25,390
California	158,175		
Kansas	151,320		

Appendix H

Estimated Number of Days Suitable for Field Work in Iowa
G. E. Ayres, C. V. Fulton (Iowa State Univ.)

Climatic Week	Dates	Av.	S.D.	Probability			
				0.24	0.50	0.76	0.88
5-8	Mar. 29-Apr. 25	13.2	4.8	17.0	14.6	10.3	8.0
9-11	Apr. 26-May 16	13.9	2.7	16.3	14.2	13.1	10.6
12-14	May 17-June 6	14.0	2.7	16.7	14.0	11.5	11.4
15-18	June 7-July 4	20.1	3.3	22.4	21.4	18.6	17.9
19-22	July 5-Aug. 1	22.0	2.9	24.8	23.2	20.5	17.9
23-26	Aug. 2-29	23.3	1.6	24.6	23.3	22.4	22.0
27-31	Aug. 30-Oct. 3	25.7	5.7	30.0	26.8	24.3	21.5
32-37	Oct. 4-Nov. 14	31.6	5.5	36.1	32.4	26.8	25.4
5	Mar. 29-Apr. 4	2.1	1.6	3.6	1.8	1.3	0.6
6	Apr. 5-11	3.4	2.0	4.9	4.1	2.4	1.0
7	Apr. 12-18	3.8	1.8	5.2	4.3	2.4	1.4
8	Apr. 19-25	4.2	1.0	5.1	4.0	3.7	3.3
9	Apr. 26-May 2	4.6	1.3	5.7	5.0	3.8	2.9
10	May 3-9	4.6	1.4	5.8	4.8	4.0	3.5
11	May 10-16	4.7	1.6	6.0	5.0	3.8	2.9
12	May 17-23	4.8	1.6	6.3	5.4	3.5	2.6
13	May 24-30	4.5	1.5	6.0	4.3	3.8	2.5
14	May 31-June 6	4.7	1.4	6.1	4.9	4.2	3.3
15	June 7-13	4.7	1.2	5.9	4.9	3.9	3.4
16	June 14-20	5.0	1.5	6.5	5.2	4.3	3.5
17	June 21-27	5.2	1.4	6.4	6.0	4.3	3.6
18	June 28-July 4	5.1	1.2	6.3	5.2	4.6	4.2
19	July 5-11	5.5	1.2	6.4	5.8	5.2	4.6
20	July 12-18	5.3	0.9	6.0	5.6	4.6	4.1
21	July 19-25	5.6	1.2	6.5	6.2	5.1	4.9
22	July 26-Aug. 1	5.6	0.8	6.0	5.8	5.2	4.4
23	Aug. 2-8	5.4	1.0	6.3	5.8	4.6	4.3
24	Aug. 9-15	5.8	0.8	6.5	6.1	5.7	5.3
25	Aug. 16-22	6.0	0.6	6.4	6.2	5.9	5.1
26	Aug. 23-29	6.0	0.5	6.7	6.0	5.6	5.5
27	Aug. 30-Sept. 5	5.7	0.7	6.6	5.7	5.3	5.0
28	Sept. 6-12	5.4	1.5	6.4	5.8	5.1	4.5
29	Sept. 13-19	4.8	1.9	6.4	5.6	3.4	2.4
30	Sept. 20-26	4.9	1.7	6.3	5.3	4.4	2.3
31	Sept. 27-Oct. 3	4.9	1.7	6.3	5.7	4.1	2.1
32	Oct. 4-10	5.0	1.8	6.6	5.6	4.3	2.6
33	Oct. 11-17	5.3	1.3	6.3	5.6	4.4	4.1
34	Oct. 18-24	5.7	1.0	6.6	5.9	4.7	4.4
35	Oct. 25-31	5.6	1.3	6.8	6.2	4.7	3.8
36	Nov. 1-7	4.9	1.8	6.8	5.5	3.6	3.1
37	Nov. 8-14	5.1	1.5	6.3	5.7	3.9	3.3
38	Nov. 15-21	5.1	1.4	6.4	5.9	4.1	3.8
39	Nov. 22-28	5.7	1.2	6.8	6.2	5.2	4.8
40	Nov. 29-Dec. 5	5.8	0.6	6.1	5.9	5.6	5.3

Expected Number of Days* Suitable for Field Operations Other than Tillage at Ames, Iowa, at Selected Probability Levels

Climatic Week	Dates	Probability Level						
		0.98	0.95	0.90	0.85	0.80	0.75	0.50
27	Aug. 30-Sept. 5	2	3	4	4	4	5	6
28	Sept. 6-Sept. 12	2	3	4	4	5	5	6
29	Sept. 13-Sept. 19	2	3	4	4	5	5	6
30	Sept. 20-Sept. 26	2	3	4	4	4	5	6
31	Sept. 27-Oct. 3	2	3	4	4	5	5	6
32	Oct. 4-Oct. 10	2	3	4	5	5	5	6
33	Oct. 11-Oct. 17	3	3	4	5	5	5	7
34	Oct. 18-Oct. 24	4	5	5	6	6	6	7
35	Oct. 25-Oct. 31	2	3	4	5	5	5	6
36	Nov. 1-Nov. 7	2	3	4	5	5	5	7
37	Nov. 8-Nov. 14	2	3	4	5	5	5	7
38	Nov. 15-Nov. 21	0	0	1	2	3	3	5
39	Nov. 22-Nov. 28	0	1	2	3	4	4	6
40	Nov. 29-Dec. 5	0	1	3	4	5	5	7
41	Dec. 6-Dec. 12	0	1	2	3	3	4	5
42	Dec. 13-Dec. 19	0	0	0	0	1	2	4
43	Dec. 20-Dec. 26	0	0	0	0	0	1	4

*Sundays included in expected number of days.

Appendix I

Number of Seeds per Mass Unit

Crop	Thousands/kg	Thousands/lb
Corn		
small kernels	4	1.8
medium kernels	3.3	1.5
large kernels	2.9	1.3
Soybeans	5.3-7.5	2.4-3.4
Cotton seed, acid delinted	6.4-9.2	2.9-4.2
Barley	19-27.5	8.5-12.5
Wheat	22-29	10-13
Oats	24-33	11-15
Sorghums	33-40	15-18
Rice	35-40	16-18
Sudan grass	79-92	36-42
Flax	132-154	60-70
Red and sweet clovers, alfalfa	484-550	220-250
Alsike clover	1450-1500	660-680
Ladino clover	1760-1848	800-840
Timothy	2200	1000
Bluegrass	4400	2000

Appendix J

SI (Metric) Units for Agriculture

Because the use of SI (Système International d'Unitès) units is approved for commercial transactions in the United States, a short description of the system as it pertains to agriculture seems appropriate. The following description, rules for usage, and conversions have been extracted from ASAE (American Society of Agricultural Engineers) Standards, ASAE EP285.7.

Units of Measure

The SI consists of seven base units, two supplementary units, and a series of derived units consistent with the base and supplementary units. There is also a series of approved prefixes for the formation of multiples and submultiples of the various units. A number of derived units are listed in Table J.1, including those with special names. Additional derived units without special names are formed as needed from base units, other derived units, or both.

Base units

For definitions refer to International Organization for Standardization ISO 1000, SI Units and Recommendations for the Use of Their Multiples and of Certain Other Units. The spellings "metre" and "litre" are those proposed by ISO.

metre (m) - unit of length
second (s) - unit of time
kilogram (kg) - unit of mass
mole (mol) - amount of substance
kelvin (K) - unit for thermodynamic temperature
ampere (A) - unit of electric current
candela (cd) - luminous intensity

Supplementary units

radian (rad) - plane angle
steradian (sr) - solid angle

SI unit prefixes

Multiples	Prefix	SI symbol
1,000,000,000,000	tera	T
1,000,000,000	giga	G
1,000,000	mega	M
1,000	kilo	k
100	hecto	h
10	deka	da
1	-base unit-	. . .
.1	deci	d
.01	centi	c
.001	milli	m
.000,001	micro	μ
.000,000,001	nano	n
.000,000,000,001	pico	p

Special Identities for Agriculture

The customary units used today in agriculture evolved with the need for practical units in both production and in commerce. The recommended SI units for travel speed, length, area, pressure, power, mass, and yield (m/s, m, m^2, Pa, w, kg, and kg/m^2) are likely to be disregarded in favor of units more nearly equal in size to customary units. Such identities are as follows

1.61 km/hr = 1 mi/hr (MPH)
2.54 cm = 1 in.
0.405 ha = 1 a
6.90 kPa = 1 lb/in^2 (psi)
0.746 kW = 1 HP
3.79 L = 1 gal
0.91 t = 1 T

The bushel is defined as a volume in customary units but is actually used as a unit of mass (weight) in commerce. A proposed replacement for the customary yield unit is the metric tonne (1000 kg or 1 Mg) per hectare. The identity between tonnes/hectare and bushels/acre is shown below where x is the bushels/acre and y is the pounds/bushel for that crop:

tonnes/hectares = xy/892

For example, 89 bushels/acre of 56 pounds/bushel grain

is $\dfrac{89 \times 56}{892}$ = 5.6 tonnes/hectare

A solidus (oblique stroke, /), a horizontal line, or negative powers may be used to express a derived unit formed from two others by division. For example:

m/s, m/s, or m × s^{-1}

Only one solidus should be used in a combination of units unless parentheses are used to avoid ambiguity.

Non-SI Units

The use of units that are not SI should be limited to those for temperature, time, and angle.

Temperature. The SI base unit for thermodynamic temperature is kelvin (K). Because of the wide usage of the degree Celsius, particularly in engineering and non-scientific areas, the Celsius scale (formerly called the centigrade scale) may be used when expressing temperature. The Celsius scale is related directly to the kelvin scale as follows:

one degree Celsius equals one degree kelvin (1 K) exactly

A Celsius temperature (t) is related to a kelvin temperature (T) as follows:

t = T - 273.15

The Celsius temperature is expressed in degrees Celsius (symbol °C). For U.S.A. usage the symbol C or °C may be used to express Celsius temperature.

Time. The SI unit for time is the second. This unit is preferred and should be used when technical calculations are involved. In other cases use of the minute, hour, day, etc., is permissible.

Angles. The SI unit for plane angle is the radian. The

use of arc degrees (°) and its decimal or minute (¢) and second (²) submultiples is permissible when the radian is not a convenient unit. Solid angles should be expressed in steradians.

Conversion from Customary Units to SI

Table J.2 lists the unit factors needed to convert from customary units to SI units. Suppose 80 acres are to be converted to hectares. Obtain the unit factor 2.471 05 acres/hectare from Table J.2 and divide into 80 acres.

$$\underline{32.375}\,\text{ha} = 80\,\text{a} \times \frac{1\,\text{ha}}{2.471,05\,\text{a}}$$

Suppose the 80 acres is to be converted to square metres. Obtain the unit factor $2.471,05\,\text{a}/10,000\,\text{m}^2$ and divide into 80 a.

$$\underline{323,749}\,\text{m}^2 = 80\,\text{a} \times \frac{10,000\,\text{m}^2}{2.471,05\,\text{a}}$$

Table J.1
Derived Units

Quantity	Unit	SI symbol	Formula
acceleration	metre per second squared	. . .	m/s^2
angular acceleration	radian per second squared	. . .	rad/s^2
angular velocity	radian per second	. . .	rad/s
area	square metre	. . .	m^2
density	kilogram per cubic metre	. . .	kg/m^2
electrical potential difference	volt	V	W/A
electrical resistance	ohm	Ω	V/A
energy	joule	J	N·m
force	newton	N	kg·m/s^2
frequency	hertz	Hz	cycle/s
magnetomotive force	ampere	A	. . .
power	watt	W	J/s
pressure	pascal	Pa	N/m^2
quantity of heat	joule	J	N·m
specific heat	joule per kilogram-kelvin	. . .	J/kg·K
stress	pascal	Pa	N/m^2
thermal conductivity	watt per metre-kelvin	. . .	W/m·K
velocity	metre per second	. . .	m/s
voltage	volt	V	W/A
volume	cubic metre	. . .	m^3
work	joule	J	N·m

Table J.2

Unit Factors for Conversion from Customary to SI Units

ACCELERATION metres/second^2 (m/s^2)

1 m/s^2 = 3.280,840 ft/s^2

Due to gravity = 9.806,650 m/s^2

AREA: metres^2 (m^2)

1 m^2 = 1550.00 in^2

1 m^2 = 10.763,91 ft^2

1 cm^2 = 0.0001 m^2 = 0.155,000 in^2

1 ha = 10,000 m^2 = 2.471,05 a

ENERGY (work): joules (J)

1 J = 0.238,892 cal

1 J = 0.737,561 ft-lb

1 kJ = 1000 J = 0.947,813 Btu

1 MJ = 1,000,000 J = 0.277,778 kW·hr

1 MJ 1,000,000 J = 0.372,506 HP·hr

FORCE: newtons (N)

1 N = 0.224,809 lb force

LENGTH: metres (m)

1 m = 39.370 in.

1 m = 3.280,840 ft

1 km = 1000 m = 0.621,371 mi

PRESSURE (force/area): 1 kN/m^2 or kilopascal (kPa)

1 kPa = 0.145,038 psi (lb/in.^{2})

1 kPa = 20.885 lb/ft^2

1 bar = 100 kPa = 14.504 psi

(1 bar is approximately 1 atmosphere)

TORQUE: newton·metre (N·m)

1 N·m = 0.737,562 lb·ft

1 N·m = 8.850,745 lb·in

VELOCITY

1 m/s = 3.280,840 ft/s

1 km/hr = 1000 m/hr = 0.621,373 mi/hr

MASS: kilograms (kg)

1 kg = 35.273,960 ounces

1 kg = 2.204,622 lb

1 Mg or tonne = 1000 kg = 2204.62 lb

1 Mg or tonne = 1.102,31 T

MASS/AREA kilogram/metre^2 (kg/m^2)

1 kg/m^2 = 0.204,816 lb/ft^2

1 tonne/ha = 1 kg/10 m^2 = 0.446,090 T/a

MASS/VOLUME (density): kilogram/metre^3 (kg/m^3)

1 kg/m^3 = 0.062,428 lb/ft^3

1 kg/L = 0.001 kg/m^3 = 8.345,39 lb/gal

1 Mg/m^3 = 1000 kg/m^3 = 62.428 lb/ft^3

(1 kg/L and 1 Mg/m^3 are approximately the density of water)

PLANE ANGLE: radians (rad)

1 rad = 57.2958 deg

POWER: watt (W)

1 W = 0.737,561 ft · lb/s

1 W = 3.412,14 Btu/hr

1 kW = 1000 W = 1.341,02 HP

VOLUME: metres^3 (m^3)

1 m^3 = 35.314,667 ft^3

1 m^3 = 1.307,951 yd^3

1 cm^3 = 0.000,001 m^3 = 0.061,024 in^3

1 L = 0.001 m^3 = 61.023,744 in^3

1 L = 0.001 m^3 = 0.264,172 gal

1 L = 0.001 m^3 = 1.056,688 qt

VOLUME/TIME: metres^3/second (m^3/s)

1 m^3/s = 35.314,667 ft^3/s

1 m^3/s = 264.172 gal/s

1 L/s = 0.264,172 gal/s

ANSWERS AND NEW PROBLEMS

SI solutions in bold type

1.1 $\mathbf{2.08\ ha/hr = \dfrac{3\ m}{} \times \dfrac{100\ m}{52\ s} \times \dfrac{3600\ s}{1\ hr} \times \dfrac{1\ ha}{10000\ m^2}}$

     $5.1\ a/hr = \dfrac{9.8\ ft}{} \times \dfrac{328\ ft}{52\ s} \times \dfrac{3600\ s}{1\ hr} \times \dfrac{a}{43560\ ft^2}$

NEW: **Find the theoretical capacity of a 6 row corn planter traveling at 5 km/hr [3.1miles/hr]. The row spacing is 0.76 m [30 in.].**

1.2 $\mathbf{a/hr = \underline{w\ m} \times \dfrac{S\ km}{hr} \times \dfrac{1000\ m}{1\ km} \times \dfrac{1\ a}{43560\ ft^2} \times \dfrac{(3.28)^2\ ft^2}{1\ m^2} = \dfrac{w\ S}{4.047}}$

     The 4.047 is estimated by 4.000

NEW: **Show that c in Equation 1.1 is 10 for SI calculations.**

1.3   a. **Theoretical capacity** $= \mathbf{\dfrac{4.8 \times 5}{10} = 2.4\ ha/hr}$ $\dfrac{2.98 \times 16.4}{8.25} = 5.92$ a/hr

     b. **Use Eq. 1.3, D = 1 unload/2 t × 2 t/ha × 4 min/unload × 1 hr/60 min = 0.06667 hr/ha**

     Effective capacity =

$$\mathbf{\dfrac{4.8 \times 5 \times 400 \times 0.95}{10 \times 400 + .06667 \times 4.8 \times 5 \times 400 \times 0.95 + 2.777 \times 4.8 \times 15} = 1.90\ ha/hr}$$

     4.69 a/hr

     c. **Field efficiency = 1.90/2.4 = 79%**

     d. **% time loss = (100 − 79) = 21%,**
        or 26.3 min. harvest, 4 min. unload, 1.315 min. turns = 31.615 min./ha, (31.615 - 25)/31.615 = 0.21

     e. **Material efficiency = 2/2.1 = 0.952**   29.69/31.17 = 0.952

NEW: **Find the effective field capacity and the percentage time loss for the above if the combine unloads without stopping.**

1.4 The bold type shows the variable changed in the table below:

S	w	L	E_w	D	t	Tank size	C
4.8	5	400	0.95	0.06667	15	2	1.90
5.76	5	400	0.95	0.06667	15	2	2.20
4.8	**6**	400	0.95	0.06667	15	2	2.22
4.8	5	**480**	0.95	0.06667	15	2	1.91
4.8	5	400	**1.00**	0.06667	15	2	1.98
4.8	5	400	0.95	**0.05333**	15	2	1.95
4.8	5	400	0.95	0.06667	**12**	2	1.91
4.8	5	400	0.95	**0.05556**	15	**2.4**	1.94

The change in all variables increased capacity, but the increased width produced the greatest effect.

NEW: Explain why the D variable was changed in the above when the tank capacity was increased.

1.5 Dimension = w/2.

Some of the crop will be backed over.

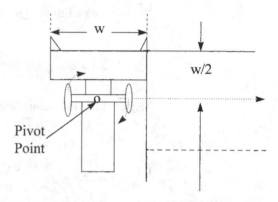

Pivot Point

NEW: Assume the drive wheels above are not reversible and have a spacing of 0.6 w. Locate the instantaneous pivot position of the right wheel to properly align the windrower after the turn. Will more or less of the crop be backed over? With the reversible design, what mounting type must the rear wheels have?

1.6 a. $[(400 - 12) \times 1.5]^{1/2} = 24.12$ m, — about 48 rows and 16 implement trips
$[(1312 - 39.36) \times 4.92]^{1/2} = 79.13$ ft, — about 48 rows and 16 implement trips

b. $n = 0.5 + [300^2/(4 \times 388 \times 1.5)]^{1/2} = 6.717$, say 7 breakthroughs (alternate lands)
$n = 0.5 + [984^2/(4 \times 1272.64 \times 4.92)]^{1/2} = 6.717$, say 7 breakthroughs (alternate lands)

$$\frac{4 \text{ tonnes}}{388 \text{ m} \times 1.5 \text{ m}} \times \frac{10000 \text{ m}^2}{\text{ha}} = 68.73 \text{ t/ha} \qquad \frac{4.4 \text{ Tons}}{1272.6 \times 4.92} \times \frac{43560 \text{ ft}^2}{\text{a}} = 30.61$$

c. metric solution = 68.73 t/ha

NEW: What are the answers if the field length is doubled?

1.7 Theoretical capacity: 450 kg/bale × 1 hr/9000 kg × 60 min/hr = 3 min/bale
Actual capacity = 3 + 1 = 4 min/bale; Field efficiency = ¾ or 75%

$$\frac{10 \text{ km}}{\text{hr}} = \frac{20 \text{ bales}}{40 \text{ min}} \times \frac{60 \text{ min}}{1 \text{ hr}} \times \frac{450 \text{ kg}}{\text{bale}} \times \frac{1 \text{ ha}}{2250 \text{ kg}} \times \frac{10,000 \text{ m}^2}{1 \text{ ha}} \times \frac{1}{6 \text{ m}} \times \frac{1 \text{ km}}{1000 \text{ m}}$$

Theoretical capacity: 1000 lb/bale × 1 hr/ 20000lb × 60 min/hr = 3 min/bale
Actual capacity = 3 +1 = 4 min/bale; Field efficiency = ¾ or 75%

$$\frac{6.19 \text{ miles}}{\text{hr}} = \frac{20 \text{ bales}}{40 \text{ min}} \times \frac{60 \text{ min}}{1 \text{ hr}} \times \frac{1000 \text{ lb}}{\text{bale}} \times \frac{1 \text{ a}}{2000 \text{ lb}} \times \frac{43,250 \text{ ft}^2}{1 \text{ a}} \times \frac{1}{20 \text{ ft}} \times \frac{1 \text{ mile}}{5280 \text{ ft}}$$

NEW: What speed would be needed to achieve the throughput if the baler did not have to stop to wrap?

1.8   **54.4 kg/round = 340kg/ha × 1 ha/10000 m² × 1 m × 800 m/trip × 2 trips/round**
119.67 lb/round = 303 lb/a × 1 ac/43560 ft² × 3.28 ft × 2624 ft/trip × 2 trips/round

   a. **Good solution. Can add 81.6 kg [180 lb] per fill: Two supply facilities would be expensive**
   b. **Acceptable. Required capacity of 108.8 kg [240 lb] per hopper is 20% above design. Might exceed frame impact strength. Variation in weight might affect depth of opener operation.**
   c. **Efficient fills. Add 88.4 kg [195 lb] (13 rounds) per fill. Extra headlands and supply facilities costly.**
   d. **Good solution, but might not be able to match fertilizer analysis to field**
   e. **Can achieve empty hoppers with 1¾ rounds but need to overfill by 5.2 kg. [10.4 lb]. Most practical for this field.**

NEW: Solve the above for a 150% rate.

1.9   a.

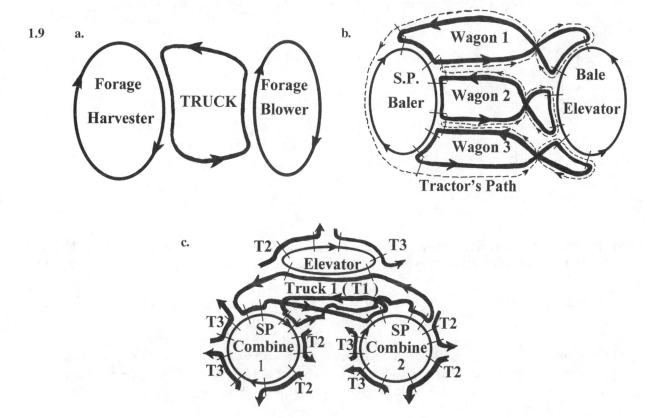

NEW: Sketch the shape of the cycle diagram for a truck supplying fertilizer and seed to three separate seeding operations. The seeder's holding capacity is 60 units of seed and fertilizer which are exhausted in 50 minutes. Five minutes are required for the truck to supply each seeder with 60 units. It takes 2 minutes for the truck to travel to the next empty seeder. The truck carries more than 180 units when filled. It takes 20 minutes for the round trip for the truck to the storage bin where it takes 8 minutes to be filled.

   a. **What is the system cycle time in minutes?**
   b. **Where is there any idle time?**
   c. **What is the system capacity in units per hour?**

1.10 There is no waiting time for the transports to be unloaded.
Let X = the required unloading time. Transport cycle time = unloader cycle time.
$21 + X = 4 \times X + 4$, $X = 5.67$ min. New system time, $T_s = 26.67$ min.

NEW: Resolve the above for just one harvester and two trucks.

1.11 Total system becomes 2 sub-systems of 1 U, 1 H, and 2 wagons with one tractor.

$T_s = 42$ min. (shuttle tractor limits) or 2.857 loads
per hour for each sub-system. Total system capacity
is 5.714 loads per hr, 1.14 loads/man-hr

Using the original loads of Fig. 1.20, as a base
capacity (6.66), the capacity of the larger capacity
wagon system is 1.5×5.714 or 8.571 loads
per hour, — or a 28.58% increase in performance.

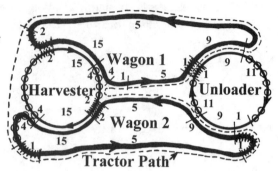

2.1 a. $(300 \times 9.807/1000)$ kN $\times$ 0.6 m/3 s $\times$ 1 kW·s/1 kN·m = 0.59 kW
660 lb/3 s $\times$ 23.6/12 ft $\times$ 1 HP·s/550 ft·lb = 0.79 HP

b. DBP = F S/c (F = drawbar pull) = $9 \times 8/3.6 = 20$ kW
$= 2025 \times 5/375 = 27.0$ HP

c. PTOP = $2 \times 3.14159 \times$ F R N/c = $2 \times 3.14159 \times 900/1000 \times 0.3 \times 1000/60 = 28.3$ kW
$= 2 \times 3.14159 \times 202 \times 1000/33000 = 38.5$ HP

d. HyP = p Q/c = $14000 \times 1.25/1000 = 17.5$ kW
$= 2030 \times 19.8/1714 = 23.45$ HP

e. EP = I E = $6 \times 12 = 72$ W
$= 0.097$ HP

NEW: What is the power requirement of a forage harvester having a rolling resistance of 3.5 kN [800 lb], and a PTO torque requirement of 270 N·m [199 lb·ft] at 1000 rpm? The forward speed is 7 km/hr [4.35 MPH].

2.2 (500 kg $\times$ 9.807/1000) kN/min $\times$ 1 min/60 s $\times$ 20 m/0.08 $\times$ 1 kW·s/1kN·m = 20.43 kW
1100 lb/min $\times$ 65.6 ft $\times$ 1/0.08 $\times$ 1 HP·min/33000 ft·lb = 27.33 HP

NEW: A 200 kg [440 lb] large hay bale is lifted from the ground surface by a fork-lift attachment on a tractor. The bale is placed on a partially loaded truck at a height of 3 m [9.84 ft]. If the power efficiency of the hydraulic power lift is 75%, calculate the power required from the hydraulic system? The lift is accomplished in 3 seconds.

2.3 Max. TE = 0.635 at 12% slip for tilled soil from Fig.2.11

a. Pull/SRAF = 0.375, $10000 \times 9.807 \times 0.75/1000 = 73.55$ kN SRAF
0.375×73.55 kN = 27.58 kN pull
Axle power = $0.96 \times 75 = 72$ kW, SRAF/AXLE POWER = 73.55/72 = 1021 N/kW
Chart speed is 6 km/hr, Calculated speed = $72 \times 0.635 \times 3.6/27.58 = 6$ km/hr
$22000 \times 0.75 = 16500$ lb SRAF
$0.375 \times 16500 = 6188$ lb pull
Axle power = $0.96 \times 100 = 96$ HP, SRAF/AXLE POWER = 16500/96 = 172 lb/HP
Chart speed = 3.7 MPH, Calculated speed = $96 \times 0.635 \times 375/6188 = 3.69$ MPH

b. Pull/SRAF = 0.445, 10000 × 9.807 × 0.65/1000 = 63.75 kN SRAF

0.445 × 63.75 kN = 28.37 kN pull,

Axle power = 0.96 × 75=72 kW, SRAF/AXLE POWER = 63.75/72 = 885 N/kW

Chart speed is 6.0 km/hr, Calculated speed = 72 × 0.635 × 3.6/28.37 = 5.80 km/hr

22000 × 0.65 = 14300 lb SRAF

0.445 × 14300 = 6364 lb. Pull

Axle power = 0.96 × 100 = 96, SRAF/AXLE POWER = 14300/96 = 149

Chart speed = 3.8 MPH, Calculated speed = 96 × 0.635 × 375/6364 = 3.6 MPH

NEW: Repeat above for a semi-mounted implement

2.4   Max TE = 0.77, slip = 9.5%, Pull/SRAF = 0.465, 70% of mass on rear

2.4 m × 7 kN/m = 16.8 kN pull, SRAF = 16.8/0.465 = 36.13 kN

36.13/0.70 = 51.6 kN or 5261 kg total mass

Drawbar power = 7 × 2.4 × 8/3.6 = 37.33 kW; then, axle power = 37.33/0.77 = 48.5 kW

PTOP = 48.5/0.96 = 50.5 kW, To accommodate load factor, 50.5/0.8 = 63.1 kW

96/12 ft × 480 lb/ft = 3840 lb pull. SRAF = 3840/0.465 = 8258 lb

8258/0.70 = 11797 lb total tractor mass

Drawbar power = 96/12 × 480 × 5/375= 51.2 HP; then, axle power = 51.2/0.77 = 66.5 HP

PTOP = 65.5/0.96 = 69.3 HP. To accommodate load factor, 68.3/0.8 = 86.6 HP

NEW: Solve above for a towed implement.

2.5   From test: 27.2 kW [36.57 HP] drawbar power, 9.12 kN [2051 lb] drawbar pull, 7.8% slip,

SRAF = 1537 kg [3388 lb], R = 420/2136 [16.5/84.1] = 2.0 (on concrete, towed load)

Drawbar Pull/SRAF = 9.12/(1537 × 9.807/1000) [2051/3388] = 0.606

Predicted slip (on concrete) = 6.3%; Actual 7.8%

Axle Power 27.3/0.922 = 29.6 kW [36.57/0.922 = 39.7 HP]

SRAF/Axle Power = 1537 × 9.807/29.6 = 509 [3388/39.7 = 85.3]

Predicted loaded speed = 9.4 km/hr [5.9 MPH]; actual 10.76 km/hr [6.7 MPH]

NEW: For the same 6th gear data, predict the performance (speed, slip) of this tractor, front wheel drive disengaged, with a towed implement in tilled soil.

2.6   a. 75/23 = 3.26 kW·hr/litre   100/6.076 = 16.46 HP·hr/gal

b. 23 × 0.84/75 = 0.26 kg/kW·hr   7 × 6.076/100 = 0.425 lb/HP·hr

c. 26 × 23/75 = 7.97 cents/kW·hr   1.00 × 6.076/100 = 6.08 cents/HP·hr

d. (75 × 3600)/(23 × 36000) = 0.32   (100 × 2545)/(6.076 × 129170) = 0.32

2.7   Ave. readings, Table 2.5 , single row, 1 m spacing: At 38 inch spacing:

Energy	4.15 kW·hr/ha		2.25 HP·hr/a
Draft	1.45 kN/row		325 lb/row

At 75 cm spacing: At 29.5 inch spacing:

Linear row length per ha: 13333 m   Linear row length per acre: 17719 ft

Energy: 1.45 × 13333 = 19333 kN·m/ha   325 × 17719 = 5758675 ft·lb/a

   or 5.37 kW·hr   or 2.9 Hp·hr

Increase: 5.37/4.15 = 1.29 or 29%   2.9/2.25 = 1.29 or 29%

NEW: What would be the change in the above if a 6-row rather than a 1-row calculation was used?

2.8 Coef. Of RR, front tires, tilled, settled soil, $0.9(2 \times 7.50 + 16) = 28$ in. diam., is **0.17**
Coef. Of RR, rear tires, tilled, settled soil, $0.9(2 \times 14.9 + 38) = 61$ in. diam., is **0.09**

RR front: $0.17 \times 333/1000 \times 9.807 = 0.56$ kN	$0.17 \times 733 = 125$ lb
RR rear: $0.09 \times 666/1000 \times 9.807 = 0.58$ kN	$0.09 \times 1467 = 132$ lb
DBP $= 1.14 \times 5/3.6 = 1.58$ kW	$257 \times 3.125/375 = 2.14$ HP
PTOP $= 1.85 \times 2 = \underline{3.70}$ kW	$2.5 \times 2 = \underline{5.00}$ HP
Total **5.28 kW**	Total 7.14 HP

NEW: Find the difference in power required if the inflation pressure in both tires was raised to 280 kPa [40 psi].

2.9 Refer to Table 13.5
Mass on tires:
$(800 + 3000)/4 = 950$ kg $(1760 + 6600)/4 = 2090$ lb
Select 7.50-16, 4-ply tire at 280 kPa 40 psi (tire section about ½ of rim diameter)
Diameter $= 0.90 \times (16 + 2 \times 7.50) = 28$ in. From Fig. 2.12 , firm soil: Coef. of RR = **0.11**
RR: $0.11 \times (3800 \times 9.807/1000) = 4.09$ kN $0.11 \times 8360 = 919.6$ lb
Power: $4.09 \times 5/3.6 = 5.68$ kW $919.6 \times 3.125/375 = 7.66$ HP

NEW: Predict the rolling resistance if a patch of loose sand must be traversed by the wagon.

2.10 a. Capacity: $2m \times 4/10 = 0.8$ ha/hr $76/12 \times 2.5/8.25 = 1.92$ a/hr
0.8 ha/hr $\times$ 24.7 t/ha $\times$ 2.25 kW·hr/t = 44.46 kW
1.92 acres/hr $\times$ 11 T/acre $\times$ 2.75 HP·hr/T = 58.08 PTO HP
Coef. of RR for harvester tires is 0.08, for wagon tires is 0.10; Fig. 2.12, (diams. are 37.8 in. and 30.6 in.,)

RR: (harv.) $0.08 \times 2000 \times 9.807/1000 = 1.57$ kN	$0.08 \times 4400 = 352$ lb
RR (wagon) $0.10 \times 5000 \times 9.807/1000 = \underline{4.90}$ kN	$0.10 \times 11000 = \underline{1100}$ lb
Total RR **6.47 kN**	1452 lb

SRAF $= 0.75 \times 3000 \times 9.807/1000 = 22.07$ kN $0.75 \times 6600 = 4950$ lb
Pull/SRAF $= 6.47/22.07 \doteq 0.29$ $1452/4950 = 0.29$
From Fig. 2.11: slip = 6.5%, DRAWBAR POWER/AXLEPOWER = 0.76
Drawbar power: $6.47 \times 4/3.6 = 7.19$ kW $1452 \times 2.5/375 = 9.68$ HP
Axle power: $7.19/0.76 = 9.46$ kW $9.68/0.76 = 12.74$ HP
PTOP: $(9.46/.96) + 44.46 = 54.31$ k $(12.74/.96) + 58.08 = 71.35$ HP

b. From Table 13.5,
for wheel loads of 1000 kg [2200 lb] for harvester, allowable is 1150 kg [2540 lb] **Permissible!**
for wheel loads of 1250 kg [2750 lb] for the wagon, allowable is 900 kg [1980 lb] **Overloaded!**
Overloaded even with a 15% increase in allowable load for speeds under 16 km/hr [10 MPH]
Raising inflation pressures to 280 kPa [40 psi] would make the wagon tire loading permissible.

NEW: Resolve for a tilled, settled soil and a 20% increase in yield.

4.1 a. 1. $190 for every year including the fifth
2. $V_6 = \$524.29$; $V_5 = \$655.36$; D = 655.36 - 524.29 = $131.07
3. $V_4 = \$1408.996$; $V_5 = \$1230.5560$; D = $178.44

b. $(2000 \times 0.36) - (2000 \times .32) = \$80$

NEW: Determine the above answers during the third year of life.

4.2   **0.195 × 60000/400 = $29.25**

NEW: **Find the costs for a $100,000 tractor used 300 hours per year.**

4.3   **a. Annual fixed costs for mower: Depr. = (3600 - 180)/20 = $171**

    Int. on Invest. (3600 + 180)/2   0.10 = $189

    Taxes, ins., shelter 3600 × 0.025 = $ 90

    Total fixed costs per year = $450

Annual fixed costs for tractor: **Depr. = (18000-900)/20 = $855**

    Int. on Invest. (18000 + 900)/2   0.10 = $945

    Taxes, ins., shelter 18000 × 0.025 = $450

    Total fixed costs per year = $2250

Hrs/year: (10 × 100)/(8 × 2 × 0.80) = 78.125 hrs   (8.25 × 247) / (5 × 6.56 × 0.80) = 77.68 hrs

Fixed costs per hour for tractor = 2250/(422 + 78) = $4.50/hr

Variable costs: (R&M)P + L + O + F + T. [Oil sump capacity is 7 L [1.8 gal] Table 2.4

(0.000875 × 3600) + (0.000083 × 18000) + 8 + (7 × 1.00/200) + (10 x 0.30) + 4.50) = $20.18/hr

(0.000875 × 3600) + (0.000083 × 18000) + 8 + (1.8 × 3.75/200) + (2.65 × 1.14) + 4.50) = $20.20/hr

Total costs: 450 + 78 × 20.19 = $2024.84, $20.24 /ha, $8.20/a

b. Est. rates for rotary weed cutting: high use, $23/ha [$9/acre] Consider hiring custom operator.

c. Custom work

Increased annual use for machinery, rent more ground, do custom work.

Reduced initial investment – used machinery, shared ownership

Longer useful life – protective shelter, paint, maintenance; repair rather than replace

NEW: **Find the (a) and (b) answers above if the use is only 50 ha [123 a] per year.**

4.4   Let H = break-even ha [acres]; Medium use rate = $53/ha [$22/a]; (R&M) = 0.000133

Actual Capacity: 4.5 × 5 × .70/10 = 1.575 ha/hr   2.79 × 16.4 × 0.70/8.25 = 3.88 a/hr

Operations Power: 10 kW·hr/ha × 1.575 = 15.75 kW   5.5 HP·hr/a x 3.88 = 21.34 HP

RR power: 0.10 × 11500 × 9.807/1000 × 4.5/3.6 = 14.1 kW   0.10 x 25300 x 2.79/375 = 18.8 HP

Total power = 15.8 + 14.1 = 29.9 kW (36.5% load)   21.4 + 18.8 = 40.2 HP (36.5% load)

Hourly costs (Use linear interpolation for fuel efficiencies)

(0.000133 × 89000) + 8 + (0.100 x 1.00) + (29.9 kW x 1 L/1.97 kW ·hr × $0.30/L) = $24.49/hr

0.000133 × 89000) + 8 + (0.027 x 3.75) + (40.2 HP × 1 gal/10.0 HP· hr x $1.14/gal) = $24.52/hr

53 H = (0.195 × 89000) + H/1.575 × $24.49 = 17355 + 15.55 H

22 H = (0.195 × 89000) + H/3.88 × $24.52 = 17355 + 6.32 H

H = 17355/37.45 = 463 hectares 17355/15.68 = 1107 acres

NEW: **Solve above for the low-use rate in Appendix C**

4.5   Use typical approximate fixed cost percentage.

    Let H = area/yr where extra hired labor costs = costs of using bale thrower

    Ave. field capacity for 4 cuttings: w/o thrower 0.3725 hr/ha   0.1500 hr/a

        with thrower 0.4139 hr/ha   0.1667 hr/a

    Extra labor costs: $8.00/hr × 0.3725 hr/ha × H = $3.04H $8.00/hr × 0.15 hr/a x H = $1.20H

    Bale thrower costs: (0.195 × 7000) + 0.4139H × (0.0004 × 7000 + 2 × 0.30) = 1365 + 1.41H

        (0.195 × 7000) + 0.1667H × (0.0004 × 7000 + 0.53 × 1.14) = 1365 + 0.57H

$$3.04H = 1365 + 1.41H$$
$$H = 837 \text{ crop ha, } 209 \text{ \underline{annual field hectares}}$$

$$1.20H = 1365 + 0.57H$$
$$H = 2167 \text{ a, } 542 \text{ \underline{annual field acres}}$$

NEW: Find the area above if the labor rate is doubled.

4.6 Use typical approximate fixed cost percentage.
Let H = breakeven area where annual rental costs equals annual ownership costs
$$8H + 2.5H + H/4 \times (5 + 0.195 \times 40000/(300 + H/4)) =$$
$$0.195 \times 6000 + 2.5H + H/6 \times (5 + 0.195 \times 40000/(300 + H/6))$$
$$3.25H + 1H + H/10 \times (5 + 0.195 \times 40000/(300 + H/10)) =$$
$$0.195 \times 6000 + 1H + H/15 \times (5 + 0.195 \times 40000/(300 + H/15))$$

Solving for H:
$$H^3 + 2863H^2 + 2304000H = 300857000 \qquad H^3 + 7158H^2 + 14355263H = 4618421053$$
$$H = 114 \text{ ha} \qquad\qquad H = 281 \text{ a}$$

NEW: Find the answer for the above neglecting the extra hours for the tractor to operate the spreader.

4.7 Annual: Depreciation = (2500-100)/6 = $400   Interest = [(2500+100)/2] x 0.12   = $156
STI   = 0.025 × 5000 = $125   R & M = 0.0005 × 5000 = $2.50/hr × 83 hrs = $208
Total cost per year = $889   Cost/ha = $8.89   Cost per acre = $3.59

NEW: Suppose a neighboring farmer shares in the fixed cost of the plow by paying according to use. What are the costs per area if the neighbor uses the plow 50 ha/yr [124 a/yr]?

4.8 a. Inflation factor 1.337, Deflated value = $22,438
(a) 6000 (b) 5557 (c) 4650 (d) 1500 (e) 4600 (g) 300 (h) $40.50 (i) 79245 (j) 1801 (k) $44.00

b. Inflation factor 1.328, Deflated value = $1130
(a) 200 (b) 201 (c) 233 (d) 400 (e) 477 (f) 600 (g) 100 (h) $1.39 (i) 5089 (j) 3837 (k) $1.34
(l) $4050 (m) 826 (n) 600 (o) $9.13 (p) $10.52

c. $10.52
d. $1.33. Repair rather than replace as costs are still going down per unit area

NEW: Resolve above when inflation is zero.

4.9 Use Eq. 4.8 x = 2 x 0.002 × 0.008 = 0.000032
$$y = 0.005 + 0.002284 - 0.00064 - 0.00264 = 0.004004$$
$$z = 0.205 + 0.0276 = 0.2326$$
$$n^3 + 125n^2 - 7269 = 0; \quad n = 7.4 \text{ years (trial \& error computation)}$$

NEW: Resolve the above if obsolescence is not a factor.

4.10 Basis is 15000 + 80000 = $95000
Section 179 deduction: $95000 – 18500 = $76500 – the new basis
Use Table 4.4, (1.5 declining balance, 7 yr.) Successive allowable depreciations for the seven years are:
8193, 14634, 11498, 9371, 9371, 9371, 9371, 4689

5.1   **25 cm** [10 in.]

NEW: **Resolve above for a 7-36 cm** [13 in.] **moldboard plow.**

5.2   **Steer to the left with adjustments provided at the plow's attachment points to the tractor.**

NEW: **What should be the adjustment if the depth is too shallow?**

5.3   **Unit draft = 0.704 N/cm²**   [0.88 lb/in.²]   **Total draft = 5.632 kN**   [1090 lb]

NEW: **If the speed above is increased 20%, what is the percentage increase in draft?**

5.4   **Solve Equation 5.2 using C_1 and C_2 values for sandy loam from Eq. 5.1 w = 1.685 m** [5.57 ft.].

NEW: **Using the answer for 5.4 above, what would be a proper operating speed and PTOP?**

5.5   **Staggered alignment means adjacent sweep passes are centered on 10 cm** [4 in.]
10 cm [4 in.] **sweeps give complete cutting if tiller is operating straight forward.**
At a 4.5° skew, the perpendicular to forward motion distance between the right edge of a sweep in the first rank and the left edge of the adjacent sweep in the second rank is 91 x sin 4.5 or 7.139 cm [36 sin 4.5 or 2.825 in.] **The skewed distance is 7.139/cos 4.5 or 7.16 cm** [2.825/cos 4.5 or 2.83 in.], **one-half is to be added to each sweep edge. Add two extensions to each sweep produces sweeps having a total width of 17.6 cm** [6.83 in.] **For a 3rd – 1st rank spacing, distance is 182 cm, width will now have to be 24.323 cm or 9.6 in.**

6.1   **Listing—places seed in the bottom of a furrow nearer to subsoil moisture**
Bedding—places the seed in the top of a raised soil ridge in a well-drained environment.

NEW: **Predict the relative soil temperature for each of the above.**

6.2   **field cultivator, or any other sub-surface tiller:**

   a. **Leaves crop residues on the surface along with the larger soil clods**
   b. **At seedbed level where the tooling operates, a fine crumbly soil condition conducive to germination should exist.**
   c. **Good control of existing weeds occurs when the sweep paths over-lap. Weed seed germination is inhibited by the rough, cloddy, and dry soil surface.**

6.3   **When the line of pull is above the line of resistance, a force couple exists to rotate the disk harrow forward and thus reducing the downward force and penetration of the rear gang.**

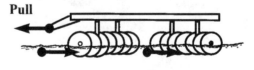

Pull

6.4   **The horizontal soil forces acting on each gang tend to rotate the disk harrow about its center of resistance. This rotation is counter balanced by an off-center drawbar pull.**

Pull

Geometric Center

8.1 3kg/30 turns × 1 turn/(3.7 m × 0.99) × 1/3.6m × 10000m²/ha = 75.8 kg/ha

6.6lb/30 turns × 1 turn/(12.14 ft × 0.99) × 1/12ft × 43560ft²/a = 66.4 lb/a

NEW: Solve the above if low air pressure and weight reduced the rolling circumference to 3.6 m [11.8 ft].

8.2 1 rev. ground drive wheel/8 seeds × 2m/1 rev. ground drive wheel × 100 cm/m = 25 cm spacing [9.84 in.].

NEW: Find the answer above for a 5% slip and an effective circumference of 1.9 m [6.23 ft].

8.3 10000m²/ha × 1/0.76m × 0.85 ha/37500 seeds × 100cm/1m = 29.82 cm/seed

43560ft²/a × 1/30in. × 12in./ft × 0.85 a/15000 seeds × 12in./ft = 11.85 in./seed

NEW: If the planter above had a 16 cell seed plate (1 seed per cell) and the squash radius is 380 mm [15 in.], how many revolution must the plate make per revolution of the drive wheel? Assume zero slip.

8.4 The seeding rate is correct only at the mid-point of the seeder's width. It is increasingly greater toward the inside of the curve and increasingly less toward the outside.

9.1

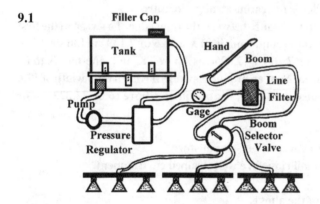

9.2 **4 kg/100 m × 1/0.76 m × 10000 m²/ha = 526 kg/ha**

8.8 lb/328 ft × 1/30 in. × 12 in./ft × 43560 ft²/a = 467 lb/a

NEW: Solve if the seeder above were PTO driven and the amount caught was during a 3 min. test.

9.3

9.4

tan 43° = 0.65 w / h

h = 0.65 w / 0.9325 = 52.9 cm [20.9 in.]

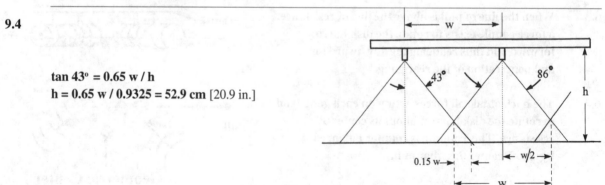

NEW: If the spraying above should be done at 200 kPa [29 psi], must the boom be raised or lowered to give the proper overlap?

9.5 20 L AI/200 L spray × 0.1 kg AI/1L AI × 10000 m²/ha × 1 hr/ 7000 m × 1/0.76 m × 2 L/min × 60 min/hr = 2.26 kg/ha

5.3 gal AI/53 gal spray × 0.83 lb AI/gal AI × 43560 ft²/a × 1 hr/4.375 miles × 1/2.5 ft × 1 mile/5280 ft × 60 min/hr × 0.53 gal spray/min = 2.0 lb AI/a

NEW: What would be the application rate if the speed were doubled and the pump output were doubled?

9.6 2 L spray/0.2 kg AI × 0.8 kg AI/ha × 1 ha/10,000 m² × 200 km/hr × 1000 m/km × 10 m × 1 hr/60 min = 26.67 L/min

2 gal/1.66 lb AI × 0.7 lb AI/acre × 1 acre/43,560 ft² × 124 miles/hr × 5280 ft/mile × 32.8 ft × 1 hr/60 min = 6.93 gal/min

NEW: What must be the pump output if the spray concentrate were to be applied without mixing with water?

9.7 1 min/2 L spray × 1L conc/50g AI × 1 kg/ha × 1000 g/kg × 1 ha/10000 m² × 1 m × 6000 m/hr × 1 hr/60 min = 0.1 ratio of conc. to spray. Mix 1 part conc. to 9 parts water

1 min/0.529gal spray × 1 gal conc/0.42lb AI × 0.89 lb AI/a × 1 a/43560ft² × 39.37in. × 1 ft/12 in. × 3.75 miles/hr × 5280 ft/mile × 1 hr/60 min = 0.1 ratio of conc. to spray. Mix 1 part conc. to 9 parts water

NEW: If the sprayer above has 16 nozzles, use the answer above and determine the flow rate in L/min [gal/min] required from a pump injecting spray concentrate into the water flowing from the main pump?

9.8 500 m × 0.25m × 1 ha/10000 m² × 1.5 kg/ha = 0.01875 kg

1640 ft × 10 in. × 1 ft/12 in. × 1 a/43560 ft² × 1.34 lb/a = 0.042 lb

NEW: What quantity of granules should be purchased per hectare [acre] covered?

10.1. Sketch the path of cut material to the cylinder =====, straw ————, clean grain x-x-x-x-x, tailings o-o-o-o-o-o, and weed seed ———————— for a combine in wheat.

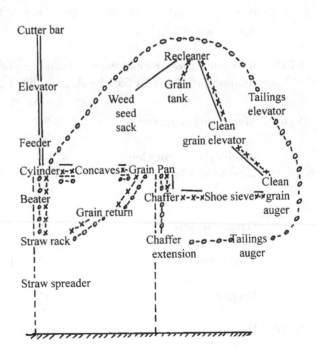

NEW: Develop a schematic flow diagram for a corn combine. Trace the flow of ears; shelled corn; cobs; and husks, silks, and other residues. Consider carefully the role of the tailings return.

10.2   **a. Open chaffer and/or shoe sieve**

        b. Reduce threshing severity; open concaves, slow cylinder speed, etc.

        c. Close chaffer and/or shoe sieves. Large quantity of straw, chaff

        d. No adjustment needed

        e. Reduce threshing severity

NEW: **Describe the conditions if the combine forward speed is excessive.**

10.3   **Annual costs of windrower = annual value of saved grain**

        [Let N = the amount of kg/ha [bu/a] lost]

        [Let A = the area in hectares [acres] harvested each year]

$$20000 \times 0.165 + 10 \times A/(7 \times 5 \times 0.85) \times (0.000183 \times 20000 + 7 + 3.5) = 0.1 \times N \times A$$

$$20000 \times 0.165 + 8.25 \times A/(4.375 \times 16.4 \times 0.85) \times (0.000183 \times 20000 + 7 + 3.50) = 3 \times N \times A$$

N = (30000 + 43.27 A)/A

$$N = (1100 + 0.638A)/A$$

NEW: **The cost of a windrow pick-up for the combine was omitted from above. Include these costs and solve the above if the pickup head's price is $9000 and has a life of 10 years.**

10.4   **Break-even slope, s, is where additional loss = additional costs**

        $0.00123s^4 \times \$0.11/kg \times 1000\ kg/ha \times 200\ ha = 30000 \times 0.195 + 200/1.5 \times (0.00005 \times 30000)$

           $27.06\ s^4 = 5850 + 200 = 6050;\ s = 3.86\%$

        $0.00123s^4 \times \$1.50/bu \times 29.7\ bu/a \times 494\ ac = 30000 \times 0.195 + 494/3.7 \times 0.00005 \times 30000)$

           $27.06\ s^4 = 5850 + 200 = 6050;\ s = 3.86\%$

10.5   **Savings = Value of cutterbar head loss—Value of row header loss 6400 seeds/kg [2900 seeds/lb]**

        Cutterbar: 200 ha $\times$ 10000 m²/ha $\times$ 80 beans/m² $\times$ 1 kg/6400 seeds $\times$ $0.22/kg =$ 5500

        Row: 200 ha $\times$ 10000 m²/ha $\times$ 26 beans/m $\times$ 1/0.6 m $\times$ 1 kg/6400 beans $\times$ $0.22/kg   =$ 2979

                                                           Savings are $2521

        Cutterbar: 494 a $\times$ 43560 ft²/a $\times$ 8 beans/ft² $\times$ 1 lb/2900 beans $\times$ 1 bu/60 lb $\times$ $6/bu = $5936

        Row: 494 a $\times$ 43560 ft²/a $\times$ 8 beans/ft $\times$ 1 lb/2900 beans $\times$ 1 bu/60 lb $\times$ $6/bu $\times$ 1/23.6 in. $\times$ 12 in./ft = $3018

        Savings are $2918

NEW: **How much greater could the row-head price be to produce break-even costs with the cutterbar head. Assume a 10 yr life for both heads and a 10 % interest rate. The operation costs are equal.**

10.6	**Missed ears at header**	**25.9%**		**Ears dropped before the combine enters the field**
				should not be charged
	Header shelled corn loss	**15.4%**		**against the machine unless**
	Total header loss	**41.3% of total loss**		**the machine capacity**
				is so low that the dropped
				ears are due to the resulting
	Separating loss	**12.1%**		**late harvest.**
	Cylinder loss	**10.5%**		
	Total threshing loss	**22.6% of total loss**		
	Dropped ear loss	**36.1% of total loss**		

NEW: **Discuss the appropriateness of including preharvest dropped ears as a factor in machine efficiency.**

10.7 The load ratings for the same size tire on a combine as on a tractor are greater:
4540 kg to 2900 kg   [10000 lb to 6400 lb]

NEW: What are the reasons why the combine tires can carry a heavier load?

11.1 **540 rev PTO/min × 3 rev mower drive/2 rev PTO × 5 cm/stroke × 2 strokes/rev mower drive × 60 min/hr × 1 m/100 cm × 1 km/1000 m = 4.86 km/hr**
540 rev PTO/min × 3 rev mower drive/2 rev PTO × 2 in./stroke × 2 strokes/rev mower drive × 60 min/hr × 1 ft/12 in. × 1 mile/5280 ft = 3.07 miles/hr

NEW: Sketch the stubble appearance if the speed above is increased substantially.

11.2 **810 rev/min × 2 cuts/rev × 2 cm/cut × 1 m/100cm × 1 km/1000 m × 60 min/hr =1.94 km/hr**
810 rev/min × 2 cuts/rev × 0.75 in./cut × 1 ft/12 in. × 1 mile/5280 ft × 60 min/hr = 1.15 miles/hr

NEW: What would be the cutter's tip speed in km/hr [MPH]?

11.3 Velocity of the hay is the resultant of the velocity vector of the rake and the velocity vector of the tooth.
Rake: 8 km/hr × 1000 m/km × 1 hr/3600 s = 2.22 m/s
5 miles/hr × 5260 ft/mile × 1 hr/3600 s = 7.33 ft/s
Tooth: 80 rev/min × 1m × 3.14159 /rev × 1 min/60 s = 4.189m/s
80 rev/min × 3.28 ft × 3.14159 /rev × 1 min/60 s = 13.7 ft/s
$(2.22^2 + 4.189^2)^{1/2} = 4.74$ **m/s**
$(7.33^2 + 13.7^2)^{1/2} = 15.54$ ft/s
Resultant angle :   A = arctan 2.22/4.189 = 27.92°
A = arctan 7.33/13.7 = 28.15°
Farthest distance: 2 m/cos 27.92° = 2.26 m
6.6 ft/cos 28.15° = 7.485 ft

NEW: What is the % reduction in hay travel if the forward speed is halved?

11.4 **Theoretical power: 30 kg × 9.807 N/kg × 4/ min × 1 W·s/N·m × 1 min/60 s × 1 kW/1000 W = 0.196 kW**
66 lb × 32.8 ft × 4/min × 1 HP·min/33000 ft· lb = 0.2625 HP
Required power: 0.196/0.15 = 1.307 kW
0.2625/0.15 = 1.75 HP

NEW: Solve the above if the rate is 6 bales/min.

11.5 Check paddle tip-to-housing clearance.

NEW: Describe how an adjustment could be made to reduce paddle tip clearance.

11.6 a. **10 t/hr × 1 ha/4.5 t × 10000 m²/ha × 1/4 m × 1 km/1000 m × 1/0.8 = 6.94 km/hr**
11T/hr × 1 a/2T × 43560 ft²/a × 1/13.12 ft × 1 mile/5280 ft × 1/0.8 = 4.32 miles/hr

b. **10 t/hr × 1000 kg/t × 1 hr/3600 s × 1 sec/1.5 strokes × 1 stroke/slug = 1.85 kg/slug**
11 T/hr × 2000 lb/T × 1 hr/3600 s × 1 s/1.5 strokes × 1 stroke/slug = 4.074 lb/slug

NEW: If 3 m [118 in.] of twine are required per bale tie, how much twine is needed per tonne [ton] of hay?

11.7 The condition occurs because the material was not placed evenly over the face of the plunger. Adjust the action of the feeder forks to produce an even distribution and/or vary the forward speed so that the volume of the material matches the performance of the feeder forks.

NEW: Describe the shape of a round bale when a narrow windrow is fed constantly into the center of the pick-up.

11.8 1 rev of cutter/6 cuts × 1 cut/0.5 cm × 62.883 cm/rev feed roll = 20.94 ratio
For cuts of 10 mm, 20 mm, and 40 mm, the ratios must be 10.47, 5.24, and 2.62
1 rev of cutter/6 cuts × 1 cut/0.2 in. × 25.13 in./rev feed roll = 20.94 ratio
For cuts of 0.4 in., 0.8 in., and 1.6 in., the ratios must be 10.47, 5.24, and 2.62

NEW: The forage harvester above has a cutting cylinder speed of 600 rpm, a throat width of 400 mm [16 in.], and is set for a 20 mm [0.8 in.] cut. What spacing between the feed roll pair would be expected if two rows of corn are being harvested at 4.8 km/hr [3 MPH]? Each stalk of corn is 2 m [6.56 ft.] long and occupies 6.45 cm² [1 in.] of throat area. The spacing of the corn stalks in the row is 20 cm. [7.87 in.].

12.1 12.75 kW·hr/t × 5.5 t/hr = 63.75 kW   15.5 HP·hr/T × 5.5 t/hr = 85.25HP

12.2 Complete a moisture calculation table:

	%	units H$_2$O	units d.m.	units total	
					a. **3.75t** 4.125T
Initial	25	**1.25** 1.375	**3.75** 4.125	**5.00** 5.500	b. **1.25/3.75 = 33.3%**
Final	20	**0.94** 1.031	**3.75** 4.125	**4.69** 5.156	1.375/4.125 = 33.3%
		0.31 0.344		**0.31** 0.344	c. **0.31 t**
					0.344 T

NEW: How much more moisture is removed when drying down to 15% WB?

12.3 Use Eq. 12.3   20 × 600/(38.7 × 0.795) = 390 kg/hr
36 × 21200/(69.4 × 12.95) = 846 lb/hr

NEW: What is the rate if the entering air above is 21°C [70°F] ?

12.4 Ignore shrinking of grain during drying

moisture %	DM	Moisture	Total		moisture %	DM	Moisture	Total
20	48	12	60		20	52.8	13.2	66
13	48	7.17	55.17		13	52.8	7.89	60.69
		4.83	4.83				5.31	5.31

a. **Bin area: 60 t × 1000 kg/t × 1 m³/770 kg × 1/1.8 m = 43.3 m²; diam. = 7.42 m**
66 T × 2000 lb/T × 1 ft³/48 lb × 1/6 ft = 458 ft²; diam. = 24.15 ft.

air flow: 43.3 m² × 4.6 m/min. = 199 m³min. 458 ft² × 15 ft/min. = 6870 ft³/min.

b. **Use Eq. 12.2, max. efficiency; Fig. 12.12 (pressure = 60% of zero flow pressure)**
199 × 42/(6000 × 0.7) = 1.99 kW   6870 × 1.66/(6350 × 0.7) = 2.56 HP

c. Assume saturated air at exit. Total moisture removal from (a)
   From psychrometric chart determine:
   0.013 – 0.012 = 0.001 kg moisture removed/kg air flow
   0.007 – 0.006 = 0.001 lb moisture removed/lb air flow
   4830 kg H$_2$O × 1 kg air/0.001 kg H$_2$O × 0.85 m^3/kg air × 1 min/199 m^3 × 1 hr/60 min = 344 hrs
   10620 lb × 1 lb air/0.001 lb × 12.95 ft^3/lb air × 1 min/6870 ft^3 air × 1 hr/ 60 min. = 334 hrs.

12.5   Use moisture calculation table and Fig. 12.17

20%	**1519** 3365		**6075** 13460		**7594** 16825	Ambient air = 0.8 m^3/kg	13 ft^3/lb
15.5%	**1115** 2469		**6075** 13460		**7190** 15929		
	404 896				**404** 896		

Air Temperature rise:
3000 MJ/hr × 1kg·°C/kJ × 1000 kJ/MJ × 0.8m^3/kg × 1 min/1000m^3 × 1 hr/60 min = 40 °C
2844141 Btu/hr × 1 lb·°F/0.24 Btu × 1 min/35315 ft^3 × 1 hr/60 min = 72.7 °F

Operating Temperature: **10 + 40 = 50 °C**
                          50 + 72.7 = 122.7 °F

Evaporative efficiency (from Fig. 12.17):   **3.95 MJ to remove 1 kg H$_2$O**
                                               1700 Btu to remove 1 lb H$_2$O

Heat requirements to raise temperature of grain to 15.5 °C [60 °F] to allow use of Fig. 12.17.

7594 kg × 1.7 kJ/kg·°C × 5.5 °C = 71004 kJ   16825 lb × 0.4 Btu/lb·°F·10 °F = 67300 Btu

Heat required for drying: **404 kg × 3.95 MJ/kg = 1596 MJ**
                              1700 Btu/lb × 896 lb = 1523200 Btu

Total heat required: **71 + 1596 = 1667 MJ**
                         67300 + 1523200 = 1590500 Btu

Total time required : **1667 MJ × 1 hr/3000 MJ = 0.556 hrs + 0.166 hrs cooling = 0.72 hrs;**
                          1590500 Btu × 1 hr/2844141 Btu = 0.559 hrs + 0.166 = 0.72 hrs

Capacity (wet basis): **7594 kg/0.72 hrs × 1 tonne/1000 kg = 10.54 t/hr**
                          16825 lb/0.72 hrs × 1 T/2000 lb = 11.68 T/hr

NEW: Solve for the drying front area in the problem above. Compute the air velocity through the grain.
         Compare the pressure drop for dry shelled corn with that from Fig. 12.9 by extending the length of the
         shelled corn curve.

13.1   Use Table 13.1:
          80 cm^2 × 2 m/s × 1 m^2/10000 cm^2 × 770 kg/m^3 × 1 t/1000 kg × 3600 s/1 hr = 44.35 t/hr
          12.4 in.^{2} × 6.6 ft/s × 1 ft^2/144 in.^{2} × 48 lb/ft^3 × 1 T/2000 lb × 3600 s/hr = 49.1 T/hr

NEW: Compute the above for oats.

13.2 Use Tables 13.1, 13.3 Use average fills for 500 and 300 rpm

Fill area in auger: $3.14159/4 \times (10^2 - 2.5^{2)} \times 0.765 = 56.3$ cm^2

$\qquad\qquad\qquad 3.14159/4 \times (4^2 - 1^2) \times 0.765 = 9.01$ in.^{2}

Capacity: 400 rev/min × 10 cm/rev × 56.3 cm^2 × 719 kg/m^3 × 1 m^3/1000000 cm^3 × 60 min/hr = 9715 kg/hr

$\qquad$ 400 rev/min × 4in./rev × 9.01in.^{2} × 44.8 lb/ft^3 × 1ft^3/1728in.^{3} × 1 bu/56 lb × 60 min/hr = 400 bu/hr

NEW: **What is the estimated power required for the above?**

13.3 a. Find type of fill; Use Eq. 13.4, Critical angle is:
$\qquad$ arc tan 100/(250-3) = 22.0° $\qquad$ arc tan 4/(10 - 0.12) = 22°
$\qquad$ - - - fill is triangular, use Eq. 13.1

$\qquad$ b. Capacity: $\qquad$ 25 × 10^2 × 1/(333.3 × 25 × tan 57.3°) = 0.193 m^3/min or 7.12 t/hr
$\qquad\qquad\qquad\qquad\qquad$ 10 × 4^2 × 3.28/ (4.8 × 10 × tan 57.3°) = 7.02 ft^3/min or 8.09 T/hr

$\qquad$ c. Power to lift grain:
$\qquad\qquad$ 7120 kg/hr × 4 m × 9.807 N/kg × 1 kW·sec/kN·m × 1 kN/1000 N × 1 hr/3600 secs = 0.078 kW
$\qquad\qquad$ 8.09 T/hr × 2000 lb/T × 1 HP·sec/550 ft·lb × 13.12 ft × 1 hr/3600 secs = 0.107 HP

Friction power:
grain: volume per cell: ½ × 10cm × 25cm × 10 cm/tan 57.3° = 802.5 cm^3
$\qquad\qquad\qquad\qquad\qquad$ ½ × 4 in. × 10 in. × 4 in./tan 57.3° = 51.4 in.^{3}
no. of cells: $\qquad\qquad$ 4 m/sin 57.3° × 1 cell/.25 m = 19
$\qquad\qquad\qquad\qquad$ 13.12 ft /in. 57.3° × 12 in./ft × 1 cell/10 in. = 18.7 or 19 cells
Total grain weight: $\qquad$ 19 cells × 802.5 cm^3/cell × 616 kg/m^3 × 1 m^3/1000000 cm^3 = 9.39 kg
$\qquad\qquad\qquad\qquad$ 19 cells × 51.4 in.^{3}/cell × 38.4 lb/ft^3 × 1 ft^3/1728 in.^{3} = 21.7 lb.
Force perpendicular to conveyor bed: 9.39 kg × 9.807 N/kg × cos 57.3° = 49.75 N
$\qquad\qquad\qquad\qquad\qquad\qquad$ 21.7 lb × cos 57.3° = 11.7 lb

49.75 N × 0.35 × 1 m/sec × 1 kW·sec/kN·m × 1 kN/1000 m = 0.0174 kW
11.7 lb × 0.35 x 3.28 ft/sec × 1 HP·sec/550 ft·lb = 0.0244 HP

$\qquad$ steel: Total chain weight: 2 kg/m × 9.807 N/kg × 4/sin 57.3° × 2 = 186.5 N
$\qquad\qquad\qquad\qquad\qquad\qquad$ 1.34 lb/ft × 13.12 ft/in. 57.3° × 2 = 41.78 lb

186.5 N × cos 57.3° × 0.57 × 1 m/sec × 1 kW·sec/kN·m × 1 kN/1000 N = 0.057 kW
41.78 lb × cos 57.3° × 0.57 × 3.28 ft/sec × 1HP·sec/550 ft·lb = 0.077 HP

Total Power: 0.078 + 0.0174 + 0.057 = 0.153 kW $\qquad$ 0.107 + 0.0244 + 0.077 = 0.208 HP

NEW: **What is the increase in capacity if the elevator is operated at the critical angle?**

13.4 a. Max. load carried at 360 kPa [52 psi] tire pressure:
$\qquad$ 6 tires x 1500 kg/tire = 9000 kg, 9 tonne $\qquad$ 6 tires x 3300 lb/tire = 19800 lb, 9.9 Tons
$\qquad$ b. 1.15 x 9 = 10.35 tonne $\qquad$ 1.15 × 9.9 = 11.39 tons

NEW: **What would be the answers to the above for a soft tire pressure of 170 kPa [24 psi]?**

13.5   M = (4000 × 1.5 + 500 × 2.5 − 0)/1 = 7250 kg (71.1 kN of force)
         M = (8800 × 4.92 + 1100 × 8.2 − 0)/3.28 = 15,950 lb.

NEW: What load is placed on each front tire in the above problem? If the load carried is only 25% of the break-out force which tires in Table 2.1 can carry the load at 8 km/hr [5 MPH]?

15.1   a. area of piston head: 3.14159/4 × (10cm)^2 = 78.53 cm^2 3.14159/4 × (4in.)^2 = 12.57 in.^{2}
         displacement: 10cm × 78.53cm^2 × 4 cylinders = 3141 cm^3 or 3.14 litre
             4 in. × 12.57 in.^{2} × 4 cylinders = 201 in.^{3}

       b. 690 kPa × 78.53 cm^2 × 1 N/(Pa·m^2) × 1 m^2/(100cm)^2 × 5 cm × 1 m/100 cm × 1000 Pa/kPa = 271 N·m
         100 lb/in.^{2} × 12.57 in.^{2} × 2 in. × 1 ft/12 in. = 209.5 lb · ft

       c. 550 kPa × 78.53cm^2 × 0.1m/stroke × 1kN/(kPa·m^2) × 1 kW·sec/kN·m × 1 m^2/(100cm)^2 × 1600 rev/
         60s × 2 power strokes/rev. = 23.03 kW
         80 lb/in.^{2} × 12.57in.^{2} × 4 in./stroke × 1 ft/12 in. × 1 HP·sec/550 ft·lb x 1600 rev/60 s × 2 strokes/
         rev = 32.47 HP

       d. Use Equation 15.1: (78.53 × 10 + 130)/130 = 7.04   (12.57 x 4 + 8)/8 = 7.29

NEW: Find the displacement, compression ratio, and theoretical air flow volume for a 6 cylinder engine with 105 mm [4.13 in.] 127 mm [5.0 in.] bore and stroke and a clearance volume of 73 cm^3 [4.5 in.^{3}] running at 2000 rpm?

15.2   Number 1 or 6 , — not both!

NEW: Explain why the valve clearance for cylinder 1 and 6 above should not be set at the same time.

15.3   3.84 rad × 57.3°/rad = 220°
         220°/opening × 1 rev/360° × 1 min/2400 rev × 60 sec/rev = 0.0152 s

NEW: Estimate the time that the valve above is at maximum lift.

15.4   0.12 m/stroke × 2 strokes/rev × 2400 rev/min × 1 min/60 s = 9.6 m/s
         5 in./stroke × 2 strokes/rev × 2400 rev/min × 1 ft/12 in. = 2000 ft/min

NEW: What is the instantaneous maximum velocity in the above?

15.6   Use combustion relation. Use atmospheric air— 0.8455 m^3/kg [13.6 ft^3/lb].
         0.74 kg fuel/litre × 800 kg O$_2$/228 kg fuel × 0.8455 m^3/kg air × 1 kg air/0.232 kg O$_2$ = 9.46 m^3 air
         6.13 lb fuel/gal × 800 lb O$_2$/228 lb fuel × 13.6 ft^3/lb air × 1 lb air/0.232 lb O$_2$ = 1260 ft^3 air

NEW: How much water would be produced in the above?

15.7   Flooded carburetor (engine may not start), excessive fuel consumption, exhaust smoke, and perhaps dripping fuel.

NEW: What corrective action should be taken for the above?

15.8 $(100)^2 \times 3.14159/4 \times 127]$ mm^3 air/stroke $\times$ 1 kg air/0.8455 m^3 air $\times$ 1 m^3 air/(100 cm)^3 air $\times$ 1 kg fuel/
(1.2×15) kg air $\times$ 1000 cm^3 fuel/0.85 kg fuel = 77.1 mm^3/stroke

$(4)^2 \times 3.14159/4 \times 5]$ in.^{3} air/stroke $\times$ 1 ft^3/1728 in.^{3} $\times$ 1 lb air/13.6 ft^3 air $\times$ 1 lb fuel/(1.2 $\times$ 15) lb air $\times$ 1 gal
fuel/7.08 lb fuel $\times$ 231 in.^{3} fuel/ gal fuel = 0.0049 in.^{3}

NEW: If the 6 cylinder engine above is turning at 2000 rpm at full load and has a 30% thermal efficiency, what power might be expected?

15.9 Low pressure injection reduces atomization, some particles don't burn, increased smoke, excess fuel consumption.

NEW: What is the effect of worn injection nozzles?

15.10 100 to 1

NEW: Explain how the primary voltage above can reach an instantaneous 250 v.

15.11 0.5 rad $\times$ 1 rev cam shaft/(2 $\times$ 3.14159) rad $\times$ 2 rev crank/1 rev cam $\times$ 1 sec/30 rev crank = 0.0053 secs.

30° $\times$ 1 rev camshaft/360° $\times$ 2 rev crankshaft/1 rev cam shaft $\times$ 1 sec/30 rev crank = 0.0056 secs

NEW: Calculate the time of fuel injection for the engine above if injection occurs over 10° of crankshaft travel.

15.12 The fluid in a low-charge battery is almost water. High charged battery fluid has a greater concentration of sulfuric acid which has a lower freezing point.

NEW: Does water or sulfuric acid have the greater specific gravity?

15.13 It senses the voltage of the battery and then varies the strength of the magnetic field in the generator or alternator. If the battery voltage is high, the generator or alternator field is weakened and the output reduced.

NEW: How does the voltage regulator above weaken the field of the generator or alternator?

15.14 The L terminal is connected to the BAT terminal of the regulator at all times. If the engine is not running, the cutout relay is open and the current from the battery passes through the ammeter in a direction opposite to that when charging.

NEW: What is the essential difference between the solid-state circuits and the earlier vibrating point circuits for charge regulation?

15.15 Use Eq. 2.3: 207/1000 $\times$ 7.57/60 = 0.026 kW   30 /1714 $\times$ 2 = 0.035 HP

NEW: Examine Fig. 15.45 and determine how the oil pump receives power.

15.16 CF for off-road engines. Continuous load does not require the extra protection provided by the other oils. Expected to be a lower-price oil.

NEW: Describe a severe duty operating characteristic for a diesel engine.

15.17 % speed droop = (no load rpm - loaded rpm)/ no load rpm   0.05 = (X·1800)/X; X = 1895 rev/min

NEW: What would an operator sense if an engine had a hunting governor?

15.18 Cooling required: 15 L fuel/hr × 45.46 MJ/kg × 0.85 kg/L fuel × 1/3 × 1 hr/60 min. = 3.22 MJ/min
4 gals/hr × 138800 Btu/gal × 1/60 min × 1/3 = 3084 Btu/min.

Coolant flow: 3.22 MJ/min × 1000 kJ/MJ × 1 kg·°C/4.18 kJ × 1/20°C × 1m³/1000 kg = 0.0385 m³/min
3084 Btu/min × 1 lb·°F/1 Btu × 1/36°F × 1 gal H₂O/8.33 lb =10.28 gals/min

NEW: Resolve the above for an anti-freeze solution having a 90% specific heat and 110 % density of water.

15.19 3.22 MJ/min × 1000 kJ/MJ × kg ·°C/1 kJ × 1/49°C × 0.82 m³/kg = 53.9 m³/min
3084 Btu/min × lb·°F/0.24 Btu × 1/80°F x 13.2 ft³/lb = 2120 ft³/min

NEW: From the above, what is the velocity of air through a 0.4m² [4.3 ft²] area radiator?

15.20 2.67 kN × 0.6 × 10cm × 1m/100cm × 1000N/kN = 160 N·m
600 lb × 0.6 × 4 in. × 1 ft/12 in. = 120 lb·ft

NEW: Resolve the above if the same size clutch is a hydraulic, wet clutch (Fig. 15.55) with a coef. of friction of 0.15?

15.21 Use torque Eq. 2.2
Engine torque: 37 × 60/(2 × 3.14159 × 2400) = 0.147 kN·m
50 × 33000/(2 × 3.14159 × 2400) = 109.4 lb·ft
Gear ratio to rear axle: 300 to 1 reduction; torque increase is 300 x engine torque 44.1 kN·m
32825 lb·ft
Radius of rear wheel: 0.9 × (19 + 15) = 30.6 in.
77.7 cm
Pull: **44.1 kN·m/0.777 m = 56.8 kN**
32825 lb·ft × 12 in./ft × 1/30.6 in. = 12873 lb

NEW: To obtain a greater theoretical pull, should the tire size be increased or decreased?

15.22 Effective area: 3.14159/4 × 10² = 78.54 cm²
3.14159/4 × 4² = 12.57 in.²
Force: 78.54 cm² × 10.3 MPa × 1 MN/1 MPa·m² × 1 m²/(100 cm)² × 1000 kN/MN = 80.9 kN
12.57 in.² × 1500 lb/in.² = 18855 lb

NEW: Calculate the maximum force available when retracting.

16.1 Use Eq. 16.2
(100 × 10 × 120 /(18.5 × 165 × 8 × 0.7) × (8 + 6))$^{1/2}$ = 9.9 m
(100 × 8.25 × 300 /(18.5 × 50 × 5 × 0.7) × (8 + 6))$^{1/2}$ = 32.7 ft

NEW: Find the least-cost width for 240 ha [600 acres] per year.

16.2 566 kg lost/60 days; k = 566/5660 × 1/60 = 0.00167   k = 9/90 × 1/60 = 0.00167

NEW: Resolve the above if there was zero loss on October 1.

16.3   Use Eq. 16.4, Table 16.2, Appendix H, Must treat each cutting individually
Week 15: U = 3.4/7 = 0.485;
$100 \times 10 \times 10/(18 \times 340 \times 9.6 \times 0.8) \times (6 + 6 + 0.01 \times 250 \times 10 /(4 \times 1 \times 0.485 \times 10)) = 2.83$
$100 \times 8.25 \times 25/(18 \times 104 \times 6 \times 0.8) \times (6 + 6 + 0.01 \times 101 \times 25 /(4 \times 1 \times 0.485 \times 10)) = 30.6$
Week 20: U = 4.1/7 = 0.586;
$0.213 \times (6 + 6 + (0.01 \times 200 \times 10 /(4 \times 1 \times 0.586 \times 10)) = 2.74$
$2.30 \times (6 + 6 + (0.01 \times 81 \times 25)/(4 \times 1 \times 0.586 \times 10)) = 29.6$
Week 28: U = 4.5/7 = 0.643;
$0.213 \times (6 + 6 + (0.01 \times 200 \times 10)/(4 \times 1 \times 0.643 \times 10)) = 2.72$
$2.30 \times (6 + 6 + (0.01 \times 81 \times 25)/(4 \times 1 \times 0.643 \times 10)) = 29.4$
Optimum $w = [2.83 + 2.74 + 2.72]^{1/2} = 2.87$ m
$= [30.6 + 29.6 + 29.4]^{1/2} = 9.47$ ft

NEW: Which modification would make the most change in the above, the mower price is doubled or the value of the crop is doubled?

16.4   Use Eq. 16.5.
$100 \times d/(FC\% \times p) = 100 \times 10/(18 \times 340) = 0.163$
$= 100 \times 10/(18 \times 104) = 0.534$
$w_{1,2} = 2.87 + 0.163/2 +/- [0.163 \times (2.87 + 0.163/4)]^{1/2}$
$= 9.47 + 0.534/2 +/- [0.534 \times (9.47 + 0.534/4)]^{1/2}$
$= 2.95 +/- 0.69; w_{1,2} = 3.64, 2.26$
$= 9.74 +/- 2.26; w_{1,2} = 12.0, 7.48$

NEW: If the price of the mower was doubled in the above, what would be the range?

16.5   Solve for A in Eq. 16.4; Use Table 1.1 to get ave. e = 0.79; k =0.01 from Table 16.2;
w = 8m [26 ft]
soybean area = corn area = A/2 ; From Appendix D, p = $640 [$197]

$8^2 = 100 \times 10 \times A/(20 \times 640 \times 5.5 \times 0.79) \times ((15 + 0.01 \times 550 \times A/2)/(2 \times 1 \times 0.6 \times 10) + (15 + 0.01 \times 600 \times A/2)/12)$
$64 = 0.018 A \times (15 + 0.229A + 15 + 0.25 A) = 0.02 A \times (30 + 0.479 A) = 0.6 A + 0.0096 A^2$
$7431 = 62.5A + A^2$
Use quadratic equation solution for $A^2 + 62.5 A - 7431 = 0$; A = 60.3 ha

$26^2 = 100 \times 8.25 \times A/(20 \times 197 \times 3.44 \times 0.79) \times ((15 + 0.01 \times 220 \times A/2)/(2 \times 1 \times 0.6 \times 10) + (15 + 0.01 \times 240 \times A/2)/(2 \times 1 \times 0.6 \times 10)$
$676 = 0.077 A \times (30 + 0.192 A) = 2.31 A + 0.0148 A^2$
Use quadratic equation solution for $A^2 + 156 A - 45676$; A = 149 acres

NEW: Use the A and other data above in Eq.16.4 and see if the w comes out be as in Prob.16.5

Problem Solving in Laboratory Exercise 1: 240 kg/min, 3.0 km/hr, 0.6 ha/hr, 2.94 ha/hr, 4.8 t/km, 0.25 km/L, 24 t/km, 16.7 c/t, 40 c/ha

INDEX